About Island Press

Since 1984, the nonprofit Island Press has been stimulating, shaping, and communicating the ideas that are essential for solving environmental problems worldwide. With more than 800 titles in print and some 40 new releases each year, we are the nation's leading publisher on environmental issues. We identify innovative thinkers and emerging trends in the environmental field. We work with world-renowned experts and authors to develop cross-disciplinary solutions to environmental challenges.

Island Press designs and implements coordinated book publication campaigns in order to communicate our critical messages in print, in person, and online using the latest technologies, programs, and the media. Our goal: to reach targeted audiences—scientists, policymakers, environmental advocates, the media, and concerned citizens—who can and will take action to protect the plants and animals that enrich our world, the ecosystems we need to survive, the water we drink, and the air we breathe.

Island Press gratefully acknowledges the support of its work by the Agua Fund, Inc., The Margaret A. Cargill Foundation, Betsy and Jesse Fink Foundation, The William and Flora Hewlett Foundation, The Kresge Foundation, The Forrest and Frances Lattner Foundation, The Andrew W. Mellon Foundation, The Curtis and Edith Munson Foundation, The Overbrook Foundation, The David and Lucile Packard Foundation, The Summit Foundation, Trust for Architectural Easements, The Winslow Foundation, and other generous donors.

Foundations of Environmental Physics

Foundations of Environmental Physics

UNDERSTANDING ENERGY USE AND HUMAN IMPACTS

Kyle Forinash

Library of Congress Cataloging-in-Publication Data

Forinash, Kyle.
Foundations of environmental physics : understanding energy use and human impacts / Kyle Forinash.
p. cm.
Includes bibliographical references and index.
ISBN-13: 978-1-59726-709-0 (cloth : alk. paper)
ISBN-10: 1-59726-709-0 (cloth : alk. paper)
1. First law of thermodynamics. 2. Second law of thermodynamics. 3. Energy conservation. 4. Energy consumption—Environmental aspects. 5. Energy consumption—Climatic factors. 6. Human ecology. I. Title.
QC173.75.T5F67 2010
333.79—dc22
2009045960

 Printed on recycled, acid-free paper

Manufactured in the United States of America
10 9 8 7 6 5 4 3 2 1

Contents

Introduction

> For a successful technology, reality must take precedence over public relations, for nature cannot be fooled."
>
> —Richard Feynman

All the accoutrements of the modern world—automobiles, vaccinations against smallpox, warm beds, and cold beer—are made possible by humanity's understanding of the physical laws of nature. This body of knowledge has been built up over centuries, and although it is occasionally modified or extended, the broad outline of our understanding of nature is unlikely to shift substantially. Physical laws often take the form of mathematical equations, but the symbols and operations in the equations represent real objects and events. Without them, current technologies, and the lifestyle they sustain, simply would not exist.

These universal laws are also the keys to understanding the origins of current environmental problems, often caused by our own technological prowess. The same laws severely limit the potential solutions available to us. Proposing solutions to central problems such as the world's looming energy crisis or toxic pollution without a complete understanding of fundamental natural laws is foolhardy and potentially dangerous. Conversely, mastery of this body of knowledge is critical to addressing today's most pressing concerns.

Most current environmental problems are a direct result of the sharp increase in population over the past 200 years and the associated technology necessary to support the 6.8 billion people alive today. The population has doubled in less than the lifetime of the author, and this fact alone has caused great stress on the environment. For portions of the last 200 years, population, energy use, food production, and pollution have increased exponentially. Although environmental problems are closely linked to population, the linkage is not inevitable. Several airborne pollutants, such as lead and chlorofluorocarbons, have been decreasing over the past 25 years, after rising in lockstep with the population for the prior 100 years. The effects of population changes on resources and the environment (pollution and its transportation, in particular) are treated first, in Chapter 1, because they have a direct and complex impact on all other environmental problems.

Of all the problems facing us, energy consumption and its consequences for the natural world may be the most dire. A core purpose of this text is to bring into focus the complexity of the current energy problem and give readers critical thinking and computational skills necessary for sorting out potential solutions. To begin to tackle energy shortages and climate change, it is critical to understand the first and second laws of thermodynamics. The first law says that energy cannot be created or destroyed but can be changed from one form to another. The implications of this law are far reaching. The first law tells us that the amount of energy in the world is limited. We can never get more energy out of a gallon of gasoline than was there initially, whether we burn it or use it in a fuel cell. The first law also tells us that energy efficiency, defined as the ratio of the energy benefit to the energy cost, is very important if energy supplies are limited, either naturally or economically. It is usually the case that the economic cost of energy conservation is less than the cost of procuring new energy supplies, but the need to determine how to conserve energy and how much can be saved necessitates an understanding of thermodynamics.

The second law of thermodynamics determines the maximum theoretical limit on the efficiency of converting one form of energy to another. Using energy always degrades it in the sense that it becomes less available to do useful work each time it is converted from one form to another. Internal combustion engines found in today's cars can never be 100% efficient, even in theory (although they certainly can be better than the current efficiency of less than 25%). The second law stringently limits the efficiency of any process that involves the conversion of heat energy (burning fuel) into mechanical energy. However, electric motors can have much higher efficiencies than heat engines (but still less than 100%) because the processes involved are fundamentally different. Chapters 2 and 3 introduce these two important restrictions imposed on us by nature and provide many practical examples of their applications.

Many other concepts from physics are also important for understanding the problems facing modern society. The basic mechanisms of heat transfer, conduction, convection, radiation, and evaporation inform how we might conserve the energy sources we do have. Heat transfer mechanisms are also important for understanding phenomena such as the surface temperature of the earth (or Venus or Mars, for that matter). Faraday's law of electrical induction is critical for explaining how credit cards, computer hard drives, electric guitar pickups, generators, and metal detectors work. Electrical losses in the power grid constitute around 8% of the total electrical energy transmitted in the modern world, and solving this problem requires an understanding of how electrical circuits work. These and other issues are also taken up in Chapters 2 and 3.

The U.S. Energy Information Administration predicts that global energy demand will increase by 60% over the next 25 years. Currently we rely on fossil fuels, a finite resource, for 85% of our energy supply. It is very likely that we have already consumed half of the easily extractable petroleum in the world. A similar phenomenon is likely to occur with coal, natural gas, and other nonrenewable resources in the next 50 to 100 years. Currently significant amounts of these resources are left in the ground, basically for economic reasons. It might be thought that we need not worry about a shortage of these resources because increasing prices will provide the economic incentive to extract more. For many other resources, such as gold or other minerals, this is the case; we do not care how much energy it takes to extract the resource because demand drives the price. Energy resources are fundamentally different, however, because the amount of energy needed to extract the resource increases as the resource becomes more difficult to remove and process. When the energy needed to extract petroleum, natural gas, or coal equals the energy content of the extracted resource, it no longer makes sense to continue mining the resource; we do not gain more energy by extracting more of it. This and other difficulties surrounding nonrenewable energy are detailed in Chapter 4.

Chapter 4 also explores the potential of nuclear energy, which some believe is an important alternative to other nonrenewable energy sources. It is very likely that the world will see a sharp increase in the number of nuclear power plants in the next 20 to 30 years. Most of this increase will occur in India and China and, to a lesser extent, other parts of Asia and Europe. France already supplies more than 75% of its electricity with nuclear power, and Japan is not far behind. China and India hope to come near that goal in the next 30 years, and plants are already under construction in those countries. At the moment, the United States seems unlikely to participate in the global "nuclear renaissance," largely because of political concerns related to safety issues such as nuclear waste, accidents, and terrorism. This is in spite of the fact that nearly all the technology being used in modern nuclear power plants around the world comes from the United States. Issues of nuclear safety are found in Chapter 8 in the broader context of risk assessment.

The "age of petroleum" will probably last less than 400 years, and we are about in the middle of that epoch. What alternatives could replace oil and other fossil fuels? Wind energy is close to being economically competitive with traditional sources such as coal and natural gas. It is for this reason that wind energy is the fastest-growing renewable resource in the world today. Although the potential amount of solar energy is enormous (and will continue to be available for centuries), this resource is still perhaps a dozen years away from being economically competitive. The limitations for solar energy are partly economic and partly technical, as discussed in Chapter 5. Hydroelectric resources will continue to be important, but much of this resource is already tapped and unlikely to expand significantly, at least in the developed world. Biofuels may provide some of our energy in the future, but the low efficiency of plant growth will probably limit this energy source to a secondary role. Even planting the entire arable land area of the United States in sugar cane (roughly eight times more energy efficient than corn) would probably not provide enough energy to meet just the current transportation needs in the United States. Again the first and second laws of thermodynamics place stringent limitations on what can be done with renewable resources. These limitations and other potential renewable energy solutions are discussed in detail in Chapter 5.

One important feature of many renewable energy resources is their intermittent nature. Although solar energy is easily capable of supplying sixty times the energy used globally today, it is available for only half of a 24-hour day on average. Some experts estimate that wind will be able to supply only 35% of our energy needs because of its intermittent nature. Energy storage is already important in large-scale applications, such as load leveling in electric power plants, and is likely to become even more important as renewable energy sources become more widespread. Transportation consumes roughly a quarter of the world's energy, and energy storage is critical for this application. Cars today are not propelled by electric motors, even though existing electric motors are at least four times more efficient than gasoline engines. This is because no one has invented an inexpensive, lightweight battery that can store sufficient energy and be easily recharged. This picture is likely to change very quickly as new research and economies of scale produce more efficient, cheaper, and lighter batteries. The issue of energy storage, on both large and small scales, is the topic of Chapter 6.

A second core element of this book, introduced in Chapter 7, is the effect of the ever-increasing use of fossil fuels on the atmosphere and climate. There is no doubt in the scientific world that carbon dioxide in the earth's atmosphere has increased over the past 200 years to a level 35% higher than at any time in the past 850,000 years. There also is no doubt that this extra carbon dioxide comes directly from human activity. It is not possible that such a significant change in the makeup of the atmosphere will have no measurable effect on the environment. In Chapter 7 we examine the scientific laws that determine the earth's climate and the evidence that informs our current understanding of how climate changes over time. Basic physical principles tell us that we should expect a change in climate, including higher surface temperatures on average, due to this 35% increase in carbon dioxide, and that is what global data collected over the past few decades appears to show.

All technology comes with associated risks, and an assessment of those risks is an essential first step in a cost–benefit analysis of any proposed solution to an environmental problem. Insurance actuarial tables probably constitute the earliest form of scientific risk assessment. These results are based on past experience and, when done carefully, are known as epidemiological studies. The use of statistics plays an important role in evaluating historical risk. Statistics applied to experiments involving laboratory animals has also been critical in deciding whether a new medical treatment is effective or a new industrial emission is toxic. These forms of risk analysis are introduced in Chapter 8. Assessing the chance of an accident in chemical plants and nuclear reactors entails a mathematical approach to risk evaluation beyond simple statistics, which is also introduced in this chapter. A scientific risk assessment is only the first step in finding and applying solutions to any environmental problem. Economic factors are also important in deciding on a plan of action for solving environmental issues, and although they are not exactly the purview of science they are treated briefly in Chapters 2 and 8.

Once published, no book can remain abreast of the rapidly changing landscape of environmental problems. However, a book can maintain a longer relevancy by focusing on the application of basic, fundamental principles to various problems while incorporating the latest information found on the Internet or in news reports. Although environmental challenges change over time, sometimes rapidly, the physical principles used to deal with these problems are not revised very often, if at all. The first and second laws of thermodynamics have been well understood for at least 200 years, and the text emphasizes a deep understanding of these two important concepts. Often a simple calculation from first principles can reveal the plausibility (or implausibility) of a proposed solution. Armed with only the first law of thermodynamics and the theoretical maximum efficiencies of various conversion processes, we can quickly discount any claim that adding some kind of water converter to your automobile engine will improve gas mileage, no matter how clever a future inventor might be. Very simple calculations based on heat transfer and efficiency quickly indicate the most effective measures to save energy in new construction, regardless of the invention of new materials with different heat transfer properties. The text emphasizes these kinds of back-of-the-envelope calculations, which give a broad answer applicable to many different proposed innovations, even those not yet in the public awareness. Many important applications appear only in the end-of-chapter problems; the reader is encouraged to work through all of them to gain a better feel for how the fundamental principles are applied in practical situations.

The first thing a student (or professor) today does when he or she sits down to work on a problem or assignment is to start up a computer and Google the question. This is true for numerical problems found in a textbook (where students hope to find the solution online) and for more open-ended discussion questions (where Wikipedia comes in handy for an overview and references). Are these resources reliable? How do we know? An important part of education today is learning how to distinguish good information from bad information found on the Internet. The text tries to lead the reader to a broad enough conceptual understanding that he or she can evaluate the reliability of online sources. About half the problems in the book ask the reader to explore the current status of relevant environmental issues using Internet or other sources and apply the principles described in the text to these questions. Some of the exercises require that the student use various Web sites (such as the United Nations population site or the U.S. Energy Information Administration site) to download and manipulate spreadsheets (e.g., population data, energy data). One exercise has the reader use downloaded software to do simple climate modeling. Although the data may change over time, the analytical process remains the same.

Scientists are by nature very skeptical, and the reader is encouraged to act likewise, especially with data found on the Internet. A careful investigation of what appears to be a legitimate counter claim against global warming often turns up faulty data sources or the support of a vested interest such as the oil industry. By digging deeper we can often determine the limits of reliability of an information source. The questions and problems at the end of the chapters are designed to encourage this kind of skeptical thinking and are an essential component of the text.

This book grew out of information collected by the author over a 15-year period of teaching a junior-level college course in environmental physics every second year. The author would very much like to thank all the students who brought up interesting articles or Web sites in class or asked thought-provoking questions over the years. I especially appreciate the students who initially did not believe my presentation of various subjects (climate change in particular), which forced me to delve deeply into the data to support my claims. My understanding of environmental issues would consist of a very small fraction of the content of this book had these students not also been interested in the subjects being discussed. The author would also like to thank Mike Moody, John Armstrong, Mike Aziz, Ben Brabson, Chris Lang, Natasha Boschen, Janardhanan Alse, Eric Schansberg, and several anonymous reviewers for comments and technical advice on portions of the manuscript. I also appreciate the help of Emily Davis and the staff at Island Press. Any errors that remain in the book are not the fault of the reviewers but are purely the result of the author's lack of expertise (or stubbornness).

Resources related to the book, including supplementary material for teaching environmental physics, can be found at www.environmentalphysics.org. This site also includes a page for submitting errata and suggestions.

CHAPTER 1

Population Growth and Environmental Impacts

It has often been asserted that a particular generation has a special place in history, and names such as *The Greatest Generation*, *Baby Boomers*, *Gen Xers*, and *Generation Y* have been given to specific groups. But although each generation has its own claim to fame, in terms of population statistics, people born in the past 50 years have a unique place in recorded history [1, 2]. For the first time, the world's population has doubled in a person's lifetime. It took 650 years for the global population to grow from a quarter of a billion to half a billion, 125 years to grow from 1 billion to 2 billion, but only 39 years to double from 3 billion to 6 billion in 2000. The world population increased in the last 5 years by half a billion, more than the total number of people living in 1600. However, if current fertility projections made by the United Nations Population Division hold, no one born after 2050 will live through another doubling of the human population. Demographic changes are not limited to the total number of people. The year 2000 marked the first time in history that the number of people older than 60 outnumbered the population younger than 10. In 2008, the number of urban people passed the number of rural people.

These facts make the current era unique in human experience and have critical implications for the natural world. As we will see in Chapter 7, the global emission of carbon dioxide is directly linked to the number of people in the world, as are the availability of clean water and the concentration of pollutants in the atmosphere and soil. In this chapter we will look at population trends, examine a simple model of population growth, and consider the environmental impacts of these profound changes.

1.1 ■ Population Growth

Figure 1.1 shows the historical rise of the global population and projections of future population growth out to the year 2150 made by the United Nations Population Division [1]. There are several things to notice in the graph. Asia, which includes India and China, has historically contained almost half the world's population, and this trend is expected to continue for the next 150 years. Sections of Asia, Africa, and to a lesser extent South America and the Caribbean

FIGURE 1.1

Global population, historical and predicted, by region. Oceania includes Melanesia, Micronesia, Polynesia, and Australasia, which is made up of Australia, New Guinea, Papua New Guinea, New Zealand, and various islands of the Malay Archipelago [1]. Northern America includes Canada and the United States.

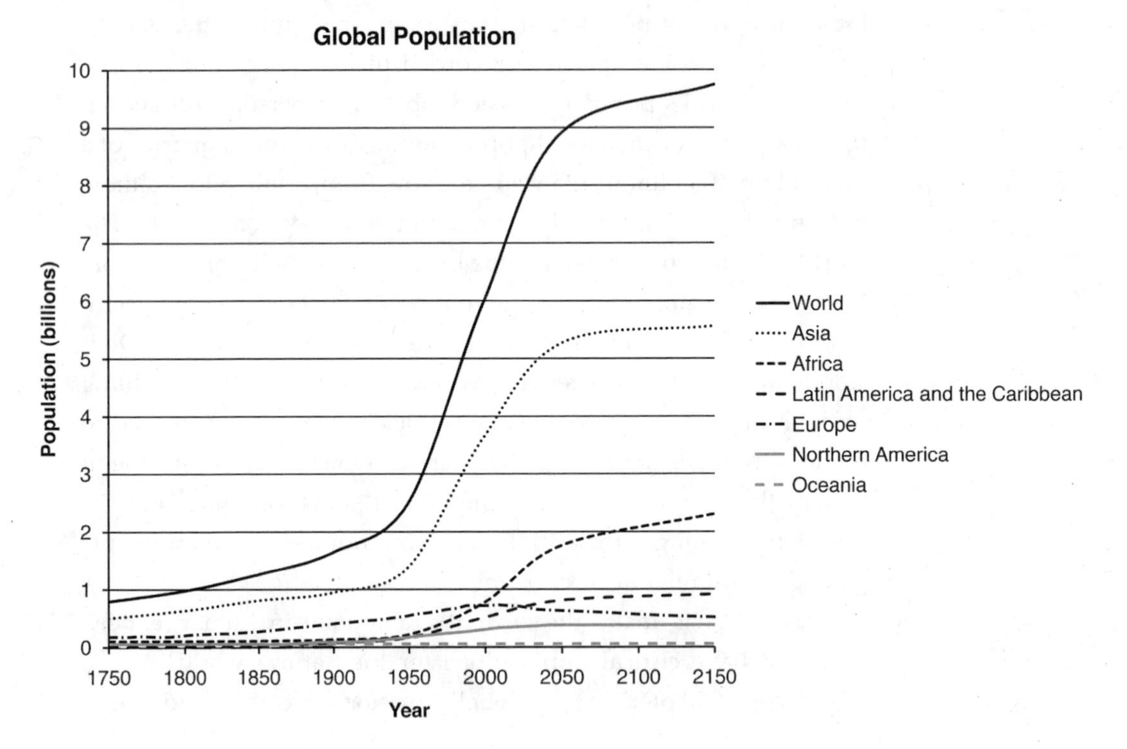

are experiencing a steep population increase, whereas Europe and northern America (not including Central America) are experiencing stable or decreasing populations. The world's population is projected to begin leveling off after 2050, reaching about 9 billion people by 2150.[1] Because of uncertainties about fertility rates, the actual global population in 2150 may be as low as 7.9 billion or as high as 10.9 billion.

Many of the terms used to describe population and demographic changes have specific definitions [3]. Population increase is the net population increase due to births, deaths, and migration in a specified period of time. The population growth rate is the percentage change per year in the population due to births, deaths, and migration. The general fertility rate is the number of live births per 1,000 women of childbearing age (assumed to be either 15 to 44 or 15 to 49, depending on the agency reporting the figure) per year. Because the yearly rate may change during a person's lifetime, another term is used more often: The total fertility rate is the number of live births per 1,000 women of childbearing age, based on age-specific birth rates during childbearing years. These rates are not equivalent to the birth rate, which is the number of live births per 1,000 people (the age of the women is not taken into account). Net migration is the total effect on an area's population due to immigration and emigration over some specified period of time. The net migration rate is the net migration per 1,000 people per year. Finally, it should be pointed out that life expectancy varies with age. Life expectancy at birth is generally lower than life expectancy in middle age, for example. This is because you are much more likely to die in the first few years of life; once this critical time has passed, your life expectancy is longer.

Table 1.1 shows a snapshot from the years 1995 to 2000 of other pertinent population data. From the table we see that approximately three quarters of the world's population lives in developing regions of the world. Average life expectancy at birth in developing regions is approximately 12 years less than in developed regions and is about 24 years lower in Africa. Fertility rates are highest in West, Middle, and East Africa and Melanesia, and these rates correspond roughly to the use of contraceptives; regions with lower fertility rates show higher use of contraceptives. This is an important indicator of how population growth rates have changed in recent history. Fertility rates decrease when women have access to contraceptives, and this tends to occur when women are better educated and have higher standards of living.

A fertility rate of two children per woman is said to be the replacement rate and, if maintained, would eventually lead to a growth rate of zero if both

TABLE 1.1

Global population growth, life expectancy, fertility, and contraceptive use, 1995–2000 by region [1].

Major Regions	Population, 1999 (millions)	Population Growth Rate (% annual)	Life Expectancy at Birth (years)	Total Fertility Rate (average number of children per woman)	Contraceptive Use, 1990s (% currently married women)
World	5,978	1.4	65	2.7	58
More developed regions	1,185	0.3	75	1.6	70
Less developed regions	4,793	1.6	63	3.0	55
Africa	767	2.4	51	5.1	20
Asia	3,634	1.4	66	2.6	60
Europe	729	0.0	73	1.4	72
Latin America and the Caribbean	511	1.6	69	2.7	66
Northern America (Canada and the United States)	307	0.8	77	1.9	71
Oceania	30	1.3	74	2.4	64

children lived to adulthood and there were no migration. Because of infant mortality, the United Nations Population Division assumes the actual replacement rate to be 2.1 children per woman, a number that varies somewhat depending on access to infant health care services.[2] Even after the replacement rate has been reached, a population will continue to grow for some time as youth born before the replacement rate is reached attain childbearing age. Only if this new generation maintains the replacement birth rate will the population level off. Fertility rates are the strongest indicator of population change, but migration can also play a significant role. In Table 1.1 we see that the fertility rate in developed countries is less than 2 (1.6 children per woman), but the annual growth rate is greater than zero (0.3). A few economically mature countries have growth rates less than zero and are therefore experiencing a decline in population. But on average, developed countries are gaining population through immigration and young women coming of childbearing age, which results in the positive growth rates shown.

The effects of regional differences in population growth rates on the local political environment can be seen by comparing recent population increases in Central America with that in the United States. The population of Mexico has increased fivefold in the past 60 years, whereas the U.S. population doubled in this time frame, a figure that includes immigration. The stress on available food, land, and water resources in Central America caused by this more rapid increase in population is one of several factors leading to increased migration, both legal and illegal, into the United States.

Whether through immigration or births, we often hear that a particular population is growing exponentially. The word *exponential* sometimes is used colloquially to simply mean a rapid increase, but in fact it has a specific meaning. When an amount increases (or decreases) over time in proportion to the current amount, the growth (or decline) is said to be exponential. Stated differently, a growth rate is exponential if, for equal time intervals, the increase is a fixed percentage of the current amount. An important example of exponential growth from economics is a fixed percentage return on an investment; each year the amount increases by a fixed percentage of the previous year's total. There are other types of growth; for example, in linear growth a fixed amount is added for each time period. Given enough time, exponential growth always leads to a larger final quantity than linear growth. In Chapter 4 we will consider a different kind of increase governed by the logistic equation; in the following we consider exponential growth.

Intuitively we expect population to increase exponentially; the more people there are, the more people are having children, so the increase in the number of people depends on the number of people of reproductive age. Obviously reproductive age does not extend over a woman's entire lifetime, so we expect population growth to be slightly different from exponential because the growth rate is a percentage of only the number of reproductive-aged women. The population figures mentioned earlier reflect total annual growth, which takes into account the fact that women do not have children throughout their lives. Despite these complications, on average the assumption holds: For the last several hundred years world population growth has been exponential although at varying rates for different time intervals.

In Example 1.1 we show how to model exponential growth. It is important to realize that growth rates larger than zero (i.e., a stable fertility rate greater than 2.1 and zero migration) if they remain constant will *always* lead to exponential growth. Negative growth rates lead to exponential decline; only a sustained growth rate of zero will lead to a stable population. From Table 1.1 we see that Europe is the only region with a stable population. All other regions in the world are undergoing exponential growth, although at significantly different rates and sometimes for different reasons (e.g., immigration).

EXAMPLE 1.1

Exponential Growth

Exponential laws in nature are fairly common. In Chapter 4 we will see examples of exponential decay; here we look more closely at the global growth rate using an exponential growth model.

The world population growth in Figure 1.1 is obviously not linear (it is not a straight line). However, we may wonder whether it is exponential with a constant rate or perhaps with a constant rate along some part of the curve. A quantity $N(t)$ is said to increase exponentially if at time t we have

$$(1.1)\quad N(t) = N(0)e^{rt},$$

where $N(0)$ is the original quantity at $t = 0$ and r is the (constant) growth rate. Here time is in years, so the growth rate represents the fraction of increase per year. The growth rate is positive for exponential growth and negative for exponential decay. A value of zero for r means an initial magnitude that does not change over time.

It is difficult to compare data with exponential curves, so a useful trick is to turn this expression into a straight line. We can do that if we take the natural logarithm (ln) of both sides, in which case we have $\ln[N(t)] = \ln[N(0)] + rt$. If we plot $\ln[N(t)]$ versus t we should have a straight line with slope r and y-intercept equal to $\ln[N(0)]$. In Table 1.2 the year, the population in billions, and the log of the global population are shown.

From Figure 1.2 we see that a plot of $\ln[N(t)]$ versus year is approximately straight. A linear equation can be fit to the data that has the form $y = 0.014x - 5.073$, as shown in the graph. The slope of this linear graph gives a growth rate constant of 0.014 or 1.4% per year, which is the value shown in Table 1.1 for the global growth rate between 1995 and 2000.

One thing that should be noted is that our model shows a constant growth rate of 1.4% between 1900 and 2000. However, closer examination of the graph reveals that the slope changed somewhat during this 100-year period. The data have a slightly lower slope (and lower growth rate) between 1900 and 1930 and a higher rate between 1950 and 1980. In fact, the growth rate was about 0.7% between 1900 and 1930 and about 1.8% between 1950 and 2000. It is also obvious that this equation cannot apply to the real population before 1900 because the y-intercept indicates a population of zero in the year 360.

TABLE 1.2

Sample population data (adapted from [4]).

Year		
1900	1.6	21.19
1910	1.7	21.25
1930	2.0	21.42
1950	2.5	21.64
1960	3.0	21.82
1970	3.6	22.00
1980	4.5	22.23
1990	5.3	22.39
2000	6.0	22.52

FIGURE 1.2

Natural log of population by year.

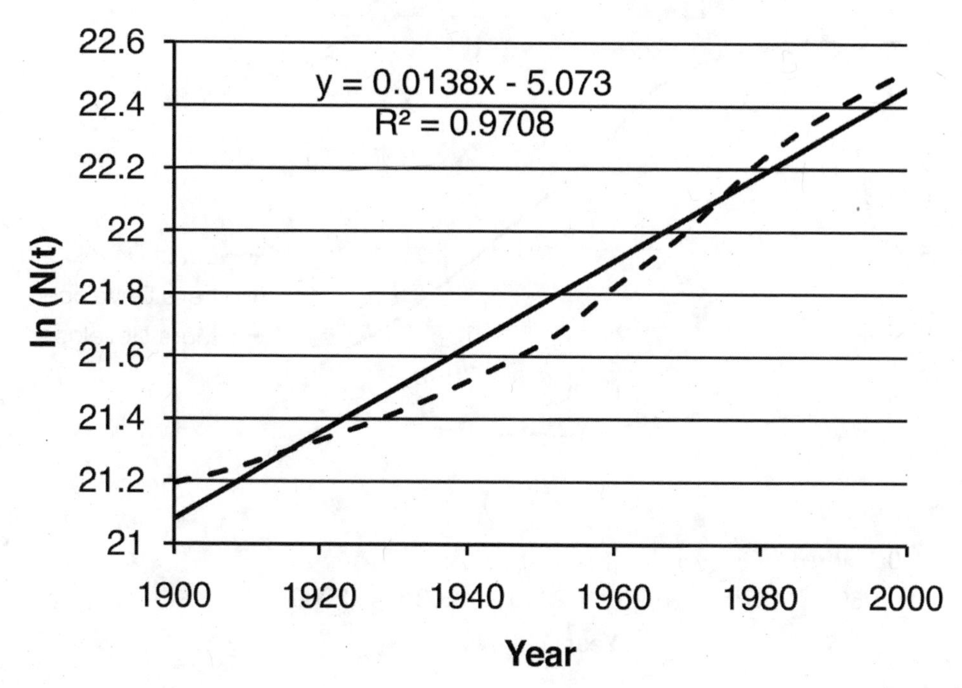

If the current fertility rates shown in Table 1.1 remained fixed, the world population would reach 244 billion people by 2150 and 134 trillion by 2300 [1]. Obviously the world population cannot continue to grow exponentially forever. It is very doubtful that a global population equivalent to that of a densely populated city, such as Calcutta, India, would be sustainable because no land would be left for growing food. However, the growth rate and subsequent population increase are very sensitive to fertility rates. As shown in Figure 1.3, fertility rates are falling and may possibly reach two children per woman by about 2050, resulting in a stabilization of the global population by about 2150. However, if the actual fertility rate falls only to two and a quarter children per woman, the global population will reach 36.4 billion by 2300. If the rate falls to 1.75 the world population in 2300 is projected to be only 2.3 billion. From this we see that because of the exponential nature of population increases, small changes in

FIGURE 1.3

Historical fertility rates and projections for various populations [1].

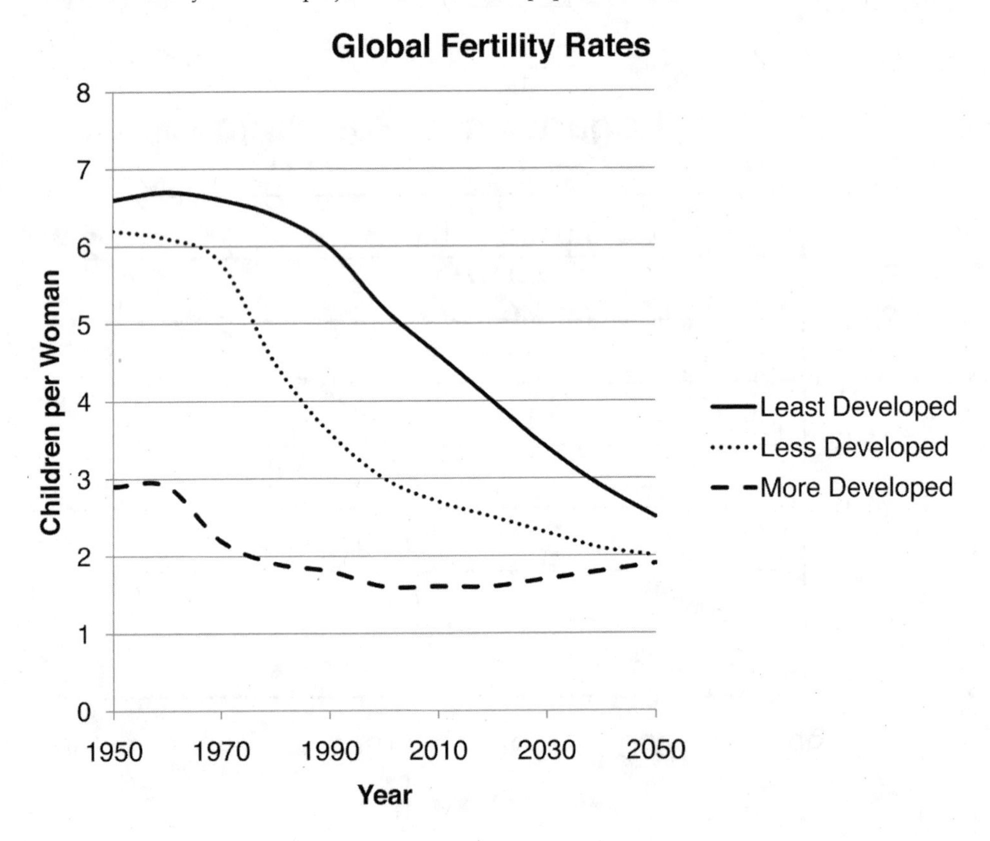

fertility rates have a large effect over time, much larger than death rates due to HIV/AIDS and other pandemic illnesses, for example.

The current decline marks the first time in history that fertility has fallen voluntarily. The only time the global population has declined was during widespread plagues such as the Black Death in Europe during the fourteenth century, when two thirds of the population is thought to have died. Despite the declining fertility rates shown in Figure 1.3, the total population is expected to continue to grow for the near future. This is because even when the fertility rate falls to two children per woman (the replacement rate), the greater-than-replacement rate experienced before this time has generated a disproportionate number of young people who will grow up to have offspring. It takes at least a generation after the replacement rate is reached for the population to stop growing. Even then, longevity often increases as countries become more developed, leading to a slower decline in the number of people living.

Declines in fertility and subsequent changes in projected population growth are occurring for several reasons. The single largest factor affecting fertility is women's education, a factor that is also positively correlated with higher economic status. In developing countries a larger part of the population has traditionally lived in rural areas where farming, much of which is done by women, is the main occupation. This mode of subsistence is changing, however, as the world's population moves into cities. Urban living often means higher economic status for women because they can work for wages, which offers the opportunity to purchase medicine, contraceptives, and better access to education, whereas an economy based solely on agriculture does not. It is also the case that farming techniques in developing countries are based on manual rather than mechanical methods, so large families are needed in order to supply workers. As societies move from rural to urban areas, there is less need for a large family; in fact, it can often be a detriment because it means more mouths to feed. Women who have moved to urban areas and have a choice and access to affordable birth control generally decide to have fewer children. This trend is widespread in developing countries and is often actively supported by government agencies in those countries. The United States is one of the few countries where subsidized family planning services and national population planning do not play a significant role in public policy [2].

Because fertility rates are generally higher in developing countries, the majority of the population growth in the next 50 years is expected to occur in the less economically developed countries of Asia and Africa. A handful of countries—India, Pakistan, Nigeria, Democratic Republic of Congo, Bangladesh, Uganda, and China—will account for most of the increase in the next 50 years. Other African countries such as Botswana, Swaziland, and Zimbabwe are also expected to grow significantly in population despite the fact that the HIV and AIDS epidemic is widespread in these countries. Because of low fertility rates in the United States, the population would level off by 2020 if there were no immigration.[3] As a result of low fertility and lack of immigration, some developed countries such as Germany, Japan, and the Russian Federation are expected to continue losing population in the next 50 years. The difference in fertility rates between developed and developing regions means that the number of people living in developing regions will increase relative to the number of people in developed areas of the world. Today the population in more developed countries makes up about 20% of the world's total. The UN Population Division projects that by 2050 the number of people living in developed regions will be only 10% of the global population.

The populations in countries where fertility rates are low tend to have a higher percentage of older people, and this trend will continue. The number of people over the age of 60 is expected to triple by 2050, growing from 20% to 33% of the global population. This is partly because people tend to live longer in developed regions. Currently in developed countries the average lifespan is about 75 years, whereas in less developed countries the average lifespan is closer to 63 years. It is also the case that higher fertility rates in developing regions mean more children have recently been born there, creating a younger population on average. In many developing countries 40% of the population is under the age of 15. Because these younger people are just now reaching childbearing age, more than half the population of these developing countries will still be under the age of 25 in 2050, whereas in developed countries up to one third of the population will be over 60 years of age [1, 2].

The exponential growth of the world's population over the past several hundred years has had significant consequences for the environment, including deforestation, overfishing, and increased pollution. Since the Industrial Revolution, those impacts have become proportionately greater as each individual's environmental footprint grows with increased use of technology. The greatest environmental damage has been done by highly developed nations, which tend to have smaller populations but more intensive industry. But as poorer countries develop and their populations move from rural to urban areas, they too are beginning to expect the same high standard of living, with the associated increase in environmental damage. On the other hand, an increase in standard of living is generally associated with greater awareness of environmental problems; developed countries spend much more on controlling environmental damage than developing countries. In the following sections we briefly discuss the impact of population growth and movement on just three factors: water resources, food supply, and pollution. Other effects of the growing global population, particularly increases in energy demand, are considered in later chapters.

1.2 ■ Effects of Population Growth

Most attempts to calculate a sustainable carrying capacity for the earth or an ecological footprint for the earth's population have been flawed. On one hand, it seems only logical that water availability, arable land for food production, or some other physically limited resource will eventually create an upper boundary to global population [5, 6]. There are also historical examples in which shortages of physical resources may have led to the collapse of local societies, such as Easter Island, early Mayan culture, and the Chaco culture in the southwestern United States. (For speculation on the causes of the demise of these cultures, see [7].) However, it is notoriously hard to predict when or whether this will occur for a particular population, given the creativity of humans in devising technological solutions. For example, the invention of fertilizer and farm machinery has increased food production far beyond what was once thought possible for a given quantity of land (although both of these innovations have led to new environmental problems). Even if physical limitations are currently not a global problem, however, they are certainly significant on local scales, and it seems sensible to examine available resources and their limitations and make plans for the future use of those resources.

1.2.1 Water

The world's growing population has put intense stress on water resources in many regions. The global per capita withdrawal of freshwater has increased slowly over the past 100 years by about 25% to approximately 600 m^3 per person per year. This trend has actually decreased by a few percent since 1980. However, because of the population increase, the total water extracted has increased by a factor of eight since 1900. Before 1900 nearly all water was used for farm irrigation. In developing countries water is still used predominantly for agriculture; India and China have the highest percentage of land under irrigation in the world. As we can see from Table 1.3, in more developed regions a higher percentage of water use is found for industry and domestic use. The percentage of domestic water use is highest in North America, partly as a result of the custom of individual houses being surrounded by a watered lawn. The average North American uses 217 m^3 of water per person per year for personal, domestic water use (e.g., taking showers, flushing toilets, washing clothes and dishes), compared with 38 m^3 for the average Indian and 26 m^3 for the average Chinese. Japanese, Italians, and Mexicans individually use about 40% less water than North Americans, the French use about half as much, and the average person in Bangladesh uses only 7% as much [8].

As shown in Table 1.3, Asia has the highest percentage of water withdrawn from locally available water resources. This reflects the heavy irrigation in India and China needed to supply these large populations with food. In other regions the withdrawal appears to be only a few percent of the total resource, but these averages hide serious problems at the local level. For example, the Near East and North Africa are now consuming 51% of the total available renewable resource. Specific countries that are much closer to withdrawing a significant proportion of their available renewable water supply include Egypt (92%), Iran (48%), Jordan (86%), and Pakistan (73%). Some countries currently import water or make freshwater from saltwater, giving withdrawal percentages higher than 100% of the available resources. Saudi Arabia uses 643%, Yemen 154%, and Libyan Arab Jamahiriya 712%, respectively, of the available freshwater found in those countries. There are also locations in the southwestern United States where water is being removed from underground aquifers faster than these aquifers are being refilled by rain and other natural sources. Even in aquifers where the amount of water is not decreasing, high withdrawal rates may result in a pressure drop, which encourages the intrusion of saltwater, as is occurring in the Upper Floridian Aquifer along the coast of Georgia and South Carolina.

Small farmers in rural areas of developing countries are most affected by water scarcities. Water shortages in Africa account for about half the world's chronically undernourished population [10]. Although modern irrigation methods such as drip irrigation are improving the efficiency with which water is being used, the capability to grow food is limited by local water availability, and this limit has been reached in several places around the globe. A now famous example is the Yellow River, which supplies an important agricultural region in China. The Yellow River has run completely dry almost every year since 1992 because of withdrawals for farm irrigation.

TABLE 1.3.

Global water resources and freshwater withdrawn in 2000. The last three columns give the percentage domestic, industrial, and agricultural use, respectively (percentages may not add to exactly 100% due to rounding). Per capita figures are given as cubic meters per person per year and include water used for industrial and agricultural uses [9, 10]; individual domestic water use per person is significantly different (see text and reference [8]).

Major Regions	Renewable Water Resource (km^3/yr)	Total Freshwater Withdrawal (km^3/yr)	Per Capita Withdrawn (m^3/person/yr)	Percentage Withdrawn of Renewable Resource	Domestic Use (%)	Industrial Use (%)	Agricultural Use (%)
Africa	5,723.5	213.24	202	3.7	24	8	68
Asia	15,436.2	2,294.84	797	14.9	15	8	76
Europe	7,507.4	392.22	557	5.2	26	50	24
North and Central America	7,620.8	622.48	260	8.2	40	36	23
South America	16,325.7	164.62	332	1.0	14	8	75
Oceania	1,669.3	26.34	363	1.6	34	19	46

Despite the apparently abundant resources shown in Table 1.3, local shortages worldwide result in more than 1 billion people not having adequate drinking water and about 2.5 billion not having adequate sanitation services—a total equal to the entire population of the world in 1950. It is estimated that 10,000 to 20,000 children a day die from water-related diseases [10]. In part this is because of a lack of local access to clean freshwater sources, but it is also the result of competition with growing industry. At the national level, it makes economic sense to use water for industry because the monetary payback is higher than that of water used in farming. However, diversion of water to industry means less is available for small farmers who cannot compete for these resources. Competition for water resources also comes from hydroelectric dams. The number of dammed reservoirs in the world has increased by nearly a factor of 100 since 1900 as a result of higher electricity demands from a growing population. This dramatic increase has meant the diversion of water from traditional channels where it was used for agriculture, the displacement of thousands of people to make way for the lakes behind the dams, and the destruction of local ecosystems [8–11]. Although dams do not use up water, the relocation of water resources and people often has the effect of making the original water source unavailable for traditional uses.

In a world where two thirds of the surface is covered with water, it might seem that this resource would never be in short supply. However 97% of the world's water is too salty for human consumption or irrigation, and of the 3% that is freshwater, two thirds is frozen in glaciers or at the poles. Desalination is still very energy intensive and therefore prohibitively expensive. Most of the large desalination facilities in the world are found in the Middle East, where there is an abundance of energy in the form of petroleum coupled with severe freshwater shortages. The global supply of freshwater from desalination plants constitutes only 0.2% of global withdrawals and is not predicted to increase significantly because of high energy costs.

1.2.2 Food

Population growth has necessitated significant changes in the way we grow our food. It is rational to ask whether food supplies might eventually limit population growth. Early estimates of how much food could be produced globally were based on the amount of land available and existing farming methods, and they were quite pessimistic. The use of fertilizers and farm equipment has changed the amount of farm land needed to support a person but has also introduced problems such as an increased reliance on fossil fuels, increased water use, reliance on monoculture crops, and an increase in pesticides in the environment, along with pesticide-resistant pests. For example, about 4,000 gallons of water and a third of a gallon of oil are needed to produce a bushel of corn using modern farming techniques when fertilizer and fuel for farm equipment are included. One result of this energy-intensive food supply system is that agriculture and livestock production account for 30% of the carbon emissions in the United States. Humans now also introduce twice as much nitrogen into the environment compared to natural processes, in large part because of heavy fertilization on commercial farms. We will further investigate the energy resources needed for farming and the diversion of food crops into energy sources in later chapters. Other issues involving food supplies are discussed here.

Estimates of how many people the earth can comfortably and sustainably support depend on what is meant by *comfortably* and *sustainably*. The world already produces enough grain and cereals to feed about 10 billion people if they all were vegetarian. However, a world with a North American diet can support only about 2.5 billion people using current technology because of the diversion of cereal and grain crops to produce beef, poultry, and other livestock. Modern industrial food production uses about 7 kg of corn to produce 1 kg of beef and about 2 kg of corn to produce 1 kg of chicken [12]. In addition to the increased demand for food caused by increased population, as societies become more industrialized and gain economic security they place greater demands on farming because diets improve. Residents of developing countries begin to switch to poultry and meat as their economies grow, and the per capita production of grain (a greater proportion of which goes to feed livestock) increases faster than the population.

Sustainable is also a controversial term. Estimates of the number of people the world can feed vary from 1,000 times the current population to figures that show we have already passed the possibility of sustainability and current growth is based on unsustainable use of resources [2]. Heavy fertilization and mechanical farming of irrigated monoculture crops are largely responsible for the current grain and cereal surplus the world now enjoys. But some experts have argued that this kind of farming is neither sustainable nor desirable [12, 13]. Modern industrialized agriculture is energy intensive in a time when traditional energy sources are becoming more limited (see Chapter 4). It also damages the land by increasing salinity and decreasing natural fertility and topsoil depth. Moreover, there are many historical examples of overreliance on monoculture crops, such as the famous Irish potato famine of the 1840s. Perhaps less well known is the collapse of the wine industry in Europe in the 1900s, when European grape vines all belonged to a single species and were therefore all susceptible to a devastating phylloxera infestation. The U.S. food industry is heavily dependent on corn and corn derivatives. For example, most nondiet soft drinks are nearly 100% corn sugar, and most processed food contains high percentages—sometimes more than 50%—of products derived from corn [12]. In any given commercial corn field, all of the plants are genetically identical, so any new pest or disease will affect the entire field, not just a few susceptible plants.

Like agronomy, livestock production has also grown appreciably over the past 100 years, as seen in Figure 1.4. From the figure we see that cattle production has increased by a factor of four, about the same as the world population for this same time span. Pig and goat production has followed a similar pattern. This is in contrast to the global horse population, which remains about the same as it was in 1900. These figures are not surprising; if the per capita consumption of a food source such as pigs stays approximately constant, the total production will increase at the same rate as the population. The rate of increase should actually be slightly higher as the standard of living in developing countries improves. On the other hand, because horses are no longer used for farming

FIGURE 1.4

Global livestock production [14–16].

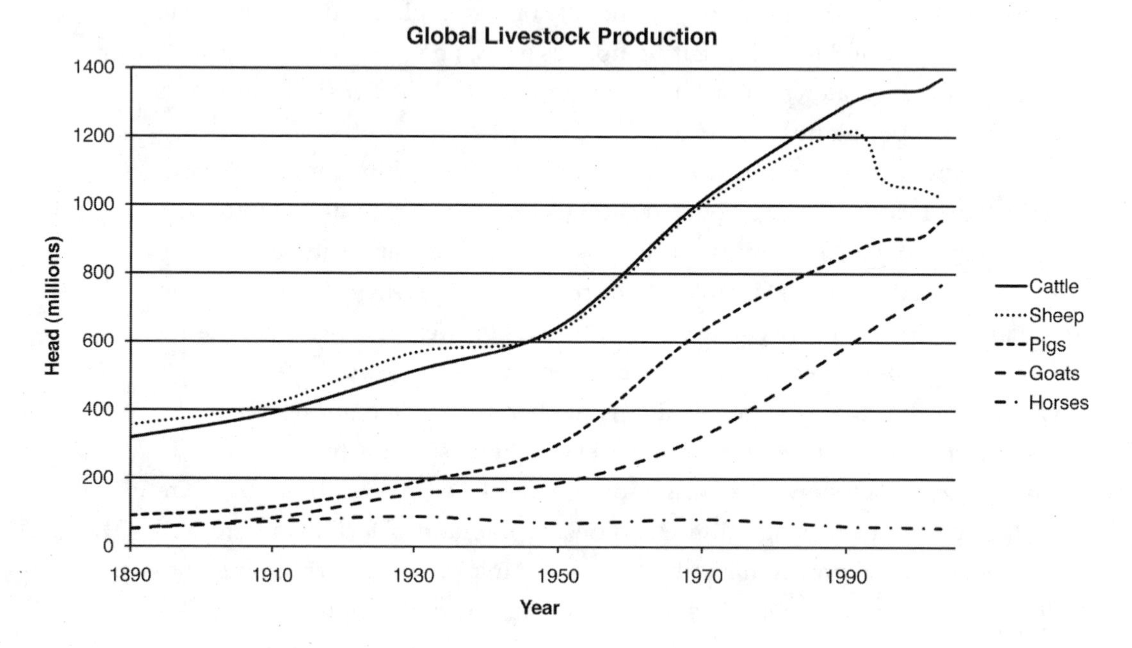

or by urban dwellers, their total numbers have remained nearly constant. The decline in sheep population since the 1990s may have to do with dwindling demand for wool rather than decreases in food production.

Fish are probably one of the few remaining wild food stocks we consume in any significant quantity. As Figure 1.5 shows, even with the rise of aquaculture more than two thirds of the fish eaten in the world today are caught in the wild rather than farmed. The downturn in catch since 2000 probably reflects global limitations of fish stock. There are many documented cases of the decline in fish populations caused by overfishing, such as the collapse of the sardine population off the coast of California in the 1940s. The population of large predatory fish such as tuna is now estimated to be one tenth of what it was before large-scale commercial fishing began [16]. Populations of oil-producing whales decreased by about 95% in the nineteenth century because of whaling and have never recovered. On the positive side, governments and international bodies are now becoming involved, and efforts are being made to manage native breeding stocks to ensure future harvests of fish. Effectively managing these resources is critical given the dire threat to the world's sustainable harvest of fish.

FIGURE 1.5

Global fish production, 1950 to 2005 [16].

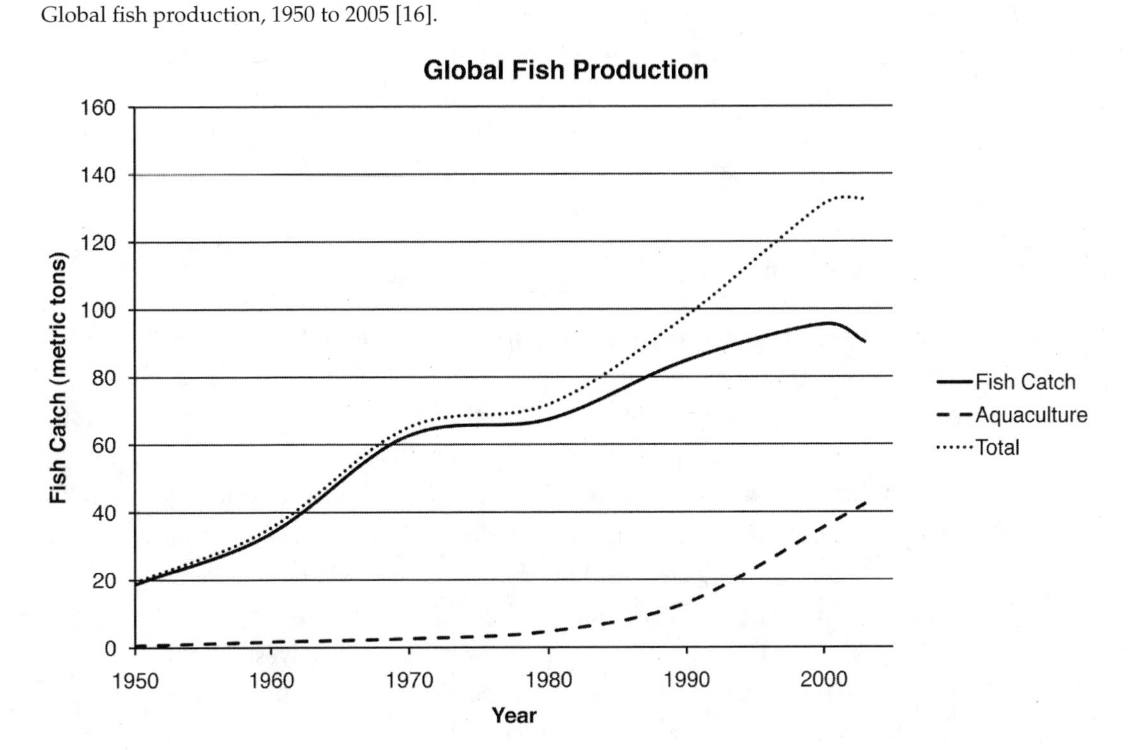

1.2.3 Municipal Waste

The United States disposes of an astounding amount of waste material each year. As can be seen in Figure 1.6, about 228 million metric tons (Mt) of waste was collected in 2006, approximately a threefold increase since 1960. This amounts to about 2 kg of garbage per person per day. About 74 Mt of the total shown for 2000 was recycled, roughly a third. Recycling rates in Europe are higher, about 60%, but even so, Europeans generate about 1.5 kg of waste per day per person. The composition of municipal waste in the United States is shown in Figure 1.7 [17]. These figures do not include industrial waste, which constituted another 200 Mt, or mineral processing waste, another 1,500 Mt, or agricultural waste, an additional 2,000 Mt.

The enormous volume of material discarded in the United States has put pressure on local governments to find new strategies for handling waste. Only so much land is available for new landfills, and incineration is a costly, energy-intensive alternative, in terms of both money and environmental damage. It is also clear that recycling will eventually reach a saturation point; automobile batteries are already recycled at a rate of 99%. More than 60% of the steel cans and yard waste are already being recycled. This figure is 45% for aluminum cans, 51% for paper, and more than 30% for plastic and tires. The rate of recycling rose very quickly between 1960 and 2000 but has nearly leveled off since then. This is partly a matter of the energy needed to recycle a given item. There is a significant energy savings in recycling steel, aluminum, and the lead in car batteries, but it takes almost the same energy to recycle paper and plastic as it does to make these products from new raw materials.

1.2.4 Pollution

The output of industrial processes has increased 40-fold over the past 100 years, in part because of population increases and more extensive use of technology. The world has also seen a shift from total use of renewable resources to a near total dependence on nonrenewable materials in the past few centuries [18]. This has affected the environment in many ways, including increased water use and toxic substances released into the air, water, and soil. Extraction of fossil fuels and the associated greenhouse gases resulting from their use will be considered in Chapters 4 and 7; here we take a brief look at the effect of an increasingly industrialized world on air pollution from heavy metals. As in the case of livestock consumption, as countries become more developed the per capita exploitation of resources used for industrialization increases at a more rapid pace than the population increase.

FIGURE 1.6

U.S. municipal waste since 1960 [17].

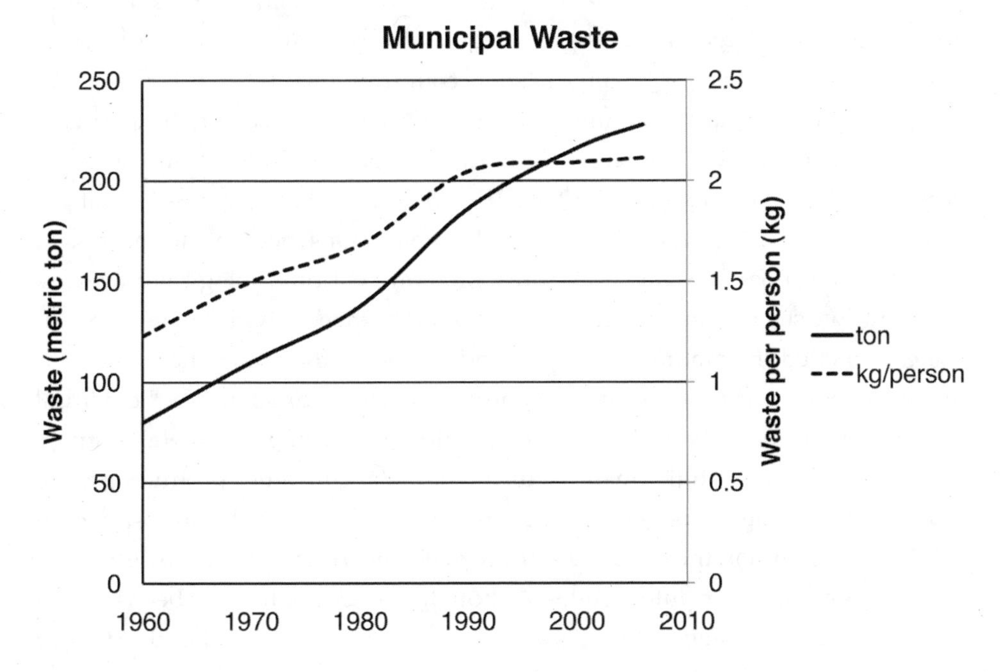

FIGURE 1.7

Approximately 228 million metric tons of municipal waste was produced in the United States in 2006 [17]. Of this, about a third was recycled.

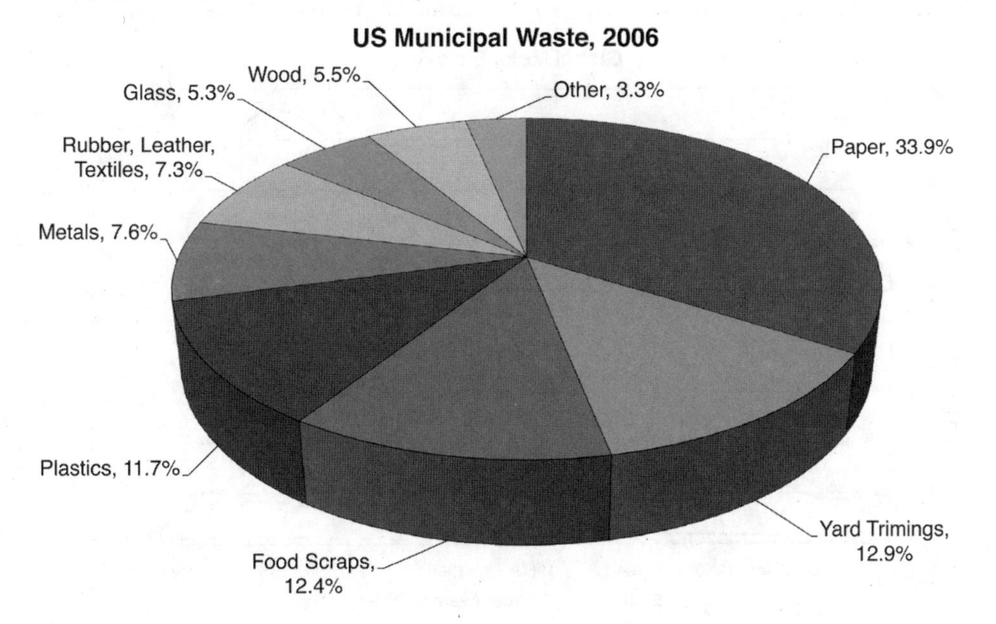

Figure 1.8 shows historical increases in human-caused emissions of five industrial metals. Again it should be clear that these increases are in step with global population increases, at least up until 1980. The decline in emissions after 1980 is caused in large part by tighter regulations on manufacturing processes and the banning of leaded gasoline in some countries. As is the case for nearly all pollutants, some metals are found in the atmosphere because of natural causes. For comparison, natural sources of cadmium, copper, lead, nickel, and zinc currently contribute approximately 12,000; 28,000; 12,000; 29,000; and 45,000 mt per year, respectively. The natural sources of these emissions include windborne soil particles, volcanoes, sea salt spray, forest fires, and various biological processes. From this we see that manufacturing and other human activities have approximately doubled the amounts of nickel and copper in the atmosphere, whereas anthropogenic contributions are currently responsible for six times the natural sources for zinc and about thirty for lead. Mining, smelting and metal refining, fossil fuel burning, and the manufacturing of metal consumer products are the major sources of anthropogenic emissions of heavy metals to the atmosphere. A significant amount of the lead in the atmosphere, nearly two-thirds, results from gasoline combustion in automobiles. Although leaded gasoline has been banned in the United States since 1986, it is still widely used in Asia and South America and some parts of Europe [19].

FIGURE 1.8

Global metal emissions of five industrially important heavy metals, not including natural sources. Each data point represents a 10-year average for the previous 10 years [20, 21].

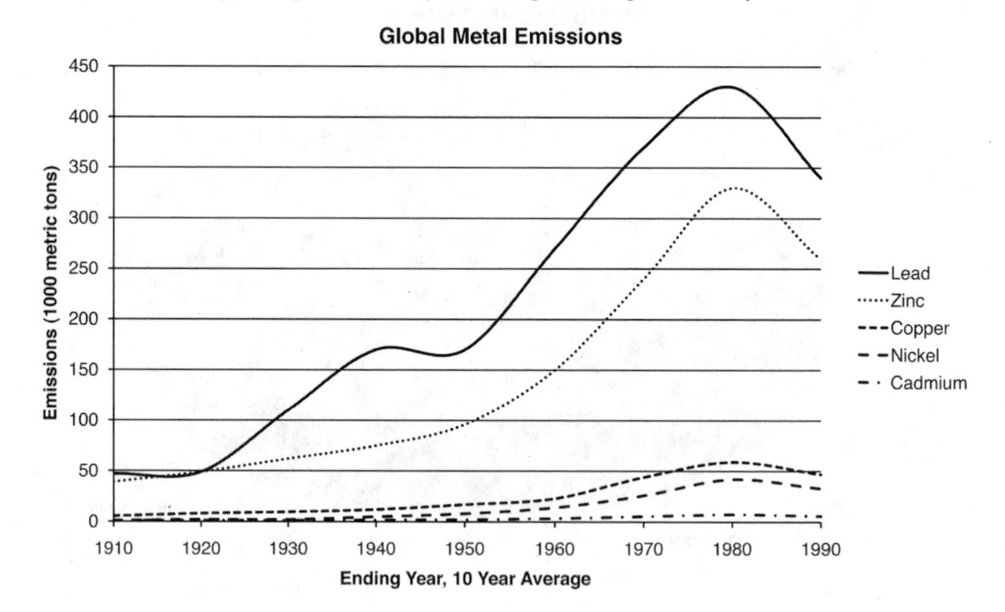

The atmosphere accounts for the largest transportation of pollutants, although this figure varies depending on the substance. For example, about 70% of the lead found in water ends up there because of atmospheric transportation, but atmospheric pollution accounts for only about 20% of the waterborne cadmium. Global transportation of pollution by air is rapid except very close to the poles, where wind patterns tend to be circumpolar and do not cross industrial areas. But even in Antarctica the level of lead found in ice core measurements has increased by a factor of four since the 1900s. At lower latitudes pollutants are more concentrated in large urban areas, but even heavy particulates such as dust and toxic metals can circle the globe in a matter of days, depending on weather conditions [22]. Example 1.2 shows a simplified calculation explaining why urban populations end up with higher pollution concentrations, even when the density of pollution sources is the same as for smaller populations.

Airborne metal pollutants eventually end up in the soil. For example, lead found in the soil of playgrounds is still a major source of lead poisoning for inner-city children in the United States, competing with lead paint as the primary source of lead poisoning in children [19]. Lead is a potent neurotoxin and is especially harmful to the developing brains of children. The high lead concentrations found in inner-city playground soil came mostly from leaded gasoline. Although leaded gasoline has been banned in the United States since 1986, most soil near highways still contains toxic concentrations of lead because, unlike the case of air transportation, heavy metals in soils do not migrate very rapidly.

EXAMPLE 1.2

An Example of Scaling

There are many examples of scaling laws (sometimes called power laws) in nature and biology in particular. In this example we will look at how pollution scales with the size of a city.

Carbon monoxide (CO) is a common pollutant from automobiles and the burning of fossil fuels. The current Environmental Protection Agency standard for air quality states that CO should not exceed 35 parts per million (ppm) averaged over 1 hour or 9 ppm averaged over 8 hours (cities with serious pollution problems such as Los Angles can reach 8-hour averages of 10 ppm) [23]. For simplicity, the sources of this type of pollution can be considered to be scattered throughout a city so that it is a uniform source per unit area, S, measured in emitted kilograms per square meter.

Most people are aware that larger cities are more polluted in general than smaller ones, but one might think that for a source that is constant per unit area the amount of pollution would not depend on the size of the city. As we will see, different scaling laws come into play, leading to a result more familiar to our experience even if the source is proportional to area.

Suppose we have two square-shaped cities, one with a length on each side of L_{small} = 4 km and the other with a side L_{big} = 40 km. We will assume both cities have the same distribution of pollution sources per area, S. The ratio of their lengths, then, is 40/4 = 10 or $L_{big} = 10\ L_{small}$; however, their areas scale as $1600\frac{\text{km}^2}{16}\text{km}^2 = 10^2$ or $A_{big} = 10^2\ A_{small}$. The rate of pollution production, $R_{production}$, for each city will be the source per area, S, times the area, or $R_{production} = SA$ so we have

$$(1.2)\quad R_{production(big)} = 10^2 R_{production(small)}.$$

This would still mean that the cities have the same pollution density because the big city is 100 times the area of the smaller one. However, the rate of removal of pollution is different for each city.

In general, pollutants cool as they rise until they are the same temperature as the surrounding atmosphere, after which they do not rise any further. This height is called the inversion height, h, and depends on the terrain, temperature, and weather patterns but for the most part does not depend on the size of the city.

Area scaling of pollution sources would lead to the same amount of pollution in each city if the rate of removal of pollutants were the same for each. However, pollution removal scales as length of one side of the city and depends on air speed, concentration, and inversion height in the following way. Wind at velocity v blows pollution away from a city through a vertical cross-sectional area of $L \times h$, where L is the length of one side of the city and h is the inversion height. The flux of pollution removal, f, in kilograms per second per square meter at one edge of a city is $f = vc$, where c is the concentration of the pollutant in kilograms per cubic meter, so the rate of removal is $R_{removal} = vcLh$ in kilograms per second.

If we now assume that all the pollution in each city eventually gets blown away, so that the rate of production equals the rate of removal for each city, we have

$$(1.3)\quad R_{production(big)} = R_{removal(big)} = vc_{big}L_{big}h$$

and

$$R_{production(small)} = R_{removal(small)} = vc_{small}L_{small}h.$$

Taking the ratio of these two equations and using the area scaling from Equation (1.2), we have

$$\frac{R_{\text{removal(big)}}}{R_{\text{removal(small)}}} = 10^2 = \frac{vc_{\text{big}}L_{\text{big}}h}{vc_{\text{small}}L_{\text{small}}h}. \quad (1.4)$$

Using the length ratio $L_{\text{big}} = 10L_{\text{small}}$ and canceling like terms, we have $c_{\text{big}} = 10c_{\text{small}}$; in other words, the pollution concentration in the big city is 10 times the amount in the small city, even if each city has the same distribution, *S*, per area of polluting sources (adapted from [23]).

1.3 ■ Ozone

It is easy to become discouraged when considering environmental problems, but there is room for some optimism. From Figure 1.8 we see that the emission of heavy metals into the atmosphere since 1980 has begun to decline. These figures signify a growing awareness of pollution problems and subsequent political changes. The decline in lead emission was largely the result of regulations prohibiting the use of leaded gasoline in cars in the United States and some parts of Europe. In this section we briefly examine the case of high-altitude ozone (O_3), which is another environmental success story [24].

There are actually two ozone problems. The first is local, ground-level, or tropospheric ozone. This problem is largely the result of the sun's action on nitrous oxides, carbon monoxide, and volatile organic compounds coming from automobile emissions, emission from industrial processes, and chemical solvents. This variety of ozone can be blown around by the wind but in general remains close to the ground, causing respiratory problems and eye irritation in humans and plays a role in the creation of smog. The second problem, involving upper atmospheric ozone, is a global problem that has improved because of world efforts to ban several classes of compounds called chlorofluorocarbons (CFCs), hydrochlorofluorocarbons[4] (HCFCs), and bromofluorocarbons (halons). The rise in the use of these chemicals is directly attributable to the increase in refrigeration used in modern society.

As we will discuss in Chapter 7, the atmosphere blocks significant portions of the electromagnetic radiation coming from the sun. For the most part, life on Earth has evolved to be sensitive to the visible portion of the electromagnetic spectrum largely because this is the part that can pass through the atmosphere largely unimpeded. Low-energy ultraviolet radiation (UVA with wavelengths,

λ, between 0.32 μm and 0.40 μm) also passes through the atmosphere and is useful to living organisms, causing the biological synthesis of vitamin D in humans. However, medium- and high-energy ultraviolet radiation (UVB and UVC) are blocked by the atmosphere, which is significant because these kinds of electromagnetic radiation are harmful to life, causing cancer and cataracts in humans and also killing phytoplankton, an important part of the ocean food chain.

High-energy radiation with wavelengths less than 0.1 μm ionize most atmospheric molecules, creating the ionosphere, which begins at an altitude of about 85 km and extends outward. Ozone is created at a lower altitude (about 25 km) when UVC radiation with wavelengths between 0.20 μm and 0.28 μm encounters diatomic oxygen, O_2, which causes the oxygen to disassociate. The production of high-altitude ozone can be summarized as two separate steps:

$$(1.5) \quad hf_C + O_2 \rightarrow O + O \rightarrow O_2$$

and

$$O + O_2 + M \rightarrow O_3 + M,$$

where $hf_C = hc/\lambda$ represents the energy of a UVC photon with frequency f_C. Here c in the expression hc/λ is the speed of light and h is Planck's constant, 6.6×10^{-34} joule-seconds. The symbol M represents any other available molecule (e.g., nitrogen or oxygen) and is needed for momentum conservation when these molecules collide.

When UVB radiation (with wavelengths between 0.28 μm and 0.32 μm) encounters an ozone molecule, it causes the ozone to disassociate into individual oxygen molecules. The resulting chemical reactions occurring are thus

$$(1.6) \quad hf_B + O_3 \rightarrow O_2 + O,$$

$$2O_3 \rightarrow 3O_2$$

$$O + O_3 \rightarrow 2O_2,$$

where hf_B is a UVB photon. These chemical reactions protect biological life on Earth from harmful UVB and UVC radiation by absorbing them in the upper atmosphere. In fact, it was only about 420 million years ago that aquatic plants increased the amount of protective oxygen in the atmosphere to the point that plants could fully colonize the earth's terrestrial surface. Before this time the atmosphere was not sufficiently shielding to permit plant life to exist except in the oceans, where water blocked this harmful radiation.

The reactions shown in Equations (1.5) and (1.6) form a cycle that generates new ozone, which eventually returns to diatomic oxygen. A third reaction involving nitrogen oxide (N_2O) is needed to fully account for the amount of ozone found in the upper atmosphere. Nitrogen oxide is created in the soil by bacteria and gradually diffuses into the upper atmosphere, where it is chemically transformed into nitric oxide (NO). Nitric oxide readily reacts with free oxygen, removing part of the oxygen available for the creation of new ozone. As a net result of this complex combination of chemical reactions, the upper atmosphere contains a largely stable layer of ozone, about 5×10^9 tons of ozone, in a region between 15 km and 40 km above the earth's surface. Or at least this layer would have remained as a stable quantity of ozone, varying only slightly with the seasons, if CFCs had not been introduced into the atmosphere by humans.

The first CFC to be created, dichlorodifluoromethane, was introduced in the 1930s as a refrigerant for use in home and commercial refrigerators. Since then dozens of compounds of carbon, fluorine, bromine, and chlorine have been introduced, which have been used as refrigerants, fire extinguisher propellants, solvents, aerosol propellants, dry cleaning fluids, and foam blowing agents. These compounds were easily manufactured, were easy to handle, had just the right physical properties to be good refrigerants or aerosols, and were thought to be harmless. In fact, the inventor of some of these compounds, Thomas Midgley, demonstrated the inertness of these gases by inhaling a lungful of one of them in front of a live audience, and indeed they are relatively inert gases. However, the fact that they are so stable also means they stay in the atmosphere for very long times, long enough to reach the upper atmosphere, where UVB photons turn them into highly reactant catalysts. It was not until the 1960s that researchers began to suspect that these compounds could have an effect on the upper atmospheric layer of ozone.

From the time of their release at the earth's surface CFCs take about 15 years to reach the upper atmosphere, at which point they act as a catalyst for the following reactions:

(1.7) $$hf_B + CF_2Cl_2 \rightarrow CF_2Cl + Cl,$$

$$Cl + O_3 \rightarrow ClO + O_2,$$

and

$$ClO + O \rightarrow Cl + O_2.$$

Notice that, once released by the absorption of a UVB photon, the chlorine molecule remains present after the last two steps. This property makes it a catalyst, a molecule that participates in the reaction but is still left at the end of the reaction so that it can continue to act. This process removes ozone from the atmosphere but does not deplete the supply of chlorine or CFC; the CFC compounds remain and can continue to turn ozone into stable O_2. One CFC molecule is estimated to be able to convert millions of ozone molecules to molecular oxygen over a lifetime of decades. It is not until the chlorine molecule interacts with some other chemical and is stabilized that it is removed from the process.

Satellite and ground measurements have shown that total ozone in the upper atmosphere has been decreasing at the rate of about 4% globally per year since the 1970s. The depletion is lowest at the equator and greatest at the poles. Because of complex chemical reactions involving CFCs and ice crystals containing nitric acid in winter clouds above the South Pole, a 70% reduction of ozone over Antarctica is seen in the spring each year (September for the southern hemisphere), when sunlight returns. This phenomenon does not occur to the same extent at the North Pole because the weather patterns there are not as stable, allowing a greater mixture of warm air, preventing the formation of ice crystals that contain other catalysts for ozone depletion. The ozone hole over the South Pole, first discovered in the mid-1980s, fluctuates in size during the year, but the spring maximum has been steadily increasing in size as more CFCs gradually reach the upper atmosphere (Figure 1.9). When the ozone hole breaks up each summer and ozone moves northward in the atmosphere, more UVB radiation reaches the earth's surface in South America, Australia, New Zealand, and South Africa, which has resulted in an increased incidence of sunburns and skin cancer in the human population of those regions.

One outcome of the discovery of the depletion of the ozone layer and the ozone hole in particular has been the banning of CFCs. Starting in the early 1980s government and international agencies have put into place regulations for the use of CFCs. These chemicals have been largely replaced as commercial refrigerants and aerosols by less harmful chemicals. This political process occurred quickly because alternatives were readily available and the cost of switching to these alternatives was not very great. As a result of the ban, atmospheric quantities of the most significant CFCs causing ozone depletion have been decreasing gradually since the early 1990s. It is now estimated that ozone levels may return to pre-1970 levels sometime in the next 50 years. As we will see in Chapter 7, carbon dioxide pollution as related to global warming is a much more difficult political problem, mainly because there is as yet no unproblematic substitute for carbon fuels. Nevertheless, the case of ozone depletion in the upper atmosphere and the subsequent banning of CFCs is an encouraging example of how science

FIGURE 1.9

The near record ozone hole of September 11, 2003. Courtesy of NASA's Astronomy Picture of the Day Web site [25].

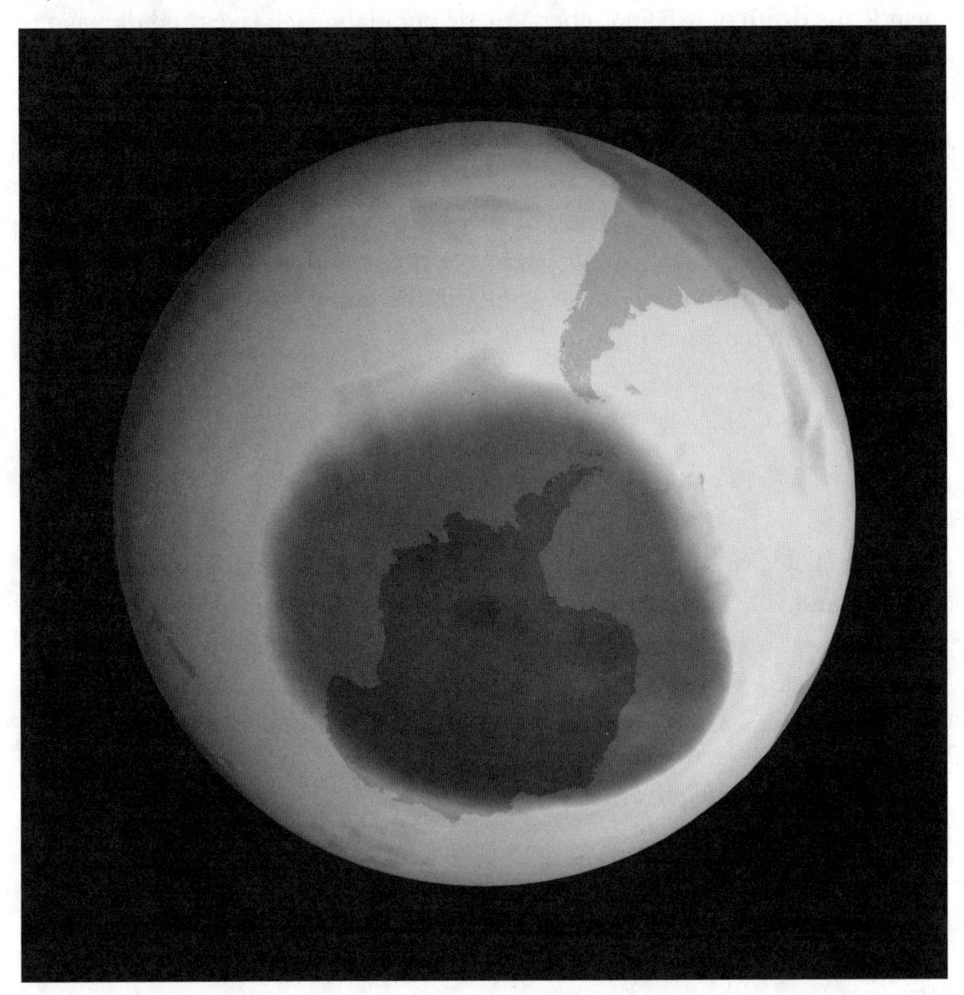

can provide sufficiently complete knowledge of an environmental problem, resulting in a solution based on political consensus and sound science.

1.4 ■ Acid Rain

Changes in acidity of rainfall associated with the burning of coal have been recognized since the mid-1800s during the Industrial Revolution in England. These effects have been locally severe in many places, including England, the northeastern United States, and eastern Canada (see Chapter 8 for a discussion of the direct health risks associated with various pollutants). Parts of Russia, Eastern Europe, and China still suffer from the effects of acid rain, which include the death of fish and other aquatic populations in lakes and damage to forests, buildings, and monuments.

The burning of coal to generate electricity in the United States is responsible for about 65% of the sulfur dioxide (SO_2) emissions and 30% of the various nitrogen compounds (NO_x, where x is a whole number) emission. The exact amount of sulfur emitted per ton of coal burned depends on the type of coal used, but generally some sulfur is always present. Once oxidized into sulfur dioxide during the burning process, the sulfur compounds undergo the following reactions in the atmosphere:

$$(1.8) \quad 2SO_2 + O_2 \rightarrow 2SO_3$$

$$SO_3 + H_2O \rightarrow H_2SO_4.$$

The H_2SO_4, sulfuric acid, is in liquid droplet form and disassociates into the ions $2H^+$ and SO_4^{-2}, which can react with plants and building materials as acid rain. The same reactions can occur for sulfate deposited directly onto the ground as particulates, a process called dry deposition.

Acidity is measured as pH, which is defined as the log of the hydrogen ion concentration:

$$(1.9) \quad pH = -\log[H^+].$$

Pure water has an H^+ concentration of 10^{-7} moles per liter, so it has a pH of 7. Liquids with pH lower than 7 are acids, and those with higher pH are bases. Some lakes have been measured to have a pH as low as 3 (about that of vinegar) as the result of acid rain.

In the 1960s it was recognized that emissions from electric utility plants in the northeastern United States were resulting in environmental problems associated with acid rain in the northeast and in eastern Canada. The increased use of tall smokestacks for coal-fired utility plants, though solving local pollution problems, had the effect of moving the acid rain problem across state and national borders. This eventually led to government intervention at the federal level.

A utility company has many choices for reducing the SO_2 emissions from a power-generating plant, including using scrubbers to remove SO_2 from the flue gas, washing the coal before it is burned to remove sulfur, building new plants that use newer technology in the burning process, using coal with lower sulfur content, and reducing demand by means of conservation and education programs [26]. However, all these mitigation strategies cost money, and the most cost-effective method depends significantly on electricity demand, the age and size of the plant, the coal source, and access to resources such as limestone or lime used in the scrubbing process. Early legislation in the 1970s (the Clean Air Act) mandated reductions in sulfur emission without regard to variances in the circumstances being dealt with at the local level. In 1990 amendments were made to the existing Clean Air Act to allow emission trading.

The emission trading scheme set a cap or ceiling on the total amount of SO_2 that could be emitted by the power industry and set up a market whereby plant owners could trade the right to emit a given amount of SO_2. So, for example, a plant operator could recoup some of the money spent on technology for cleaning up that plant's emissions by selling the unused permits to emit SO_2 to another operator. The permits could also be saved for use in the future should the utility need to build new generating plants. The caps and trading were phased in over several years, starting with the largest, most polluting plants.

The cap-and-trade system for the reduction of SO_2 in the United States has been very successful by all accounts. Not only has the amount of sulfur emission been cut by roughly 50%, but the estimated cost of this reduction is far below what was estimated for a purely regulatory method (a full discussion of the economics involved can be found in [26]). The cap-and-trade concept has been implemented in several other countries around the world for other types of pollution. The Kyoto Protocol of 1997 calls for a cap-and-trade approach for greenhouse gas emissions, which, as discussed in Chapter 7, are responsible for global climate change. The United States is the only developed country that has not ratified the Kyoto Protocol.

1.5 ■ Summary

Today, the world is a fundamentally different place than it has ever been in the history of humankind. The population explosion of the last hundred years, coupled with changes in standards of living, has created unprecedented demands on the environment. In particular, there have been radical increases in water and energy use, food production, and pollution. Some have doubted that humans could significantly change the earth's environment, but as we have seen, small changes at the personal level result in global transformations. Reference [14] has many more examples of humanity's impact on the environment over the past 100 years.

Many of the scientific problems facing society today—from finding new energy resources and reducing pollution to increasing food and water supplies to curtailing human-caused climate change—originate in the exponential growth of the world's population. Even as population growth slows, we will continue to confront the complex side effects of our numbers and levels of consumption. However, as the examples of acid rain and upper atmospheric ozone show, when motivated with enough conclusive evidence and economically viable alternatives, people can solve seemingly intractable challenges.

Questions, Problems, and Projects

For all questions,

- Cite all sources you use (e.g., book, article, Wikipedia, Web site). Before using Wikipedia as a source, please read the policies of research and citing in Wikipedia: http://en.wikipedia.org/wiki/Wikipedia:Researching_with_Wikipedia and http://en.wikipedia.org/wiki/Wikipedia:Citing_Wikipedia. For Wikipedia you should verify the information in the article by reading the sources listed.
- For all sources, briefly indicate why you think the source you have used is a reliable source, especially Web resources.
- Write using your own words (except for direct quotes, which should be enclosed in quotation marks and properly cited).

1. Suppose you are investigating two investment strategies. In the first you start with $1, and the value increases at a rate of 10% per month (exponential growth rate of 0.1 per month). In the second scheme you start with $100 and add $100 each month (linear growth). What are the results of the two investment schemes after 1 year? After 2 years? Which investment scheme is better? Explain. (*Note*: This problem assumes dollars that retain the same value over time. However, because it can be invested, money changes value over time. As a result, formulas for compound interest found in accounting books will be different from a simple exponential function. See Section 8.7 for more details.)

2. Using the data in Example 1.1, verify the population growth rate of 0.7% stated in the text for the years 1900 to 1930 and the rate of 1.8% stated for the period 1950 to 2000.

3. Go to the United Nations Population Web site and find the data shown in Figure 1.1. Apply the exponential curve analysis to the individual populations of Asia, Africa, and northern America (the United States and Canada). Perform the least squares fit for the past 100 years. Do your calculated growth rates match those given in Table 1.1 for each of these regions? Explain any discrepancies.

4. According to the United Nations Population Division of the Department of Economic and Social Affairs, given current fertility rates, the world population would reach 244 billion people by 2150 and 134 trillion by 2300. Apply the exponential curve analysis to verify these figures using the fertility rates shown in Table 1.1.

5. Densely populated urban areas have something like 10^4 people per square kilometer (e.g., Mexico City). The total land area of the earth is about 1.5×10^8 km^2. Verify these two numbers using reliable sources on the Internet or in the library or by making an approximation using some simple estimates (show your work). If we use this total land area and the data for 1980 in Table 1.2, we find that the population density was $4.5 \times 10^9/1.5 \times 10^8$ km^2 = 30 people/km^2 in 1980. If the world growth rate remains at 1.3%, when will the population density for the whole earth reach that of a densely populated urban area? (*Hint*: The population density is $\frac{N(t)}{\text{area}} = \frac{N(0)}{\text{area}} e^{rt}$, where $r = 0.013$. You are solving for t when $\frac{N(t)}{\text{area}}$ equals a density of 10^4/km^2. You can pick $\frac{N(0)}{\text{area}}$ to be the density in 1980, in which case t is the years from 1980.)

6. Assume that the average mass of a human is 70 kg. If the world growth rate remains at 1.3%, when will the mass of people living on the earth equal the mass of the earth? (*Hint*: The mass of the population is $mN(t) = mN(0)e^{rt}$, where $r = 0.013$. You are solving for t when $mN(t)$ equals the mass of the earth. You can pick $N(0)$ to be the present population, in which case t will be the time from now when this occurs.)

7. Read Chapter 2, "The Bottleneck," from *The Future of Life* by E. O. Wilson (2002, Alfred A. Knopf). You may be able to find a version of it online. a) List the arguments given by the Economist in support of her

view. b) List the arguments given by the Environmentalist in support of her view. Make a list of the kinds of additional data (facts and figures) you would need to locate in order to support each side. In other words, what else would you like to know about each side before you made a decision as to who was right? c) Write a one-page essay supporting one view or the other. In your argument, say why the other view is wrong and give further reasons why your view is right.

8. Investigate how access to family planning, girls' education, and economic opportunities for women can reduce fertility and slow population growth. Present a specific example if you can find one. (*Hint*: Start with http://www.populationaction.org/Publications/Fact_Sheets/FS7/Summary.shtml).

9. Go to the U.S. Geological Survey Web page called Earthshots (http://earthshots.usgs.gov/tableofcontents). Read the explanation of what the maps represent given in the selection on Garden City, Kansas. Pick images for two other locations and investigate how the landscape has changed over time. Briefly describe your conclusions in a paragraph or two for each site. Be prepared to share your results with the class.

10. Using a reliable source, find a discussion of an example of overdrawing water from an aquifer and the resulting effects (e.g., in the United States the Upper Floridian aquifer, Biscayne aquifer, New Jersey Coastal Plain aquifer, and Ogallala aquifer are all under stress). Write a short summary, starting with a definition of an aquifer, and include specific examples of problems with overdrawing.

11. Using a reliable source, find a discussion of two of the following water problems, write a short summary, and try to find photographs of the affected areas before and after the problem was noted: (a) the Yellow River in China, (b) the Aral Sea, (c) the Three Gorges Dam project, (d) Lake Chad in Africa, (e) Lake Nakuru in Kenya, or (f) some other regional water resource problem of your choice.

12. Using a reliable source, find a discussion of the allegations of recent wrongdoing against one of the two giant water companies, Suez or Vivendi Environment, and write a short summary. Be sure to list your sources and state why you think the source is reliable (the first section of reference [11] may be a good place to start).

13. Using a reliable source, find a discussion of one of the various new water-saving methods of irrigation and write a short summary. Include pros and cons, success stories and failures, and future prospects.
14. Go to the Water Footprint page (http://www.waterfootprint.org) and calculate your water footprint. Summarize your findings and list ways you could reduce your annual water consumption.
15. Using a reliable source, find a discussion of the Asian brown cloud problem and write a short summary.
16. Using a reliable source, find a discussion of the Great Pacific Garbage Patch problem and write a short summary.
17. Find a discussion from a reliable source of the desalination plant at Jebel Ali in Saudi Arabia and write a short summary of how the plant works. Include the energy needed per liter of fresh water produced. Extrapolate that amount to find how much energy would be needed to supply all the water used in a medium-sized city (or choose the city you live in).
18. Membrane reverse osmosis and distillation are two methods of removing salt from saltwater. Using reliable sources, write a short summary of how each works and the advantages and disadvantages of each. Comment on the amount of energy needed for each process. Make a short list of other means of desalination and how they work.
19. Older reverse osmosis plants use about 5 kWh (1 kWh = 3.6×10^6 J) to make 1 m^3 of freshwater from seawater [27]. (a) Find the amount of energy in quads (1.05×10^{18} J) to supply the entire annual freshwater withdrawn by North America (Table 1.3) from seawater by the reverse osmosis process. (b) Newer, more efficient plants use less than 2 kWh to make a cubic meter of water. Find the energy needed to supply all of Asia's freshwater using this process. (c) Comment on your results. How do these amounts of energy compare to other energy needs (see Chapter 4)?
20. Using reliable sources, write a definition and give examples of the concept of virtual water. Reference [27] has a readable discussion of virtual water.
21. Find a discussion from a reliable source of an example of a crop disaster that was the result of overdependence on a monoculture crop (e.g., the

Irish potato famine of the 1840s, the Vietnamese rice failure of 2008, the phylloxera infestation of grapes in Europe) and write a short summary.

22. Using reliable sources, find out how much corn it takes to create the following food products (i.e., the percentage of the following that comes from corn): french fries, Chicken McNuggets, Coca-Cola, a McDonald's milkshake, your favorite salad dressing, frozen pizza, Tang, Cool Whip, Cheez Whiz, toothpaste, disposable diapers, trash bags, and charcoal briquettes. Use the Internet or the label of your favorite cereal, bread, lunchmeat, cookie, or other processed food and see whether you can determine approximately what percentage of the product comes from corn. (Note: The following are or can be made from corn: cornstarch; monoglycerides, triglycerides, and diglycerides; lecithin; vegetable shortening; hydrogenated corn oil; and high-fructose corn syrup).

23. Go to the Union of Concerned Scientists Web site on sustainable agriculture (http://www.ucsusa.org/) and search for information on sustainable agriculture. Summarize the discussion of one of the topics found there. Be prepared to share your findings in class.

24. Find a discussion from a reliable source of a specific case of overfishing and write a short summary. Cod fishing in the North Atlantic, sardine fishing off the coast of California, and salmon fishing near the Alaskan coast are all interesting examples, as is the whaling industry of the last century.

25. Find a discussion from a reliable source of methods being used by various governments to prevent overfishing and write a short summary. The Norwegian management of cod fishing is a particularly interesting example.

26. Pick some product, such as one of those mentioned in the graphs for food in this chapter (Figures 1.4 or 1.5) or the water data graph. You can also look at manufactured goods if they have existed for a few decades (e.g., cars, refrigerators). Graph the growth in the number of these items over time using reliable sources. Using the method demonstrated in Example 1.1, determine whether this product increased exponentially during some period of time. Does your curve have the same growth rate as the growth rate of the population during the same period of time? Give reasons why the growth rates are not exactly the same.

27. Using reliable sources, write a brief report on the amount and type of pollution associated with a large-scale commercial farm operation raising one of the following animals for consumption: (a) pigs, (b) cattle, (c) chickens, or (d) farmed fish. Using the global consumption of these food sources mentioned in the text, estimate the total world pollution produced by these industries.

28. Suppose the population of a town triples and the town decides to triple the size of a water tower to be used in case of emergency. The strength of many common building materials scales as the cross-sectional area; doubling the radius increases the strength by four. However, mass scales as volume, so doubling the radius increases the mass by a factor of eight. Imagine a model for a water tower that has a concrete column with radius r_c and a spherical tank on top with radius r_T. Concrete can support a compression force of about $20 \times 10^6 \frac{\text{N}}{\text{m}^2}$ or 20 MPa (but only about half that if it is stretched).

    a. The original tank has a radius of 10 m and is filled with water of density $1{,}000 \frac{\text{kg}}{\text{m}^2}$. What minimum radius was needed for the concrete support column for this tank?

    b. Suppose the town hires a builder to triple the size of the tank. The builder decides to just triple the radius of both the tank and the support column. Do the calculation (using three times the minimum radius found in part (a) for the column) and state why this is not a good way to proceed.

    c. The strength of bones scales as cross-sectional area, the same way concrete does. Picture a person as a spherical tank (torso) supported by two columns (legs). What has to happen to the thickness of the legs compared to the thickness of the torso if we want to make a person 10 times bigger (10 times the mass)?

    d. Make an argument based on scaling laws against ever seeing terrestrial animals much bigger than the dinosaurs. Do some sample calculations and assume dinosaur bones are about as strong as human bones, which can support a compression stress of $1.5 \times 10^8 \frac{\text{N}}{\text{m}^2}$.

29. For the following items, using reliable sources, find out how many are discarded per year in the United States (or in the world or other countries if you can find these figures) and what proportion are recycled: (a)

cell phones, (b) computers (monitors and the computer itself), (c) automobiles, (d) televisions, (e) furniture (e.g., sofas, beds, chairs, tables). As an alternative, it might be interesting to search for the number of other items that are disposed of each year, such as ballpoint pens, clocks, toys, or Barbie dolls (your choice). Be prepared to share your findings with the class.

30. For one of the following types of pollution, find reliable sources of information (cite your references) and summarize the scientific understanding of the problems listed, examples of the problem, proposed solutions, the feasibility of the solutions, and the progress made to date on the solutions.
    a. Air pollution: acid rain, indoor air quality, particulates, smog.
    b. Water pollution: eutrophication, hypoxia, marine pollution, ocean acidification, oil spills, ship pollution, surface runoff, algal blooms, thermal pollution, wastewater, waterborne diseases, water quality, water stagnation.
    c. Soil contamination: bioremediation, herbicides, pesticides, overfertilization of lawns.
    d. Other types of pollution: invasive species, light pollution, noise pollution, radio spectrum pollution, visual pollution.
31. Using a reliable source, report on the global nitrogen pollution problem. List sources of nitrogen pollution, the effects of nitrogen pollution, and how these have changed over the last 50 years. (*Hint*: A recent article in *Scientific American* discusses this issue.)
32. Using reliable sources, report on the London smog disaster of 1952.
33. Using reliable sources, report on the effects of noise pollution on humans and other animals. In particular, find information regarding the masking of animal signals by noise of human origin.
34. Find the energy (in electron volts) and frequency ranges for the UVA, UVB, and UVC wavelength ranges given in the text.
35. Find a discussion from a reliable source of how ozone in the upper atmosphere is measured and write a summary. (*Hint*: There are ground and satellite measurements.)
36. Find a discussion from a reliable source of the current status of the ozone hole above Antarctica and write a summary.

REFERENCES AND SUGGESTED READING

1. United Nations Population Division (http://www.un.org/popin/functional/population.html).

2. J. E. Cohen, "Human Population Grows Up," *Scientific American* 293(9) (Sept. 2005):48.

3. Population Reference Bureau, *Population Handbook*, 5th ed. (http://www.prb.org/).

4. J. Harte, *Consider a Spherical Cow* (1988, University Science Books, New York).

5. J. E. Cohen, *How Many People Can the Earth Support?* (1995, W. W. Norton, New York).

6. E. O. Wilson, "The Bottleneck," in *The Future of Life* (2002, Alfred A. Knopf, New York).

7. J. M. Diamond, *Collapse: How Societies Choose to Fail or Succeed* (2005, Penguin, New York).

8. A. Y. Hoekstra and A. K. Chapagain, "Water Footprints of Nations: Water Use By Peoples as a Function of Their Consumption Patterns," *Water Resource Management* 21 (2007):35. (http://www.waterfootprint.org/Reports/Hoekstra_and_Chapagain_2007.pdf).

9. Food and Agriculture Organization of the United Nations, Water Report 23, "Review of World Water Resources by Country" (http://www.fao.org/nr/water/aquastat/main/index.stm).

10. P. H. Gleick, "Making Every Drop Count," *Scientific American* 284(2) (Feb. 2001):41.

11. B. McDonald and D. Jehl, eds., *Whose Water Is It?* (2003, National Geographic, Washington, DC).

12. M. Pollan, *The Omnivore's Dilemma* (2006, Penguin, New York).

13. Union of Concerned Scientists, *Sustainable Agriculture* (http://www.ucsusa.org/).

14. J. R. McNeill, *Something New Under the Sun: An Environmental History of the Twentieth-Century World* (2000, W.W. Norton, New York).

15. Food and Agriculture Organization of the United Nations (http://faostat.fao.org/).

16. R. A. Myers and B. Worm, "Rapid Worldwide Depletion of Predatory Fish Populations," *Nature* 423 (May 15, 2003):280–283.

17. Environmental Protection Agency (http://www.epa.gov/epawaste/nonhaz/municipal/msw99.htm).

18. M. F. Ashby, *Materials and the Environment: Eco-Informed Material Choice* (2009, Butterworth-Heinemann, Burlington, MA).

19. H. Mielke, "Lead in the Inner Cities," *American Scientist* 87(1) (Jan.–Feb. 1999):62.

20. J. O. Nriagu, "A History of Global Metal Production," *Science* 272(5259) (1996):223.

21. J. O. Nriagu, "Global Metal Pollution," *Environment* 32(7) (1990):7.

22. D. W. Griffin, C. A. Kellogg, V. H. Garrison, and E. A. Shinn, "The Global Transport of Dust: An Intercontinental River of Dust, Microorganisms and Toxic Chemicals Flows through the Earth's Atmosphere," *American Scientist* 90(3) (2002):228.

23. D. Hafemeister, *Physics of Societal Issues* (2007, Springer, New York).

24. A. Irwin, "An Environmental Fairy Tale," in G. Farmelo, ed., *It Must Be Beautiful: Great Equations of Modern Science* (2002, Granta Press, London).

25. For more information on the ozone hole, see http://antwrp.gsfc.nasa.gov/apod/ap031006.html.

26. J. M. Deutch and R. K. Lester, *Making Technology Work* (2004, Cambridge University Press, Cambridge).

27. V. Smil, "Water News: Bad, Good and Virtual," *American Scientist* 96(5) (Sept.–Oct. 2008):399.

Notes

[1] These are projections made by the United Nations Population Division based on current and historical fertility rates and plausible assumptions about future regional changes in populations. They are merely projections and should not be considered inevitable facts.

[2] The replacement rate could be as high as 2.4 children per woman if mortality of children under 5 years old is high.

[3] Immigration currently accounts for about a third of the increase in the U.S. population each year. The additional population increase is due to population momentum, the rise due to women already born as they reach childbearing age.

[4] HCFCs were initially proposed as a substitute for CFCs because they break down faster, but because they also contain chlorine they do cause some ozone depletion. A different set of compounds, hydrofluorocarbons, have also been proposed as a substitute for CFCs, and they do not cause ozone depletion in the upper atmosphere. Unfortunately, they are implicated in global warming (see Chapter 7).

CHAPTER 2

Efficiency and the First Law of Thermodynamics

Because of the increasing global population and an expanding use of technology, there is a growing demand for energy in the world. But as we learn from the first law of thermodynamics, energy is not something that can be created or destroyed; it is a conserved, finite quantity in the universe. This is only one of several physical limitations that constrain how the world's supply of energy will respond to economic demand. In this chapter we review the first law, methods of energy transfer, and thermodynamic efficiency and give several examples of applications to environmental problems. The reader is assumed to be somewhat familiar with the first law of thermodynamics and the various forms of energy transfer, so only a brief review is given here (an excellent introductory book on the general principles of thermodynamics is [1]). The second law of thermodynamics is mentioned in this chapter, but a full treatment is delayed until the next chapter. Once we have looked at some of the fundamental principles related to the law of conservation of energy, we will look at energy conservation, the least expensive way to provide more energy given the fact that energy supplies are not infinite. The chapter closes with a discussion of economic and social issues related to energy efficiency and conservation.

2.1 ■ The First Law of Thermodynamics

The first law of thermodynamics is equivalent to the law of conservation of energy: Energy cannot be created or destroyed, but it can change forms. In mechanics we are often interested in systems where the total energy remains fixed, but energy can change from one type to another. A typical example is the change from kinetic energy to gravitational potential energy for a projectile moving upward in the earth's gravitational field. A closed system in mechanics is defined to be a system or process in which energy is not added or subtracted but rather only changes form. In contrast, for a thermodynamic process energy may flow into or out of the system, although the total energy is again conserved. A closed thermodynamic system is defined to be a situation in which the physical material making up the system does not increase or decrease, but energy is allowed to enter and leave. In order to apply conservation of energy to thermodynamic processes, the flow of energy into and out of the system must be accounted for, unlike purely mechanical processes, for which we generally choose to define systems so that the total energy remains fixed.

In place of the definitions of energy found in mechanics, it is often more useful in thermodynamics to classify energy into three basic types: work (W), heat (Q), and changes in internal energy (ΔU). The quantities Q and W are positive if energy is added to the system and work is done on the system, and they have the opposite signs for heat leaving the system and work done by the system. There are several units of energy to choose from, but in this chapter we will most often use joules. (See Appendix A for a list of alternative units and conversion factors.) The definitions provided here allow us to write a simple form for the first law of thermodynamics for processes in which only energy flows into or out of the system as

$$\Delta U = Q + W. \tag{2.1}$$

The symbol Q, often referred to colloquially as heat flow, refers to an amount of energy transferred as the result of a temperature difference. We cannot measure a certain amount of heat in an object, so strictly speaking it is incorrect to talk about heat flow from one object to another. However, when there is a temperature difference between two objects we can measure changes in quantities such as temperature, work done, and (in some cases) changes in internal energy in each object. The energy change due only to a temperature difference is Q, in joules, and occurs by one of the three mechanisms of convection, conduction, or radiation. Evaporation also transfers energy to or from an object, but in this case the system is not closed, and a mass transfer is involved that must also be taken into account.

Work is defined in mechanics as $W=\int_{x_i}^{x_f}\vec{F}\cdot d\vec{s}$, where $\vec{F}$ is a force that acts over a distance between some initial location, x_i, and a final location, x_f. In thermodynamics, however, we are often interested in work done by an expanding or contracting gas. For a gas or fluid pushing a piston inside a cylinder, the expansion will generally be in the same direction as the force, so we may omit the dot product and vector notation. Because pressure is defined to be force per area, or $P = F/A$ (in pascal = N/m^2) we can write the thermodynamic work equation as

$$W=\int_{V_i}^{V_f} PdV, \tag{2.2}$$

where we have used the fact that a surface area A, moving through a distance dx sweeps out a volume dV, or $Adx = dV$ (Figure 2.1). We remind the reader that, analogous to the case for mechanical work, the area under a curve in a pressure versus volume (P–V) graph is the work done.

Internal energy (U), is the sum of all the types of energy the individual molecules may have. It includes mechanical energy of the individual molecules such as rotational, vibrational, and kinetic energy at the microscopic level. Internal energy also includes energy absorbed or released by chemical reactions between molecules. As a result, energy changes related to heat capacity, latent heats, and chemical reactions such as burning are included. In an ideal gas the molecules are noninteracting point particles and so do not have rotational, vibrational, or other types of internal energy. The only kind of internal energy an ideal gas can have is the random kinetic energy of its molecules (temperature), and as a result they can never condense into a liquid or solid.

FIGURE 2.1

A piston with area A moving through a distance dx will sweep out a volume dV.

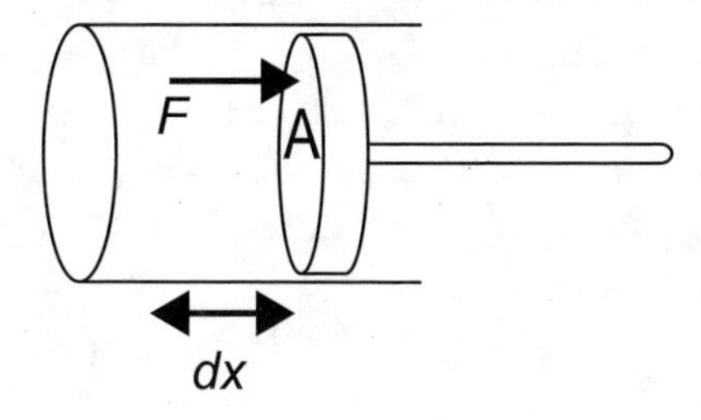

In this chapter and the next we will see several examples of heat engines. A heat engine is any device that changes a heat flow caused by a temperature difference into mechanical work. A gasoline engine is an example of a heat engine; it burns gasoline to form a hot gas, which expands to do mechanical work. A heat engine is an example of a cyclic process, a combination of thermodynamic processes that returns the system to the same internal energy state. In this process work is done, heat flows, but the net change of internal energy at the end of the cycle is zero. Most technological applications of thermodynamics involve cyclic processes. In the remainder of this book, we will generally use the word *process* to mean *cyclic process*. Example 2.1 applies the aforementioned concepts of heat, internal energy, and work to a simple hypothetical heat engine undergoing a cyclic process.

EXAMPLE 2.1

A Simple Heat Engine

Suppose a simple heat engine undergoes the cycle shown in Figure 2.2 using an ideal gas. $P_a = 100$ kPa, $V_a = 1.00$ L. (1 L = 10^{-3} m^3.) At b the pressure and volume are three times as high as at a. During process $a \rightarrow b$ 420 J of heat is absorbed, and in process $b \rightarrow c$ 160 J is expelled. Find the work done in each step, the total work done, the internal energy changes in each step, and the heat transfer during $c \rightarrow a$, assuming no energy is lost to friction or other sources.

The work done in the process $a \rightarrow b$ is the area under the line, which is a combination of a triangular and a rectangular area, so we have $W_{a\rightarrow b} = \frac{1}{2}(200 \text{ kPa} \times 2.00 \text{ L}) + (100 \text{ kPa} \times 2.00 \text{ L}) = 400$ J.

The work done from b to c is zero because the volume does not change.

The work done from c to a is negative because the final volume is smaller than the beginning volume. The area under the line gives $W_{c\rightarrow a} = -100 \text{ kPa} \times 2.00 \text{ L} = -200$ J.

The total work for the cycle is the sum of the work in each step, which is also the work inside the triangle: $W_{\text{cycle}} = 400 \text{ J} - 200 \text{ J} = 200$ J.

Using the first law we have $\Delta U_{a\rightarrow b} = Q_{a\rightarrow b} + W_{a\rightarrow b} = 420 \text{ J} + 400 \text{ J} = 820$ J.

Also using the first law, $\Delta U_{b\rightarrow c} = Q_{b\rightarrow c} + W_{b\rightarrow c} = -160 \text{ J} + 0 = -160$ J.

In a cyclic process using an ideal gas the system returns to exactly the same state, which means the internal energy also returns to the same value. So $\Delta U_{\text{cycle}} = \Delta U_{a\rightarrow b} + \Delta U_{b\rightarrow c} + \Delta U_{c\rightarrow a} = 0$, and we have $0 = 820 \text{ J} - 160 \text{ J} + \Delta U_{c\rightarrow a}$, so $\Delta U_{c\rightarrow a} = 660$ J.

Again using $\Delta U_{\text{cycle}} = 0$ and the first law, we have $0 = W_{\text{cycle}} + Q_{\text{cycle}}$ or $0 = 200 \text{ J} + Q_{a\rightarrow b} + Q_{b\rightarrow c} + Q_{c\rightarrow a}$, from which we have $Q_{c\rightarrow a} = -460$ J.

FIGURE 2.2

The *P–V* diagram for a theoretical heat engine.

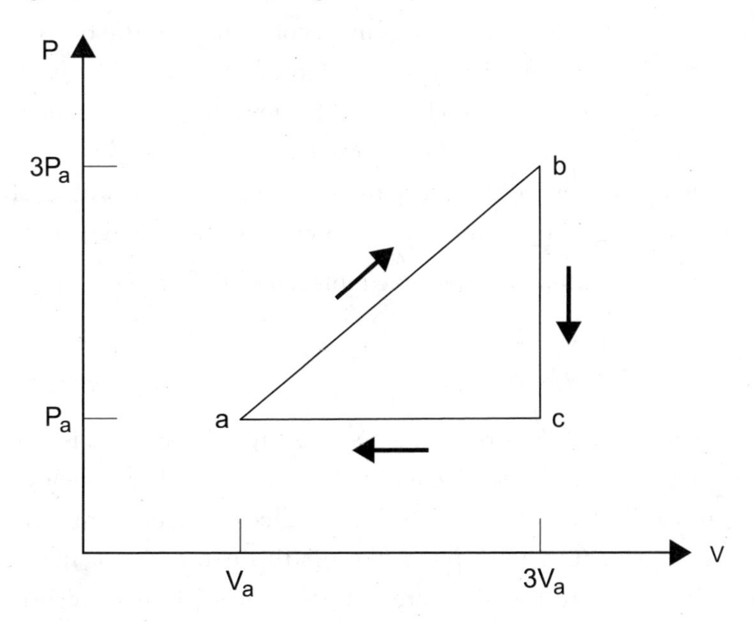

The first law says that the total energy of a closed system would remain fixed if there are no energy exchanges with the surroundings. If we add heat energy to a substance via a temperature difference and do work on it (change the volume), the internal energy will change by an amount equal to the added energy. The principle also works in reverse: Energy can flow out of a substance or it can do work but only in an amount equal to the change of internal energy. Knowledge of this law makes any claims found in the popular media of creating more output energy than input energy highly suspect. For example, proposals seen on the Internet by various inventors claiming to use ordinary water as a fuel for automobiles, assemblies of flywheels or magnets as generators of power, energy from space, or other such schemes are all impossible according to the first law of thermodynamics; you cannot end up with more energy than you started with.

2.2 ■ Some Other Useful Thermodynamic Quantities

In this section we define several other thermodynamic quantities that we will find useful later in this chapter and later in the book. This is intended as a review and is not to be considered a comprehensive treatment. The reader is referred to reference [1] for a more in-depth discussion of these quantities.

2.2.1 Temperature

Temperature scales are defined empirically, with the result that there are several such scales. For scientific purposes it is most convenient to use the Kelvin scale, which defines absolute zero kelvin as the temperature at which in theory the volume of an ideal gas would reach zero as the temperature is decreased. Based on this concept and experimental measurements, we have 0K = –273.15°C. For many purposes it is also useful to keep in mind the classical relation between the macroscopic quantity temperature, T, in kelvin and the average kinetic energy, $\left\langle \frac{1}{2}\mathrm{mv}^2 \right\rangle$, of the individual, microscopic molecules of a sample:

$$(2.3)\quad \frac{3}{2}k_B T = \left\langle \frac{1}{2}\mathrm{mv}^2 \right\rangle,$$

where $k_B = 1.38 \times 10^{-23}$ J/K is Boltzmann's constant. The average is important in this expression because we are dealing with a large number of randomly moving molecules. In fact, temperature can be thought of as a measure of the random kinetic energy present in a sample. Although this expression is not quite right because it predicts zero kinetic energy at zero temperature (which is not true due to quantum mechanics), it will be useful later for understanding evaporation and several other processes. A more sophisticated definition of temperature based on changes in entropy is found in Chapter 3.

2.2.2 Power

Power is the rate at which energy is used,

$$(2.4)\quad P = dW/dt,$$

and is measured in watts = joules/second. The power unit called the horsepower, commonly used for motors in the United States, is approximately 746 W. A kilowatt-hour, used in measuring electrical energy, is actually an energy unit (equal to 3.6×10^6 J) because it has watts multiplied by time.

2.2.3 Specific Heat Capacity

The energy gain by an object due to any of the various transfer mechanisms mentioned earlier can have different effects. Added energy could cause the temperature to increase:

$$(2.5)\quad Q = mC\,\Delta T,$$

where Q is the total energy transfer (in joules); m is the mass of the substance being heated; C is the specific heat capacity, which is a property of the substance; and ΔT is the change in temperature. The specific heat capacity is the heat capacity per unit mass. For materials that change volume at different pressures, such as gases, there are two values, one for constant pressure applications, C_p, the other if the volume is kept constant, C_V. For an ideal gas this is the only possible outcome of adding energy; the added energy increases the average kinetic energy of the molecules because there is no other way to store it. For ideal gases, the only internal energy is kinetic.

2.2.4 Latent Heat of Fusion and Vaporization

If energy added to a substance weakens or breaks bonds between molecules rather than increasing their average kinetic energy, the substance could change phase (e.g., from a solid to a liquid). If the substance changes from a solid to a liquid at the melting temperature, the energy added is

$$(2.6) \quad Q = mL_f,$$

where L_f is the latent heat of fusion for a mass m of that substance. For water the latent heat has the value of 334 kJ/kg. If the added energy breaks the bonds between molecules in a liquid, we have a change of phase from liquid to vapor and

$$(2.7) \quad Q = mL_v,$$

where L_v is the latent heat of vaporization, which is 2,260 kJ/kg for water. The inverse of this process, condensation, releases energy. It is this latent heat of condensation that gives tropical storms their energy; water evaporated from a warm ocean surface releases energy when it condenses into rain.

Generally the temperature of a substance rises as energy is added until it reaches its melting or boiling point. At that temperature any additional energy begins to break bonds, melting or vaporizing the substance, in other words, causing a change of phase while the temperature remains constant. In the case that a solid changes directly to a gas (sublimation), another latent heat value is used, L_s, the latent heat of sublimation.

The conditions under which evaporation will cool an object and how much energy will be transferred due to evaporation given a particular surrounding humidity will be treated in a later section. Here we ask a simpler question:

How much energy is removed if a given quantity of liquid is evaporated? The following example looks at the case of energy loss by sweating.

EXAMPLE 2.2

Sweating

Suppose a jogger loses 1.5 L of sweat per hour. How much energy is needed to evaporate this sweat? What power is needed?

The heat of vaporization of water is 2.26×10^6 J/kg Sweat is mostly water and has a similar heat of vaporization of ~2.4×10^6 J/kg. A liter has a volume of 10^{-3} m^3, and water has a density of 1,000 kg/m^3, so the mass of water evaporated is 1.5 L × 10^{-3} m^3/L × 1,000 kg/m^3 = 1.5 kg. Using Equation (2.7), we have the energy loss as Q = 1.5 kg × 2.4 × 10^6 J/kg = 3.6×10^6 J per hour, or 1,000 W.

2.2.5 Thermal Expansion

Added energy may also cause a substance to expand or, in rare cases, contract. For most substances the expansion coefficient is constant over a reasonable range of temperatures, and we may write

(2.8) $\Delta L = \alpha L \, \Delta T,$

where ΔL is the change in length, L is the original length, ΔT is the change in temperature, and α is the thermal expansion coefficient. A similar equation may be written for area or volume expansion but with different coefficients. If the material is isotropic, meaning that it expands the same amount in all three directions for a given temperature change, the volume expansion coefficient is three times the linear coefficient. These coefficients will generally be positive for normal temperatures and pressures but may be temperature dependent over larger ranges of temperatures and pressures and can occasionally be negative. Water is an important example of a material that has a temperature-dependent coefficient of expansion. Water above 4° Celsius expands as it is heated, but it also expands as it is cooled between 4° and 0° Celsius. When ice forms at 0° Celsius it is less dense than water at 0° Celsius and therefore floats.

2.2.6 Efficiency

For many applications, such as the use of heat engines, we are interested in converting a given amount of energy into useful work using some cyclic process.

In most applications we want to get the most output work for any given input energy. In general, thermodynamic efficiency can be defined as the ratio of the energy benefit to the energy cost [1]. For a heat engine undergoing a cyclic process, the energy benefit is the work done and the cost is the input heat flow, and we have

$$(2.9)\quad \eta = \frac{\text{benefit}}{\text{cost}} = W_{\text{cycle}} / Q_H,$$

where Q_H is the heat flow produced from some hot reservoir (e.g., burning gasoline). It is customary to convert this fraction to a percentage. The first law limits the efficiency of heat engines to 100%; you cannot get more energy out of a cyclic process than input energy (however, see Chapter 3 for a definition of coefficient of performance, which can be larger than 100%). Using Equation (2.9), we calculate the efficiency for the heat engine in Example 2.1 as $W_{\text{cycle}}/Q_H = \frac{200\text{ J}}{420\text{ J}} \times 100\% = 48\%$.

For a connected sequence of cyclic energy conversion processes, the total efficiency is the product of the efficiencies of each of the individual steps. As a result, the total efficiency of a series of energy conversions is limited by the least efficient step. The following example shows how to calculate the total efficiency of a series of conversions.

EXAMPLE 2.3

Efficiency of Combined Processes

The conversion of electricity to thermal energy is very efficient; it is possible to convert 100% of the electrical energy in a circuit into thermal energy. Looking at things purely from an efficiency point of view, should we use electric heat or gas heat in our homes?

The efficiency of converting coal to electricity is about 35% in a conventional generating plant in use today, whereas newer gas turbines can achieve up to 65% efficiency. (This is one reason why power plants being built today are often gas turbines. Natural gas also burns cleaner than coal and in many places is less expensive.) The transmission efficiency of electricity through high-tension power lines is about 90%. So the total efficiency of burning natural gas to make electricity, transmitting it to your house, and then using the electricity as a heating source is, at best,

$$\eta = 0.65 \times 0.90 \times 1.00 = 0.58 \text{ or } 58\%.$$

For a traditional coal power plant with 35% efficiency, the total efficiency drops to only 31%. Most modern home gas furnaces have efficiencies greater than 90%. Burning the gas directly in your furnace is much more efficient than burning natural gas to make electricity and then heating your home with electricity.

However, should be noted that cost also plays a role in making the decision to heat by gas or electricity. In many locations the local power plant can provide a given number of joules of electrical energy for a price much lower than an equivalent number of joules in the form of natural gas (e.g., in a region where coal is cheap and plentiful but natural gas is not). In this case the monetary cost of electric heat is lower, even though the thermodynamic efficiency is lower than that of gas heat.

In the case of heat engines undergoing a cyclic process, the energy not converted into useful work is considered waste energy and is generally given off to the environment as heat. As noted earlier, traditional electrical power plants turn only about a third of the input fuel energy into electricity, discarding two thirds in the form of heat. In other words, for every dollar of fuel purchased in the form of coal or natural gas, we only get 35¢ of electricity from a traditional power plant. If, in some combination of cleverly arranged steps, we could put the waste energy to use, we can then include it as part of the useful work, raising the efficiency for the combined process. In this case the total efficiency of a combination of processes can end up higher than the individual steps, unlike in the example of electric heat, where the processes occurred sequentially.

Plants that supply both electricity and thermal energy from the same source are called combined heat and power (CHP) plants, and the practice of using waste energy from one process for other purposes is called cogeneration. It is estimated that waste heat from electrical power generation could supply as much as 50% of the heating needs in the United States. In fact, as much as 30% of the discarded energy from power plants in the United States was used to heat homes and businesses in the early 1900s [2]. As newer power plants were located as far away from population areas as possible because of esthetic and pollution considerations it became more difficult to use the discarded energy effectively, and the practice was abandoned. Another significant obstacle to wider use of CHP plants today is conflicting regulations, which often are different for heat generation and electrical generation. It is also the case that the typical time scale for replacing existing power plants is 40 years and about 80 years for buildings

[3]. This means that CHPs can only be phased in gradually over the next 40 to 80 years as old factories and buildings are replaced by newer ones. Example 2.4 shows how waste heat can be recycled to achieve higher efficiencies than those of isolated processes.

EXAMPLE 2.4

Cogeneration

The kraft process is a chemical method used in modern paper manufacturing whereby wood is converted into wood pulp (predominantly cellulose), which is used as raw material in paper mills. The process has several stages, but for the purposes of this example we may imagine two separate phases. First, wood chips are cooked for several hours at temperatures of about 180°C in the presence of sodium hydroxide, sodium sulfide, and other chemicals. At the end of the process these compounds can be extracted from the wood pulp for reuse. However, before these chemicals can be reused they have to be further processed at high temperature and pressure.

Suppose the initial cooking stage for a pulp production operation requires 25,000 GJ of thermal energy, and the reprocessing of the catalytic compounds requires an additional 45,000 GJ. If natural gas is used for these processes and the furnaces are 80% efficient, the input energy needed is 87,500 GJ, of which 17,500 GJ (20%) is given off to the environment as waste energy. Suppose an additional 15,000 GJ of electricity is needed to run the plant. If this was generated separately using natural gas and generators that are 50% efficient, we would need 30,000 GJ more energy input (and have 15,000 GJ additional wasted thermal energy). The total energy input for the entire operation would be 117,500 GJ, and the total efficiency is thus $\frac{85{,}000\text{ GJ}}{117{,}500\text{ GJ}} = 72\%$. Can we improve the overall efficiency of the plant?

It turns out that the wasted energy of the kraft process is hot enough to produce steam, which can be used to generate electricity. Suppose 85% of the 17,500 GJ of wasted thermal energy can be recaptured and used in an electric generator that is 50% efficient. This would supply 7,400 GJ of electricity, so we need to generate only 7,600 GJ more from natural gas. Again assuming 50% efficiency, we would now need 15,200 GJ input energy for electricity instead of 30,000 GJ. Our total input energy is now 87,500 GJ + 15,200 GJ = 102,700 GJ, which gives an improved efficiency of $\frac{85{,}000\text{ GJ}}{102{,}700\text{ GJ}} = 82\%$, a 10% gain in efficiency.

It should be emphasized that we have not broken the first law of thermodynamics. All we have done is combine processes in such a way that energy normally given off from one process as wasted thermal energy was put to use for other purposes. In this sense a careful combination of processes can end up with a higher efficiency than the product of each individual process. However, individual cyclic processes can never have efficiencies higher than 100% because of the first law of thermodynamics.

2.2.7 A First Look at the Second Law

As a final observation in this section we should point out that the first law of thermodynamics tells us energy is conserved but does not provide any restrictions about how energy may be converted from one type to another. The second law of thermodynamics restricts the kinds of energy conversions that are possible. For example, it says we cannot convert all the thermal energy directly into work; other changes in internal energy will have to occur at the same time. As we will see in Chapter 3, the second law determines the maximum theoretical efficiencies of energy conversions and transfers; these restrictions are limitations imposed by nature and cannot be thwarted by any new technology or invention.

The efficiencies of some interesting energy conversion processes are shown in Table 2.1. There are two things to notice about these processes. First, any time energy is converted from one form to another, some energy is lost; none of these processes is 100% efficient. As we will see later, at least some of these losses are unavoidable because of the second law of thermodynamics. It should also be noted that some conversion processes are much more efficient than others. For example, electrical conversions (motors and batteries) are in general more efficient than internal combustion processes. Note also that lighting is, in general, an inefficient process, as is overall plant photosynthesis. Plants are very efficient (nearly 90% in some cases) at capturing light but very inefficient at turning this energy into a stored form that would be useful for other purposes [5].

2.3 ■ Types of Energy Transfer

As we have discussed, energy may be transferred from one object to another through a temperature difference. There are four such mechanisms of transferring energy, and it is important to realize that these are the only available means of transporting thermal energy between objects at different temperatures. The

TABLE 2.1

A few conversion processes and their efficiencies [4]. The last column lists the efficiencies of prototypes being developed in laboratories. It should be noted that most sources list lighting efficacy (lumens of light produced per watt of electricity used) rather than efficiency (light energy radiated per electrical energy used) as an attempt to compensate for the fact that our eyes have varying sensitivity to light at different frequencies. Here we list efficiency, the energy converted into electromagnetic waves, regardless of whether they are visible. Additional interesting examples of energy conversions can be found in [4].

Conversion Process	Type of Process	Typical Efficiency (%)	Laboratory or Theoretical Maximum Efficiency (%)
Batteries	Chemical to electrical		
Lithium ion		99	
Nickel cadmium		66	
Lead storage		45	
Electric motors	Electrical to mechanical		
Large (10,000 W)		92	
Small (1,000 W)		80	95
Internal combustion	Chemical to mechanical		
Diesel		50	60
Gasoline		20	37
Steam		25	50
Gas turbine	Chemical to electrical	60	70
Plants	Radiant to chemical		
Light capture		95 (visible only)	
Total efficiency (most plants)		1–2	6–8 (see [5])
Sugar cane		1.2	
Solar electric cells	Radiant to electrical	25	30
Solar thermal	Radiant to thermal	40	
Fuel cells	Chemical to electrical		
Proton exchange		30–50	60
Methanol		25–40	83
Hydroelectric	Gravitational to electrical	75	
Lighting	Electrical or chemical to radiant		
Candle		<1	
Incandescent		3	5
Fluorescent		10–12	15
Light-emitting diode		4–10	22
Sodium discharge		22–27	

first three (conduction, convection, and radiation) do not involve an exchange of substances and so apply to closed systems. Evaporation involves the exchange of material and is an example of the more general process known as mass transfer. We consider simple examples of each mechanism in this section.

2.3.1 Conduction

If two objects are at different temperatures, the molecules of the hotter object will have a higher average kinetic energy. If the two objects are in physical contact with each other, the surface molecules will transfer this energy to the molecules of the cooler object by collisions at the boundary between the two. This kind of energy transfer is called conduction. For applications we generally are interested in energy transfer through a slab of material, such as an insulating wall, the skin, or a fur coat. In this case we may write an empirical equation for the energy transfer per time as

(2.10) $dQ/dt = kA\Delta T/L,$

where dQ/dt is the rate of energy transfer or power (in joules per second or watts) through a slab of material of area A and thickness L. Here ΔT is the temperature difference from one side of the material to the other. The parameter k is the thermal conductivity, a constant that depends on the insulating material and is given in units of watts per kelvin per meter. Some values of thermal conductivity for common substances are given in Table 2.2.

TABLE 2.2

Thermal conductivities of common substances.

Material	Thermal Conductivity, *k* (W/K/m)
Copper	401
Gold	314
Aluminum	237
Iron	79.5
Concrete	0.9–1.3
Glass	0.8
Water	0.6
Wood	0.05–0.36
Fiberglass insulation	0.04
Air (still)	0.02
Styrofoam	0.01

Intuitively, the meaning of Equation (2.10) should be clear; for example, the larger the area a window has and the larger the difference between inside and outside temperatures, the more energy it leaks. Making the glass thicker increases L and so reduces the rate of energy transfer. Using two or three layers of glass with a thermally insulating gas in between decreases the net thermal conductivity, k, while increasing the thickness, L. In the case of infinitesimal temperature differences, we can replace $\Delta T/L$ with dT/dx in Equation (2.10), but for practical applications the empirically derived equation is sufficient.

EXAMPLE 2.5

Thermal Conduction

Equation (2.10) applies to many interesting circumstances. Suppose we are interested in energy transfer through clothing for a human (or perhaps through the fur coat of an animal). On a cold day a person might be wearing layers of clothing 1.5 cm thick. Dry clothing has a k value of about 0.045 W/K/m, but wet clothing has a thermal conductivity closer to 0.65 W/K/m. What is the difference in energy transfer when the person has dry clothes as compared to wet clothing, assuming only energy transfer due to conductivity through the clothes, other factors being equal?

Let us approximate a human as a cylinder h meters tall with a radius r. For a 1.6-m-tall person with a radius of 15 cm this gives a surface area of $2\pi r \times h + 2\pi r^2 = 1.6\ \text{m}^2$. Surface temperature of the skin is about 37°C. On a cold day the outside temperature might be 0°C. If we further assume that the entire body is covered with the same thickness of clothing, the rate of energy transfer through the dry clothes would be $\frac{dQ}{dt} = \frac{kA\Delta T}{L} = \frac{0.045 \times 1.6\ \text{m}^2(37°\text{C} - 0°\text{C})}{0.015\ \text{m}} = 178\ \text{W}$. In the case of wet clothing we get 2,565 W.

In 1 hour of activity the person in dry clothing will need to replace 178 W × 3,600s = 639,360 J of energy or 152,229 calories, which is about 152 food calories, about the amount in a power bar. If the person is wet, the same calculation shows they will have to replace about 2,200 food calories or consume about 15 power bars per hour. Obviously, being stranded in the rain for any length of time is going to take significantly more food resources to stay warm if the person is wearing wet clothing. This could quickly develop into a situation where the person could not metabolize food rapidly enough to stay warm and is the mechanism behind hypothermia.

2.3.2 Convection

For conduction it is generally assumed that the bulk material of the hot and cold objects does not move. If the hot object stays in one place but a cool fluid (liquid or gas) moves past the object, the fluid will first be heated by conduction and then transport the heat away by a process known as convection or advection. Convection can be quite complicated, depending on the fluid transporting the heat, the buoyancy of this material as it changes temperature, the shape and size of the object, its initial temperature, and whether the fluid is forced to move or moves only under the effects of buoyancy and gravity. In most cases a thin boundary layer of fluid remains stationary near the surface of the cooling object so that convection actually involves an initial conduction through an insulating boundary layer. In many cases we may use a simplified equation in which the energy transfer is proportional to the exposed surface area, A, of the object, the temperature difference and the properties of the material. This law is known as Newton's law of cooling and is given by

(2.11) $dQ/dt = hA\,\Delta T.$

Here h, is called the heat transfer coefficient in units of $W/m^2/K$, and is highly dependent on the particular situation. In general h is found empirically, but it can be calculated for a few special cases. An excellent discussion of the various factors involved and some interesting biological cases (e.g., cooling of plant leaves, birds, insects, and other animals) can be found in reference [6].

Natural convection occurs when a fluid near an object is heated and expands, thus becoming more buoyant. The heat transfer coefficient in this case depends on the viscosity of the moving fluid, the thermal expansion coefficient, the size of the object, and the temperature difference between the object and the surrounding fluid. If the fluid moves smoothly past the surface, the convection is said to be laminar flow. The fluid flow may also be turbulent, in which case there are eddies and swirls in the fluid. The difference between these two regimes depends on the shape of the object, the viscosity of the fluid, and the temperature difference.

Two interesting cases of natural convection are air carrying heat away from cylinders and flat surfaces. For laminar flow, which occurs when the temperature difference is not too large and for objects roughly a meter in size, the heat transfer coefficient is approximately given by

(2.12) $h \sim 1.4(\Delta T)^{1/4},$

where ΔT is the temperature difference between the object and the surroundings. Here the units of h will be W/m²/K. For turbulent flow (larger temperature differences, in kelvin, and larger objects) we have

(2.13) $h \sim 1.3(\Delta T)^{1/3}$.

In part the difference in powers in Equations (2.12) and (2.13) (and similar equations later in this chapter for forced convection) have to do with the boundary layer, which acts as an insulating layer clinging to the outside surface of the object. Turbulent flow reduces the thickness of the layer of still air sticking to the surface of the cooling object.

In the case of forced convection, the heat loss will obviously depend on the speed of the air, v. We also expect to find different values of the heat transfer coefficient depending on whether the air movement is laminar (streamline) or turbulent. For slow, laminar air movement over a flat plate we have

(2.14) $h \sim 3.9\sqrt{v}$,

and for turbulent flow over a flat plate, when air speeds are faster, we have

(2.15) $h \sim 5.7v^{0.8}$.

For forced convection of air over a cylinder at speeds high enough to be turbulent, we have

(2.16) $h \sim 4.8v^{0.6}$.

For all these cases the units of h will be W/m²/K if v is in m/s. The equations shown here are approximately the same for vertical and horizontal surfaces and also work reasonably well for heat transfer in the opposite direction from a hot, convecting fluid to a solid object at a lower temperature. If the flowing material is a fluid different from air, the coefficients given earlier will be different [6], but the divisions between turbulent and laminar flow remain.

EXAMPLE 2.6

Convective Energy Transfer

In addition to the energy loss by conduction for a human calculated in Example 2.5, there will also be heat loss by convection. Let us compare the two cases, again assuming a cylindrical person of area 1.6 m². With a temperature difference of 37°C, if there is no wind, using Equation

(2.7) we expect to have $h \sim 1.4\ (\Delta T)^{1/4} = 3.4\ \text{W/m}^2\text{/K}$. The heat loss due to natural convection, then, is $dQ/dT = hA\ \Delta T = 3.4\ \text{W/m}^2\text{/K} \times 1.6\ \text{m}^2$ (37°C – 0°C) = 204.2 W. Comparing this to Example 2.5, we see that the conductive heat flow through wet clothes due to the high thermal conductivity of water is the larger of the two forms of heat loss.

Suppose the wind picks up to 15 m/s (about 30 mph). This would be a case of forced convection, and we have $h \sim 4.8v^{0.6} = 24.4\ \text{W/m}^2\text{/K}$. The heat loss is now 1,442.8 W, substantially more than the rate of heat loss by natural convection or by conduction through dry clothes. This extra heat loss due to forced convection is the basis of the wind chill factor given in weather reports; the rate of heat loss from a human is higher as the air speed increases.

Heat may also flow in a fluid or gas in a slower form of convection where individual molecules of one substance with higher average kinetic energy gradually penetrate a fluid of a different substance with low average kinetic energy. This process is called diffusion and mixes both the type and the kinetic energy of molecules of the two substances. Like evaporation, diffusion is an example of heat transport accompanied by a mass transfer.

It should be noted that the word *convection* is also used in cases of fluid flow where there is no thermal energy transfer. For example, convective flows in the ocean often depend on the salinity of the water rather than temperature differences. Water with a higher salt content is denser and, by Archimedes' principle, will sink when surrounded by less salty water. This may occur in some cases even if the less salty water is cooler. The relative humidity of two parcels of air is another property that may cause convection in the atmosphere without a heat transfer.

2.3.3 Radiation

A complete description of the behavior of the electrons associated with an atom or molecule requires quantum mechanics. From these calculations, verified by empirical results, we find that the energy levels available to electrons in an individual molecule are discrete; an electron trapped in the electrical potential of a nucleus inhabits fixed energy levels. These levels are determined by the number of protons in the nucleus (and other nearby protons and electrons in the case of complex molecules) and so will be different for each different type of molecule or atom. If the electron changes from one energy level to another, it must give off (or absorb) a discrete amount of energy, exactly equal to the difference in energy between the two energy levels. This energy may turn up as thermal (i.e., random molecular kinetic) energy, but often it is given off as an electromagnetic wave with a fixed frequency. The relation between the energy given off (or absorbed) and the frequency of the wave emitted is

$$E = hf = hc/\lambda, \tag{2.17}$$

where E is the energy in joules or electron volts (eV), f is frequency in hertz, c is the speed of light (2.999×10^9 m/s), and h is Planck's constant, 6.63×10^{-34} Js (= 4.14×10^{-15} eVs). Each different atom or molecule has a unique set of energy levels, with the result that when electrons change from various levels in that atom it will emit a unique set of frequencies of light. The unique collection of all available discrete frequencies is called a discrete spectrum and can be used to identify a substance, by examining either the set of frequencies emitted or the unique set that is absorbed when the substance is exposed to a continuous spectrum of electromagnetic waves. Because the energy is emitted (or absorbed) in discrete packets, $E = hf$, which have particle-like properties, they are often called photons instead of electromagnetic waves.

The picture of discrete energy levels for individual molecules can change somewhat when there is a large number of molecules close together in a gas, liquid, or solid. In a gas, collisions with other molecules can cause an atom to emit a photon with a little more or a little less energy, shifting the photon frequency a bit. The result is that, depending on the temperature and density of the gas, the discrete spectrum may be smeared out, and a continuous spectrum of waves is emitted. Free electrons, such as those in a metal sample or ionized gas, may also lose energy by emission of electromagnetic waves in a continuous spectrum.

In a solid each atom contributes energy levels that overlap with those of its neighbor, resulting in a nearly continuous set of energies available to the electron. Because the molecules (and electrons) of any object above absolute zero have a range of possible kinetic energies because of their temperature, we would expect these oscillating electrons to give off a broad spectrum of electromagnetic waves. This does occur, but the calculation is again a quantum mechanical one that takes into account the fact that an electron trapped in a solid still has some energy levels that are forbidden to it. Broadly speaking, for solids, liquids, and gases the thermal motion of their atoms and their associated electrons results in the emission of a broad spectrum of electromagnetic waves called a blackbody spectrum. The loss of this energy in the form of emitted photons will cause the radiating object to cool, even if it is not in direct contact with any other material. In the reverse process an object absorbing photons as a blackbody will gain energy, which may result in an increase in temperature. The earth receives almost all its energy from the sun by means of this mechanism (radioactive decay in the earth's core also supplies heat energy to the earth).

In general, blackbody radiation, unlike a discrete spectrum, is independent of the exact material but does depend on the temperature, as we have noted. The energy transfer due to radiation from any object above absolute zero kelvin is given by the Stefan–Boltzmann equation,

$$\frac{dQ}{dt} = \sigma \varepsilon A T^4, \tag{2.18}$$

where A is the surface area of the radiating object, T is the temperature in kelvin, and σ is the Stefan–Boltzmann constant, $\sigma = 5.669 \times 10^{-8}$ W/m²/K⁴. The energy transfer is also proportional to a unitless constant called emissivity, ε, which provides information about the nature of the surface of the object. Some objects are more efficient at absorbing (or emitting) photons, and emissivity is an empirical measure of this property. Objects that are efficient at emitting radiation are also efficient absorbers of photons. The emissivity indicates the fraction of absorbed (or emitted) radiation from the surface and varies between zero (perfectly reflecting) and one (perfectly absorbing). The term *blackbody* comes from the fact that an object that absorbs all energy (emissivity equal to one) and reflects nothing would appear black.

As defined, Equation (2.18) gives the energy emitted per second (the power emitted, $P(T) = dQ/dt$ by an object at temperature T_o. However, it should be noted that the same equation can also be used to calculate the energy absorbed by an object from surroundings that are at some other temperature, T_s. In many important applications, such as layers of the earth's atmosphere, the object in question is absorbing and emitting at the same time, in which case it is useful to know the net energy transfer from the object, $P(T_o) - P(T_s)$. The following gives a simple illustration.

EXAMPLE 2.7

Radiative Heat Loss

Estimate the power output by means of radiation of a person inside a house that is at a temperature of 20°C. The emissivity for a human is about 0.50. As before, we assume an area of 1.6 m² for the individual.

If we suppose the person has a temperature of 35°C (308 K), then the radiation emitted per second is $\frac{dQ}{dt} = \sigma \varepsilon A T^4 = 410 \text{ W} = P(T_o)$.

However, this calculation does not take into account the fact that the room also radiates energy back to the person. The energy absorbed by the person from a room at 293 K will be $\frac{dQ}{dt} = \sigma \varepsilon A T^4 = 336 \text{ W} = P(T_s)$ so that the net heat flux per second from the person into the room is 410 W – 336 W = 74 W.

Because the radiating molecules of a hot object do not all have the same kinetic energy (recall that temperature is related to the *average* molecular kinetic energy), we cannot expect the photons given off to all have the same frequency. Instead, as mentioned earlier, a broad spectrum of frequencies called the black-body spectrum is given off. Max Planck was the first person to figure out the correct equation for the irradiance, $I(T,\lambda)$, of a blackbody, which is the energy per second per square meter given off at a fixed temperature and a given wave-length. Stated differently, the irradiance is the energy given off at a particular wavelength by a blackbody at a given temperature. The irradiance of an object at temperature T for a particular wavelength λ is

$$(2.19) \quad I(T,\lambda) = \frac{2\pi hc^2}{\lambda^5\{e^{hc/k_B \lambda T} - 1\}}$$

where $k_B = 1.38 \times 10^{-23}$ J/K is Boltzmann's constant. Here $I(T,\lambda)$ is in units of W/ m^2 per wavelength. Equation (2.19) is plotted in Figure 2.3 for four identical objects at four different temperatures. Notice that hotter objects give off much more total energy. Integrating Equation (2.19) over wavelengths from zero to infinity gives the total energy radiated (the power) at a certain temperature, in other words the Stefan–Boltzmann law given in Equation (2.18).

From Figure 2.3 we also see that the peak of the spectrum occurs at different wavelengths for objects with different temperatures. Setting the derivative of Equation (2.19) equal to zero and solving for λ gives the peak wavelength as a function of temperature in Kelvin, known as Wien's law:

$$(2.20) \quad \lambda_{\max} = \frac{2.898 \times 10^{-3}}{T} \text{K} \cdot \text{m}.$$

For very hot objects the peak of the spectrum might fall in the visible part of the spectrum, in which case the object will visibly glow. For example, a tungsten

FIGURE 2.3

Blackbody curves for four identical objects at different temperatures.

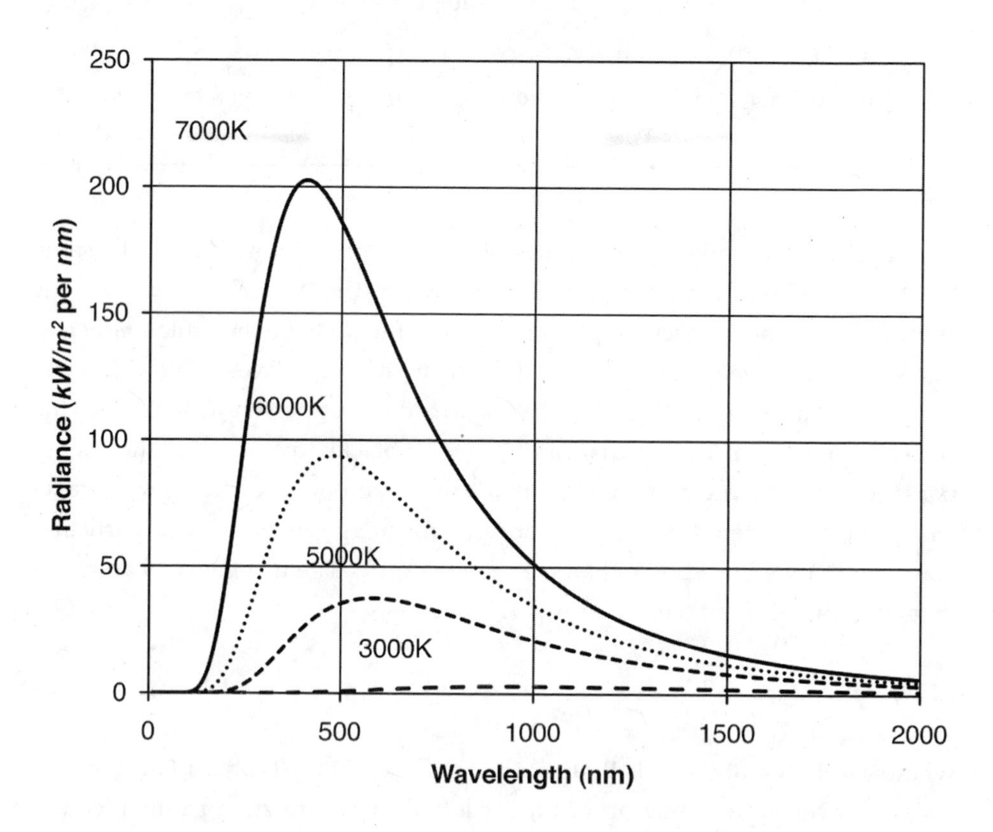

light bulb filament is hot enough to emit some of its blackbody radiation in the visible spectrum. This peak may be used to determine the temperature of objects for which a thermometer is impractical, such as a very hot object such as molten steel or remote objects such as the surface of a star. Cooler objects also give off electromagnetic waves but not in a part of the spectrum we can see with our eyes. This blackbody spectrum can be detected, however, and this is the basis of some kinds of night vision technology. Living creatures are warmer than their surroundings and so emit a different blackbody spectrum, allowing them to be detected against a cooler background, even when there are very few visible photons. As we will see later, blackbody radiation is also part of the mechanism by which various atmospheric gases cause surfaces of Mercury, Mars, Venus, and Earth to be warmer than they would be otherwise.

2.3.4 Evaporation

As mentioned previously, the three means of heat transfer mentioned earlier do not involve the transfer of substances and so are applicable to closed systems. Evaporation is a heat transfer that necessitates the transport of atoms or molecules. This type of process, like diffusion and bulk mass transfer, is important in energy transfers between open systems.

As we saw in Equation (2.3), temperature is proportional to the average of the random molecular kinetic energy; $\frac{3}{2}k_B T = \left\langle \frac{1}{2}\mathrm{mv}^2 \right\rangle$. Suppose a liquid sits in a container with an exposed surface. Near the surface of the liquid the higher-energy molecules have a greater chance to escape the surface tension of the liquid than do the lower-energy molecules. As the liquid loses these higher-energy particles, the average kinetic energy of the remaining molecules has to decrease. But a decrease of average kinetic energy means the temperature has decreased. This process is called evaporative cooling and can occur even when there is no temperature difference between the liquid and its surroundings (although once a temperature difference is generated by this process, energy will begin to transfer back into the liquid by one of the other three means of energy transfer).

The first derivation of Equation (2.3) is credited to Einstein, who was trying to find a theoretical explanation of Brownian motion. Brownian motion is the random motion of small, inanimate objects (e.g., pollen, dust) when viewed under a microscope and was first noticed by biologist Robert Brown in 1827. In 1905 Einstein showed that the motion could be explained statistically by the random bombardment of these small objects by smaller, invisible particles: atoms and molecules. This was considered to be the first direct proof of the existence of molecules and greatly advanced the field of physics known as statistical mechanics. This random jiggling motion is also responsible for diffusion, the slow movement of one set of molecules as they infiltrate into another substance. An example of diffusion is the slow movement of perfume molecules through the air after a bottle of perfume is opened.

Heat loss by evaporation for the case of a person who is sweating can be approximated by an equation similar to those for convection and conduction and depends on vapor pressure. Vapor pressure, which is temperature dependent, is the pressure at which the liquid (or solid) and gas phase of a substance comes into equilibrium. At this pressure (and temperature) equal amounts of the substance are evaporating and condensing, so the system is in equilibrium. For example, in a closed container of water and air at a given temperature and pressure, a fixed amount of water vapor will remain in the air. In the case of sweat evaporation from the skin we have

$$(2.21)\quad \frac{dQ}{dt} = l(P_s - P_o),$$

where l is the evaporative heat transfer coefficient with units of W/Pa, P_s is the water vapor pressure adjacent to the skin, and P_o is the water vapor pressure in the surrounding air. As in the case of convection, the coefficient, l depends on the air velocity and in general is determined empirically. The water vapor pressure close to the skin's surface, P_s, will depend on the temperature of the person's skin and the rate at which the person is sweating. The vapor pressure of the surrounding air depends also on the temperature and on the humidity. From the equation we see that on a hot humid day (high values of P_o) the rate of heat loss due to sweating will be lower. As in the case of the closed container mentioned earlier, impermeable clothing is uncomfortable (and even dangerous) because evaporation (sweating) will cease once the layer of air trapped by the clothing becomes saturated.

2.4 ■ Applications of Energy Transfer

In the engineering world of real objects, the aforementioned concepts are often simplified to a more narrow set of problems where some of the details are known (or assumed to be known) in advance. In these cases, empirical data for standard material (e.g., the *R* factor for building insulation) are given. Here we look at three common applications of the principles of energy transfer discussed previously.

2.4.1 Heat Transmission through a Composite Wall

For a more general case of energy transfer in which there is both conduction and convection we may combine Equations (2.10) and (2.11) into a single equation, ignoring radiation and evaporation for the time being. The most general case would be heat loss between two reservoirs separated by a barrier of thickness L. In this case we may write the empirical equation

$$(2.22)\quad dQ/dt = UA\Delta T,$$

where the constant U, the total heat transfer coefficient per unit thickness, accounts for the possibility of convection on either side of the barrier and conductive energy transfer through the barrier. It should be pointed out that whereas thermal conductivity, k (in W/K/m), is an intrinsic property of the material, independent of the size of the sample and surrounding conditions, the U-factor is not an intrinsic property of the material and is empirically derived under standardized conditions. The units of U are also different from those of conductivity; it is measured in W/K/m² rather than W/K/m. Although U-factors are generally derived by direct measurement, an approximate value for U may be calculated as

$$\frac{1}{U} = \frac{1}{h_C} + \frac{1}{h_H} + \frac{L}{k} + H, \tag{2.23}$$

where h_C is the convection coefficient on the cool side of the boundary, h_H is the convection coefficient on the hot side, k is the thermal conductivity for the material separating the two substances, L is the thickness of the boundary, and H is an empirical constant that depends on the shape and type of surface of the boundary.

In the commercial sector one often finds the insulating properties of a particular material given as the R-value, called the thermal resistivity, typically given in units of ft² °Fh/Btu in. The R-value is the inverse of the U-factor, converted to the appropriate units. Here the British thermal unit (Btu) is a measure of energy equal to 1,055 J. R-values, then, summarize the net energy transfer through 1 inch of a particular insulating material as measured under a standard set of conditions. It should also be noted that these values are only approximate and can change depending on the conditions. For example, brick that has absorbed rainwater has a slightly lower R-value than brick that has dried out in the sun. Table 2.3 gives a list of typical building material conductivities and their R-values per inch. For the case of a composite wall or barrier, adding the R-values may give a sufficiently accurate value for calculating energy transfer, even though gaps between layers and differences in adhered boundary layers of air may change the effective R-value of the combination.

EXAMPLE 2.8

Conduction through a Composite Wall

Suppose the wall of a house is composed of 0.5 inch of drywall, 3.5 inches of fiberglass batting, and 3 inches of brick. A layer of still air clings to both the inside and the outside; this layer provides insulation with an R-value of 0.2 on the outside and 0.7 inside (convection has more of an effect on the boundary layer outside than the air layer inside on average because of wind). The wall is 2.0 m high and 4.0 m wide, and the temperature difference between inside and outside is 20°C. What is the total energy loss through this wall?

First, using values from the table we find the total R-value, which is $0.5 \times 0.9 + 3.5 \times 2 + 3 \times .02 + .02 + 0.7 = 8.95$ where, as noted previously, R has units of ft^2 °F h/Btu in. The total thickness of the wall (allowing 0.5 inch for the air layers) is 7.5 inches, or about 0.18 m. The conversion factor to Km^2/W is 1 ft^2 °F h/Btu = 0.176 Km^2/W, so we have R = 1.575 Km^2/W, which gives a U-factor of 1/1.575 = 0.635 W/K/m^2. Using Equation (2.22), we have a energy transfer of $\frac{dQ}{dt} = UA\Delta T = 0.635 \times 8.0\ \text{m}^2\,(20.0°\text{C}) = 101.6$ W.

In a calculation of energy loss for an entire building several other factors should be included. A totally sealed building with people living inside it would eventually have serious condensation problems, not to mention a buildup of carbon dioxide from respiration. All buildings need ventilation. The number of times the air in a building is replaced per hour is called the exchange rate and is federally mandated in the United States for public buildings. The recommended exchange rate of air for a room is about 1.5 exchanges per hour or, alternatively, about 8 L of fresh air per person per second. Recommended ventilation rates for commercial kitchens (30 exchanges/h), restaurants (10/h), classrooms (4/h), and offices (4/h) are higher. These ventilation requirements result in energy losses of about 15% for a typical building. An additional 25% of its energy is lost through the roof, 10% through windows, 35% through the walls, and 15% through the floors.

Most buildings also have heat sources in them. A human generates about 100 W if seated, up to 140 W if active. Lighting typically provides 20 W to 40 W per square meter, depending on the type of lighting, computers about 150 W, televisions 100 W, and photocopiers up to 800 W. With these sources plus captured solar energy and better insulation it is possible to make buildings that do not need central heating units, even in cold climates (see Problem 2.22).

2.4.2 Newton's Law of Cooling (Again)

In Examples 2.1 and 2.2 we calculated the energy transfer from an object assuming the temperature difference between the two objects remained constant. However, if the hot object cannot maintain a constant temperature (by metabolism in the case of living organisms or a furnace in the case of buildings), then

TABLE 2.3

Approximate *R*-values for various common substances. These values are only approximate because they may depend on how the material is used. For example, *R*-values for double- and triple-pane windows depend on the size of the space between the windows and the gas used to fill that space. A more comprehensive list and links to other tables of values can be found in [6].

Material	*R*-Value in ft^2 °F h/Btu inch
Vacuum	30
Thinsulate (clothing)	5.75
Air (no convection)	5.75
Soil	0.1
Water	0.24
Stainless steel	0.01
Aluminum	0.0006
Copper	0.00036
Polyurethane foam	3.6
Wood panel	2.5
Bale of straw	1.45
Snow	1
Fiberglass, loose	2.2
Fiberglass batting	2
Polyisocyanurate panel with foil	6.8
Cellulose, loose	3
Cardboard	3
Brick	0.2
Poured concrete	0.1
Drywall	0.9
Windows	
Single pane	1
Double pane	2
Triple pane	3
Double pane with coating	3

the temperature will begin to decrease because of the loss of energy. Newton's law of cooling (Equation 2.11) still applies, as do the equations before it, but here we will use them in a different way.

If we imagine very small energy transfers occurring in short periods of time, we may write Equation (2.5) as

$$dQ/dt = mC\,\Delta T/dt \quad \text{or} \quad dQ/dt = mCdT/dt. \tag{2.24}$$

We may now combine Equations (2.22) and (2.24) to get

(2.25) $dQ/dt = UA\,\Delta T = -mCdT/dt,$

where the minus sign indicates a heat loss by the hot substance (Equation (2.22) was originally applied to heat flow through a barrier from hot to cold). Notice that in this equation ΔT is not a constant because now we assume the hot object is going to cool off. Writing temperature as a function of time, $T(t)$, and the surrounding temperature as T_S, we may now write Equation (2.11) as the differential equation

(2.26) $$-\frac{UA(T(t) - T_s)}{mC} = \frac{dT(t)}{dt},$$

a second version of Newton's law of cooling.

The solution to Equation (2.26), which may be verified by direct substitution, is

(2.27) $T(t) = T_s + (T_o - T_s)\, e^{-(\lambda t)},$

where $\lambda = \frac{UA}{mC}$, and the initial temperature of the cooling object is T_o Recall that U, the total heat flow coefficient, is empirically determined so the value $\lambda = \frac{UA}{mC}$ is only approximate and can usually be determined directly from experiments or indirectly calculated from other information, as shown in Example 2.9. From the solution we see that an object will cool exponentially over time.

A more useful version of the solution can be found by writing it down twice, once at time t_1 and again at a later time t_2. If we then divide the two versions, we have

(2.28) $$\frac{T(t_1) - T_S}{T(t_2) - T_S} = e^{-\lambda(t_1 - t_2)}.$$

Taking a natural logarithm of both sides gives

(2.29) $$\ln\left(\frac{T(t_1) - T_S}{T(t_2) - T_S}\right) = -\lambda(t_1 - t_2),$$

which is a straight line on a semi-log plot with slope $\lambda = \frac{UA}{mC}$.

EXAMPLE 2.9

Newton's Law of Cooling

Suppose a corpse is discovered at a murder scene and we want to establish the time of death. We happen to know that room temperature was maintained at 20°C between the time of death and the discovery of the corpse and that the temperature of the corpse at discovery was 32°C. We also know that in the 2 hours from the time of discovery until the corpse is removed from the scene, its temperature drops by another 2°C while the surrounding temperature stays constant. When did the person die?

First, using the temperature change from 32°C to 30°C we can find the constant λ (this is easier than trying to find the overall energy transfer coefficient, U, the mass, m, area, A, and heat capacity, C, of the victim). Using Equation (2.29) with $t_1 - t_2 = -2$ h, we have

$$\ln\left(\frac{32^\circ\text{C} - 20^\circ\text{C}}{30^\circ\text{C} - 20^\circ\text{C}}\right) = -\lambda(-2\text{ h}) \text{ or } \lambda = 0.182/\text{h}. \quad (2.30)$$

Now apply the same equation to the unknown time of death, assuming an initial temperature of 37°C:

$$\ln\left(\frac{37^\circ\text{C} - 20^\circ\text{C}}{32^\circ\text{C} - 20^\circ\text{C}}\right) = -0.182\Delta t, \quad (2.31)$$

in which case the time since death is $\Delta t = -1.9$ h, or 1.9 h before the corpse was discovered.

An application more closely related to environmental issues than Example (2.9) is the rate of cooling of a thermal storage device at night. As we will see in Chapter 6, solar thermal energy can be stored during the day to keep a building warm at night. The rate at which the stored energy is given off can be determined by Newton's law of cooling, Equation (2.27).

2.4.3 Window Coatings

Before the 1970s, about 5% of the total energy consumed in the United States could be attributed to energy transfer through windows, either as energy lost through them during the winter or as additional cooling costs due to unwanted energy gain during the summer [7]. Improved window design has made it possible to save nearly 40% of the heating loss in winter in northern climates and 32% of the cooling loss in summer in southern climates, lowering the energy loss to 3%. Three basic changes in window design have allowed these improvements: windows with multiple glazing (several layers of glass), various coatings on the glass, and changes in the chemical makeup of the glass itself [8].

As mentioned in Section 2.3.1, most conducting surfaces have a thin layer of insulating air attached to them. Because still air (without convection) has a low thermal coefficient, this layer plays a significant role in the insulating properties of a window, accounting for 20% of the insulation of a single pane of glass. Forced convection in the form of wind reduces the effectiveness of the outside layer, but making multi-glazed windows or putting a storm window on the outside allows multiple air layers to be trapped, decreasing the overall thermal conductivity of the window. The advantage of multiple glazing over storm windows is that the gas trapped between layers of glass experiences less convection and can be chosen to be something with a lower thermal conductivity than air. Filling the space between glazing with aerogel, fused nanometer-sized particles of silica, which has very low conductivity, has also been proposed [9].

Two thirds of the energy loss through a double-glazed window is lost as infrared (IR) radiation with wavelengths greater than about 1,000 nm because of the blackbody radiation of objects inside the building. Normal, uncoated glass transmits visible light but absorbs infrared wavelengths from inside and then re-radiates the energy to the outside. Ideally we would like the window for a building in a cool climate to admit visible radiation from the sun but reflect energy in the infrared spectrum back into the building. Initial improvements in window coatings involved simply blocking all frequencies, both visible and IR, which also had the advantage in the case of office buildings of reducing glare and preventing overheating during the summer. Newer coatings are designed to have low emissivity, which in this application means they allow high transmission rates for visible wavelengths but high reflectance for wavelengths in the IR band, reducing the effective U-value by 30%. A further innovation is to use a coating that blocks all IR in warm climates but use a different coating that admits IR between 1,000 nm and 3,000 nm (called near infrared [NIR]) for windows in cooler climates. Admitting NIR allows solar heating during the winter at high latitudes, where the sun is at a much lower angle in the sky.

Changing the properties of the glass itself affects the conduction of heat through a window. Initially this was a permanent change to the makeup of the glass, such as tinting or any of the coatings mentioned earlier. It is now possible to make glass that actively responds to the surroundings. Although the glass that changes tint on exposure to sun, used in some eyeglasses, is currently too expensive for use in windows, it is expected that this technology will eventually be available for construction at a reasonable price. It is also now possible to make electrically controllable (e.g., electrochromic) windows that can be changed from being highly reflective to all wavelengths to being selectively transparent to various ranges of wavelengths [9]. These developments make it possible to change a window's properties on demand or automatically match changes in the environment.

2.5 ■ Lighting Efficacy

We defined thermodynamic efficiency in Section 2.2.6 as the energy benefit divided by the energy cost (Equation (2.9)). The intensity of electromagnetic waves is measured in watts per square meter (W/m^2), so one way to define lighting efficiency would be to divide the total output energy at all frequencies by the energy used to create that output. However, it is usually the case that a significant portion of the light given off by a lamp is not in the visible portion of the spectrum and so is not useful for lighting purposes. In comparing different kinds of commercial lighting it is more practical to use the unit of luminous flux called the lumen to measure output from a lamp. The lumen is a subjective measure of light intensity as seen by the human eye, adjusted for frequency (the human eye does not have the same sensitivity at all wavelengths). Efficiencies of commercial lighting are usually given as luminous efficacy, which is the ratio of lumens per watt or visible light output per input electrical power. Because approximately 20% of the electricity generated in the United States is used for lighting, the topic of lighting efficacy is significant. The efficacies for various types of commercial lighting are listed in Table 2.4; these values should be compared with the efficiencies for lighting given in Table 2.1.

2.5.1 Incandescent Lamps

As we saw in Section 2.3.3, any object above absolute zero kelvin will give off a continuous spectrum of electromagnetic radiation. The peak wavelength of this spectrum is given by Equation (2.20): $\lambda_{max} = 3 \times 10^{-3}/T$ K·m. For the surface of the sun, with a temperature of about 6,000 K, this wavelength is about 500 nm, which is in the center of the visible spectrum. This also is the mechanism for incandescent lamps; a filament is heated using an electric current to a high

TABLE 2.4

Typical efficacies for various types of commercial lighting.

Type of Light	Efficacy (lm/W)
Candle	0.3
Incandescent	5–12
Halogen	16–24
Fluorescent, compact	45–60
Fluorescent, tube	50–100
Xenon arc	30–50
Sodium discharge	150–200
Sulfur discharge[a]	100
Light-emitting diode, white	26–70
Light-emitting diode prototypes	150

[a]In a sulfur discharge lamp the sulfur is excited by microwaves rather than directly by electrical currents.

enough temperature so that it gives off at least part of its light in the visible spectrum. One problem in making an incandescent lamp is that most materials vaporize if raised to a temperature of 6,000 K. The compromise is to use a lower temperature and enclose the filament in a low-pressure inert gas in order to prevent the filament from oxidizing. Tungsten is a metal with the valuable properties of not being very brittle while having a very high melting temperature, and so it is typically used in incandescent applications. Incandescent lamp filaments generally reach temperatures of about 3,000 K, with the result that the peak wavelength is outside the visible spectrum, in the infrared. However, the spectrum given off is broad enough that some radiation emitted is in the visible spectrum, which is the light we see coming from the bulb. This means the majority of the energy given off by the bulb is not useful for illumination but appears instead as heat (in fact, the surface of an incandescent light bulb can reach 300°C). Incandescent lights convert only about 5% of the input energy into visible light, and the rest is given off as heat energy or other frequencies that are not visible to the human eye.

2.5.2 Fluorescent Lamps

Table 2.4 lists several lighting choices that have higher luminous efficacies than incandescent lamps. Halogen lamps achieve a slightly greater efficiency by enclosing the filament in a small amount of a halogen gas such as bromine or iodine, which allows a higher operating temperature. In these bulbs atoms from the tungsten filament do vaporize and are deposited on the inside of the glass bulb, but the halogen atoms return the tungsten to the filament, thus preventing the filament from disintegrating. A way to avoid heating a filament altogether is to use an electric current to heat a gas to a high enough temperature to give off light. Because the gas is already vaporized, it can be heated to higher temperatures than a tungsten filament, thus making them more efficient. Various arc lamps use this method to create light.

Fluorescent lamps are examples of gas discharge lamps, which operate at much lower temperatures than arc lamps. To understand the mechanism, first recall from Section 2.3.3 that in addition to a blackbody spectrum, an element can also give off a discrete spectrum because of its particular electron configuration. In this case only certain wavelengths are given off, rather than a continuous spectrum, as occurs for blackbody radiation. For example, when an electric current passes through neon gas it excites the electrons in such a way that they move to higher energy levels. When they eventually return to the lower energy levels, these electrons give off characteristic wavelengths of light, most of which are in the orange–red part of the spectrum. A neon sign is easily identified from its distinguishing reddish orange color. Likewise, sodium and mercury vapor lamps, commonly used for streetlights, have distinctive yellowish (for sodium) or bluish (for mercury) tints because of their discrete spectrums.

In commercial fluorescent lights a mixture of mercury, argon, and neon is excited by a current so that the mercury atoms give off a discrete spectrum in the ultraviolet range. An uncoated mercury vapor lamp would not emit visible light because ultraviolet cannot be seen by the human eye. However, if the bulb is coated with material that will absorb ultraviolet and re-emit the energy in the visible range, the bulb can be used as a source of light. This process of absorption at high frequencies (low wavelengths) and re-emission at lower frequencies (higher wavelengths) is called fluorescence, a phenomenon that can be seen in so-called black lights, which are in fact ultraviolet lights without a fluorescent coating. Black lights are sometimes used in bars or nightclubs, in which case the ultraviolet light is absorbed by clothing or other material that fluoresces in the visible range, making the material appear to glow with visible light. Depending on exactly which elements are used for a fluorescent lamp coating, the visible spectrum given off can have slightly different shades of white, giving rise to "cool white," "natural white," or other colors available commercially. Because fluorescent lights depend on discrete radiation rather than blackbody radiation, they can operate at lower temperatures and be more efficient. It should be noted that, like any object above zero kelvin, they also have a blackbody spectrum but with a peak wavelength for a much lower temperature, one that is not useful for producing much visible light. For this reason they do have temperatures higher than their surroundings and are not 100% efficient in converting electrical energy into visible light.

2.5.3 Light-Emitting Diodes

Light-emitting diodes (LEDs) offer yet another mechanism for turning electrical current into visible light. Recall from Section 2.3.3 that in a solid the energy levels available to an electron are smeared out into bands of energy levels. The exact structure of these bands and the number of available electrons in them depend on the chemical properties of the atoms making up the material and can be calculated using quantum mechanics. It is often the case that these bands of available energy levels are separated by bands of forbidden energy levels called band gaps. As shown in Figure 2.4, there are four basic arrangements of bands and band gaps for any solid. Electrons may fill and be limited to a lower band, called the valence band, with no chance to change their energy, in which case the solid is an insulator. A partially filled band, known as the conduction band, allows electrons to change energy, in which case the material is a conductor. If a filled lower band overlaps with an empty band (i.e., no band gap) so that plenty of free energy levels are available to the electrons, they can move around in the solid, and the material is also a conductor. A large number of free electrons in a conductor can easily move through the conductor if an electrical potential (a voltage) is applied because there are many available energy levels. In the case where the valence band is filled with electrons but is close enough to the empty conduction band that at least a few electrons can jump across the band gap (e.g., due to thermal kinetic energy), the material is an intrinsic semiconductor.[1]

When an electron moves into the conduction band it leaves behind a vacancy or unfilled energy state in the valence band. These vacancies or holes act as if they are positive charge carriers. An electron that moves into a hole is equivalent to the hole moving in the opposite direction, much like an empty seat in a row of chairs would appear move to the left if the occupants in the surrounding chairs each moved to the right.

New energy levels can also be produced in the band gap for a particular substance by addition of trace elements to the material, a process called doping. If the doping leads to extra filled energy levels in the gap just below the bottom

FIGURE 2.4

Dark shading represents energy levels full of electrons (the valence band), light shading represents energy levels that are empty and available for electrons to move to (the conduction band). For the material represented on the left, the band gap is so large the electrons cannot move from the valence band to the conduction band, so the material is an insulator. If there is a partially filled band, the material is a conductor. The third figure shows a substance with overlapping bands, so the electrons also have empty energy levels to move to, and the material is a conductor. The last drawing shows a semiconductor; some electrons may be able to jump across the band gap if they have sufficient thermal energy because the gap is small.

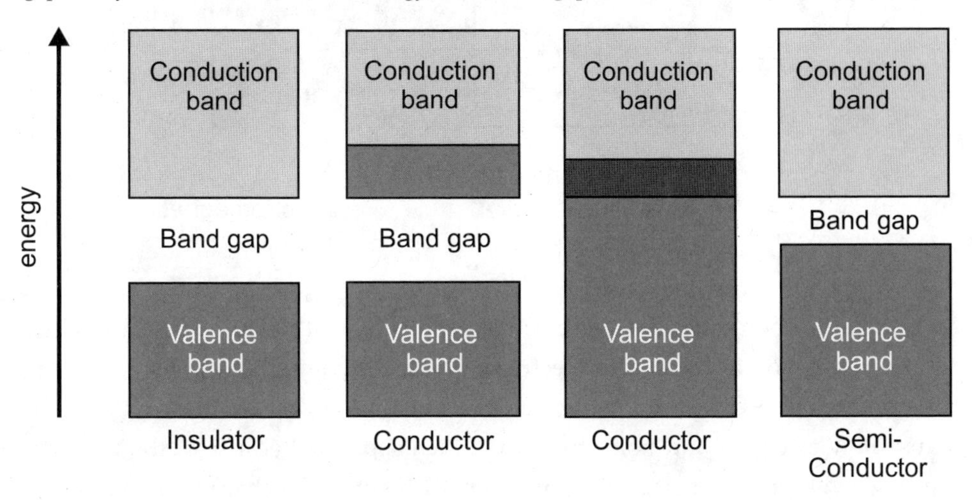

of the conduction band, the material is called an n-type semiconductor (*n* for *negative charge carriers*). For n-type semiconductors, thermal energy is enough to promote the electrons in the new energy levels into the conduction band, where they can act as charge carriers in a current flow. If the doping process ends up creating unfilled energy levels just above the top of the valence band, electrons will move to these energy levels from the valence band, leaving behind vacancies. These holes in the valence band act as positive charge carriers, and the material is said to be a p-type semiconductor.

The basis of the behavior of LEDs, solar cells, and several other electronic devices such as transistors and diodes is the combination of two or more differently doped semiconductors. Most solar cells and LEDs are based on joining of a piece of n-type silicon with a piece of p-type silicon, a combination called a p–n junction. Because of the different available energy levels on either side of the junction, electron flow occurs preferentially in one direction. A simple p–n junction acts as a diode in an electric circuit, allowing current to flow through it in one direction but blocking current trying to flow in the other direction. A schematic diagram for a p–n junction acting as an LED is shown in Figure 2.5. In this

FIGURE 2.5

At the junction between p-type and n-type semiconductors under an applied voltage, V, holes will recombine with electrons resulting in energy given off as light. Conventional current, here labeled I, actually indicates electrons traveling in the opposite direction.

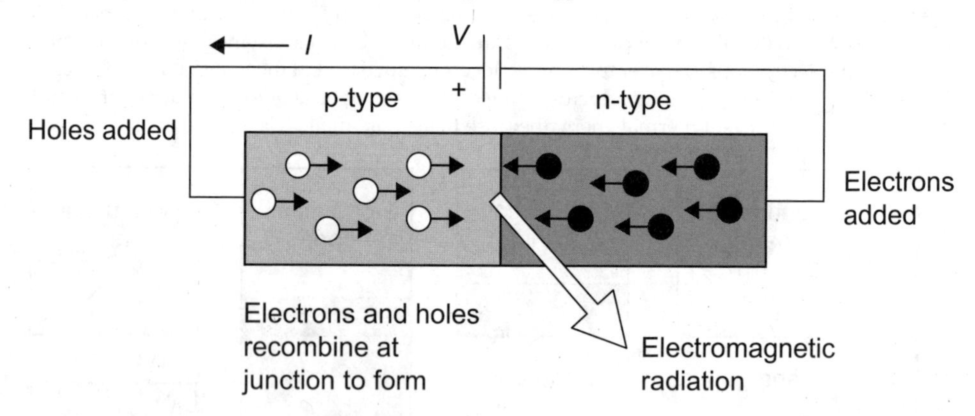

application, when the holes and electrons recombine at the junction, the energy of recombination is emitted as electromagnetic radiation. Solar cells, discussed in more detail in Chapter 5, work in a reverse manner: Incoming solar photons cause new electrons and holes to form, which separate because of the different energy levels on either side of the junction. This supply of charge carriers can be used to power an external electric circuit.

The band gap determines the energy of the photons given off. For a band gap of energy E, the frequency of photon given off is $E = hf = hc/\lambda$, where E is the energy in joules or electron volts (eV), f is frequency in hertz, c is the speed of light, and h is Planck's constant, as in Equation (2.17). This means that normally an LED will produce only a narrow range of colors, with photon energies approximately equal to the band gap energy. Designing a material with a band gap that gives off visible light is difficult because of limitations of the material making up the LED. The first commercial LEDs produced light in the infrared range. Monochromatic LEDs followed this development. More recently, blue LEDs have become available.

By careful modification of the physical arrangement of the material, LEDs can be made into solid state lasers, which are used to read and write CDs. The ability to create DVDs, which hold more information than CDs, had to await the development of blue LED lasers, which have a higher resolution because of their smaller wavelengths. An LED can produce white light in much the same way as fluorescent bulbs, by producing ultraviolet light that is re-emitted in the visible range by various coatings. Layering various colors of LEDs can also be used to produce white light.

Like fluorescent lights, LEDs do not depend on blackbody radiation to give off light and so do not have to operate at high temperatures with associated loss of efficiency. In monochromatic applications where only one color is needed, such as car tail lights or traffic signal lights, LEDs have much higher efficacies than other types of lamps that start with white light and filter out the unwanted colors. White LEDs have recently come on the market that use only a third of the electricity for the same light output as fluorescent tubes. LEDs are also used in some types of flat panel television and computer screens. In these applications, red, blue, and green LEDs are combined in different amounts of each color to produce the other colors and hues needed. In theory these monitors should use less energy than an equivalently sized TV using older technology, although often other factors, such as the amount of electricity used in standby mode and larger screen sizes, can actually make newer technology less efficient.

LEDs built with organic compounds (OLEDs) are currently under investigation. Laboratory prototype versions of white OLEDs have very high efficacies, and it is expected that white LEDs and OLEDs that surpass the efficacies of fluorescent lights will become commercially competitive in the next few years [10]. Greater efficacies, longer lifetimes, resistance to shock, small size, and the fact that they do not contain mercury make LEDs an attractive light source for the future [11].

2.6 ■ Electrical Energy Transfer

Although electrical current flow is not a heat transfer mechanism, it is an important means of moving energy from one place to another, which is a central theme of this chapter. As noted previously, traditional coal-fired electric generation plants are about 30% to 40% efficient. Because of these conversion losses, every 1 kWh (3.6×10^6 J) of electrical energy consumed in the United States requires the use of approximately 3 kWh of energy from primary sources such as coal or natural gas. Of the 1 kWh of electricity generated, more than 7% is lost in transmission from where it is generated to where it is used. In this section we will review a few concepts of basic electricity and then explain why alternating current is used for electrical transmission.

Current, *I*, is the amount of charge passing a point in a wire per second. Charge is measured in coulombs (an electron carries a charge of 1.6×10^{-19} C). An ampere of current is defined to be the flow of 1 coulomb per second and thus constitutes a very large flow of electrons. Electrons are the actual charge carriers in most cases, but electrons were not proven to exist until long after many engineering applications. As a result, the conventional current shown in circuit diagrams flows from the positive terminal of a battery to the negative, opposite the direction the electrons actually travel.

The energy per charge of the charge carriers is measured in volts, where a volt is defined to be equal to a joule per coulomb. An applied voltage gives the electrons potential energy, much like lifting a mass in a gravitational field gives a mass gravitational potential energy. In a conductor the electrical potential causes the electrons to start moving. Positive charges leaving a resistor have a lower potential energy than when they enter, and we say that conventional current will flow from regions of high voltage to low voltage if a path is available, much like a mass falling in a gravitational potential field. The amount of current flow through a resistor depends on the voltage and the resistance and is given by Ohm's law:

(2.32) $V = IR,$

where R is the resistance measured in ohms. Voltage is analogous to pressure in a pipe, and it should be remembered that there can be a voltage (a potential for current flow) where there is no current flow, just as there is water pressure in a pipe even when the tap is closed.

While traveling through a circuit, the current meets with some resistance. Many electrical devices, such as lights, toasters, and other appliances, can be approximated as a series of resistances, measured in ohms. Resistance occurs when electrons interact with the atoms that make up the material through which they are flowing. A full explanation requires quantum mechanics, but a simple picture to have in mind (known as the Drude model) is to think of the electrons as small particles passing through a maze of closely spaced atoms that are oscillating back and forth in place because of their thermal energy. The kinetic energy lost by the electrons as they strike atoms in the conductor becomes random kinetic (or thermal) and electromagnetic energy that causes the light bulb to light, the toaster to heat up, and so on. The same amount of current (the same number of electrons) flows out of a resistor as flows into it, but the charge carriers have less energy when they leave than when they started with because of collisions with the stationary atoms.

From Equation (2.3) we see that the temperature of the wire will rise because of an increase in the average kinetic energy of the atoms in the material as they absorb the kinetic energy of the electrons. Power lost by a resistor due to heat flow out of it is proportional to current squared,

(2.33) $P = RI^2.$

Electric currents are linked to magnetism by Ampere's law and Faraday's law. Ampere's law says current flow will produce magnetic fields. Faraday's law says a changing magnetic flux will cause an electrical potential:

$$\varepsilon = -\frac{d}{dt}\int \vec{B} \cdot d\vec{A}, \tag{2.34}$$

where ε is the electromotive force (emf), measured in volts, and A is some area where there is magnetic field. Here the term *emf* is used in place of *voltage*, mainly by convention; for calculation purposes *emf* and *voltage*, V, are interchangeable. The integral with the dot product counts how much magnetic field penetrates the area and is called the magnetic flux. If the flux changes over time, either by a change in magnetic field, a change in the area, or a change in the angle between the area and the magnetic field, then a voltage (emf) will appear around the edge of the area. It is important to note that something must change over time; static conditions do not produce emfs. If an emf is present, it can be used to do useful work, making current flow via Ohm's law, emf = $V = IR$.

Faraday's law is the basis of many modern technological devices. Credit card readers, magnetic tape players, the read head of a computer disk drive, electric guitar pickups, traffic sensors embedded in the highway, transformers, and electric generators all depend on Faraday's law. In all these cases a changing magnetic field causes a voltage, which then causes a current flow. So, for example, when you swipe a credit card the magnetic field embedded in iron particles on the back of the card passes by a coil in the reader, which responds by generating a current that is read by a computer circuit. The current pulses mirror the coded magnetic signature embedded in the card.

A transformer consists of a block of iron (which intensifies the magnetic field) with two separate loops of wire wound around it. Each of the two loops has a different number of turns, which are electrically insulated from each other. Current enters the transformer through one set of loops, called the primary, and leaves the transformer from the second coil, called the secondary. It is the changing magnetic field produced by the primary that transfers the electrical energy to the secondary. If the current in the primary were constant, there would be a constant magnetic field in the transformer, and no current would flow in the secondary. If a current that alternates in one direction is flowing in the primary, then the magnetic field will fluctuate. This fluctuating magnetic field produces a current flow in the secondary by Faraday's law. The voltage on each side of the transformer is proportional to the number of turns of wire on each side:

$$\frac{V_S}{V_P} = \frac{N_S}{N_P}, \tag{2.35}$$

where V_P and V_S are the voltages across the primary and secondary and N_P and N_S are the number of turns in the primary and secondary, respectively.

The use of transformers in the power grid requires the use of alternating current because transformers do not work with direct current. Alternating current generated at a power plant is stepped up to high voltage by means of a transformer to send the current to the end user. This may appear at first to break the first law of thermodynamics, but as the voltage increases, the current decreases proportionally. Because electrical power in watts is $P = VI$, increasing the voltage while proportionally decreasing the current results in no energy gain. The best case scenario is that all the energy in the primary part of the transformer gets transmitted to the secondary (in the real world there will be some loss of energy because the transformer gives off electromagnetic waves as well as heat). For a perfect transformer we have

$$(2.36)\quad P_P = V_P I_P = V_S I_S = P_S.$$

From Equation (2.33) we see that power loss in a transmission line will be proportional to current squared. It is thus advantageous to use high-voltage, low-current electrical transmission to reduce the energy lost in the wires linking the power plant to the user. In the neighborhood where the current is to be used, the voltage is stepped back down and the current increases by the same proportion. This provides a higher current for use in the home, where the energy can be put to useful work. It should be noted that there are other energy losses in a current-carrying wire, particularly if the current is alternating, but resistance loss is the largest loss.

EXAMPLE 2.10

Electrical Transmission

Estimate the energy savings for burning a 100-W light bulb continuously for a period of 1 week if the energy is transmitted to your house via alternating current at a voltage of 10,000 V as compared with transmitting it at a voltage of 110 V.

House voltage is 110 V, so the current used in the bulb, using $P = VI$, is $I = 0.9$ amp. The energy used by the bulb during the week is 100 W × 7 days × 24 h × 3,600 s = 6.0×10^7 J.

A typical power transmission line might have a resistance of about 0.4 Ω/km. If we assume the power plant is 5 km away, this is a total resistance of 2.0 Ω. If we transmit the current at 110 V, the power loss is $P = RI^2 = 1.6$ W, or for the week the energy loss is 9.8×10^5 J.

If we transmit the current at 10,000 V we first have to use a transformer to step up the voltage, which steps down the current. Using Equation (2.36) we have 10,000 V $\times I_s =$ 110 V $\times$ 0.9 amp, so the current in the transmission wire is now $I_s = 0.0099$ amp and the power loss is now $P = RI^2 = 2.0 \times 10^{-4}$ W. Multiplying by the seconds in the week gives an energy loss of only 118.6 J.

2.7 ■ Energy Efficiency and the Economy

As we will explore further in Chapter 4, the world is facing a growing problem of increased energy demand. The first law of thermodynamics limits the output energy of any process to be equal to or less than the input energy, which puts further constraints on how much energy is actually available. Additional efficiency limitations imposed by the second law of thermodynamics are the topics of the next chapter. Given the constraint of finite supplies of energy, achieving higher efficiencies by cogeneration and reducing unwanted energy flows are examples of ways to reduce energy demand. A well-insulated home using efficient appliances uses a third of the energy of an uninsulated building using older technology. Savings for larger buildings can be substantially more, sometimes enough to circumvent the need for central heating (see Problems 2.22 and 2.25).

Some have argued that government mandates for energy efficiency would have unwanted economic and social consequences. In this section we briefly take up the connection between energy conservation, government policy, economics, and standards of living. Further discussion of the economics of energy is found in Chapter 8.

2.7.1 Corporate Average Fuel Economy

The average refrigerator built in the United States today uses one fourth the energy of one built in 1974 [7]. For the United States this is equivalent to a savings of 40×10^9 W of electrical power that does not have to be generated.[2] This improvement in efficiency and similar improvements for other appliances resulted in part from increases in energy costs and subsequent government policies put into place as a result of the oil crisis of 1973. One such policy was the

Corporate Average Fuel Economy (CAFE) standards for cars and trucks. Figure 2.6 shows the standards and the resulting fleet fuel average. Notice that from 1975 to 1985 the standard was raised from approximately 17 mpg to about 27 mpg for cars. During this same time period the average fleet economy increased to about 28 mpg. From 1985 to 2007 the standards were not changed, and the average gas mileage also changed very little.

One of the first so-called muscle cars made in the United States was the 1949 Oldsmobile Rocket 88, which in one early incarnation had about 175 hp, much higher than the average car on the road at that time. The average light truck (which includes SUVs) today has about 220 hp. In effect, the average car today has the power of a high-end sports car of 50 years ago. Average horsepower declined from 137 hp in the early 1970s to 107 hp in 1981 but has risen to more than 200 hp today. Plotted in Figure 2.6 is also the ratio of vehicle horsepower to weight, which has increased steadily since 1980. It should be apparent from the graph that between 1985 and the present, technological improvements were used primarily for making cars bigger and more powerful rather than more fuel

FIGURE 2.6

Corporate Average Fuel Economy (CAFE) standards and actual fleet average fuel economy for cars and trucks. The ratio of engine horsepower to vehicle weight is also plotted (times 500).

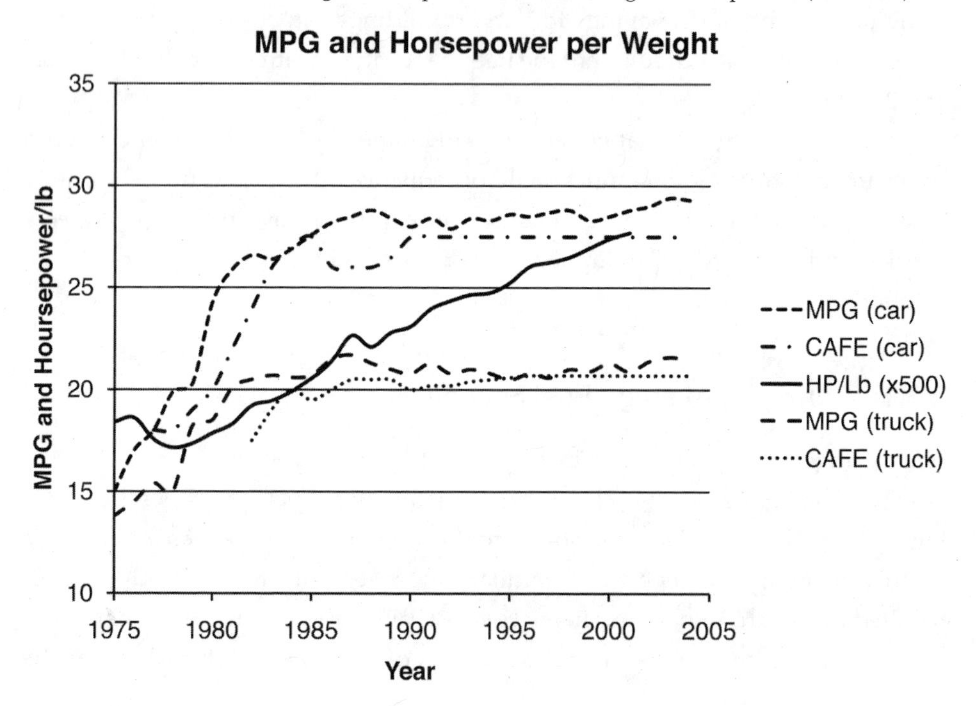

efficient. On average, cars today are about 9% heavier and have 40% more horsepower than in the 1980s. It is estimated that an additional 50% improvement in average gas mileage is possible in the next 10 years if we reverse this trend and apply existing technology for better gas mileage [3]. This trend has been driven primarily by economic demand by consumers for larger, faster vehicles.

2.7.2 Energy Efficiency and the Economy

It is often argued that if the government mandates efficiency, the economy will suffer because the economic burden of better efficiency will slow economic growth. One such argument assumes that economic growth, as measured by the gross domestic product (GDP), is directly proportional to energy use. It is true that energy use and GDP for both the United States and the world have steadily increased over the past 40 years (Figure 2.7). However, other arguments indicate that energy and GDP are not strongly linked. Figure 2.7 shows that the ratio of energy to GDP has been declining for the past 40 years, which indicates that the energy needed for a given increase in GDP is not a constant number. Some

FIGURE 2.7

World energy use (in Quad = 1.055×10^{18} J) versus GDP (in constant year 2000 U.S. dollars times 10^{11}). The ratio of energy use to GDP (times 100) is also shown. Data from the World Bank [12].

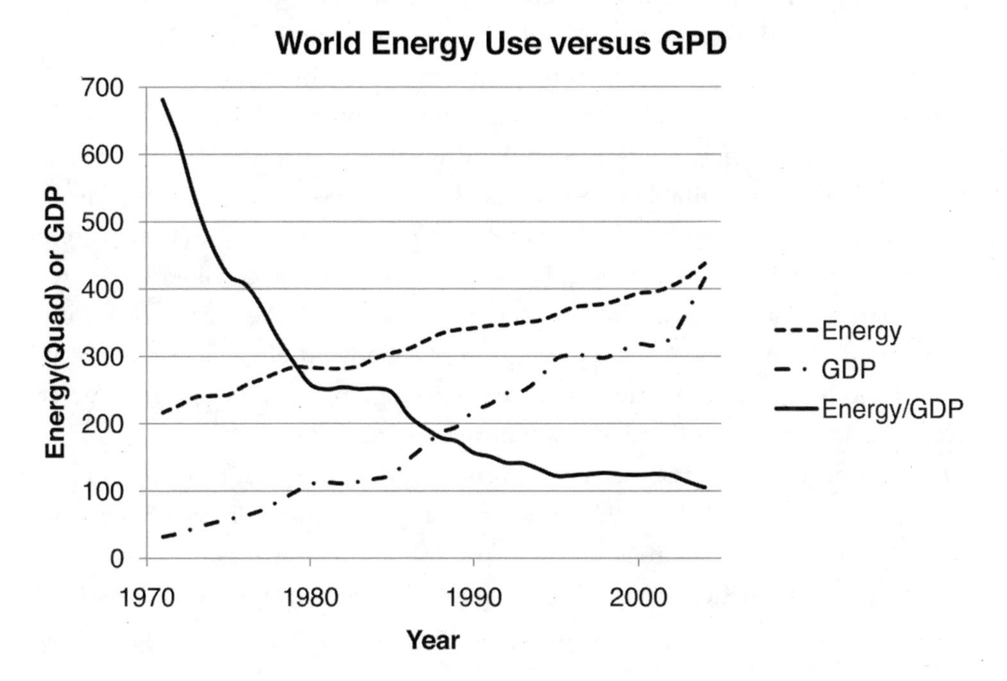

of this trend for the United States can be explained as a slow shift away from heavy manufacturing to service industries, but this explains only a portion of the decreasing ratio of world energy per GDP.

A second argument against regulating efficiency is that the increased burden of higher efficiency, if mandated by the government, will make industries less competitive in the global market. For example, the additional cost of making a more fuel-efficient car might raise the price of the vehicle so that it cannot compete against cars with lower fuel economy that cost less. There are several industry-specific counterexamples to this perception, however. For example, Asian and European car manufacturers produce cars with higher fuel efficiencies (mainly as a reaction to high gasoline and diesel fuel prices in Asia and Europe). Despite the additional manufacturing costs involved in making a car with higher fuel economy, these cars are very competitive in U.S. and other markets. Other examples include refrigerators, which have increased in efficiency by 75% since the early 1970s while dropping in price by 60% and increasing in volume by 20% on average.

Tax incentives, low-interest loans for development, consumer education programs, and mandated efficiencies are possible government interventions, all of which have different effects (see Section 8.7.1). Tailoring the regulation to the industry and the desired effect allows a flexible approach that preserves an industry's competitive edge. So, for example, a carbon tax on the national steel industry designed to reduce fossil fuel use could be coupled with a carbon tariff on steel imported from noncompliant foreign manufacturers, thus leveling the playing field. The degree of government intervention should also be considered. It has been suggested that direct funding is more appropriate to the early development phase of a new energy technology, whereas indirect measures such as tax breaks are more suitable for encouraging industries to introduce new technology into the marketplace [13]. From an economic perspective it seems reasonable to reduce all forms of regulation (aside from safety considerations) for new energy technologies as they reach the point at which they can successfully compete in the market. A slow phasing in of mandated efficiencies over long time periods allows industries to make long-range plans, an important factor considering the 40-year lifetime of a typical power plant.

Barriers to increased efficiencies in construction include the fact that buildings, which consume 39% of the primary energy used in the United States, are not built by the same entities that will use the building, so there is an incentive to cut construction costs at the expense of incorporating energy-saving technologies. Often the occupant of the building does not notice the additional energy expense because it is low compared with mortgage or rent expenses, even though in the long term the energy needs for operating the building are generally 10 times higher than the energy needed to construct the building [14]. Tax incentives and mandated efficiencies are ways to address this problem.

Another way to think about the question of energy conservation and its economic cost is to compare standards of living and energy use, as shown in Figure 2.8. The United Nations created the Human Development Index (HDI) as a way to include not only economic well-being (consumption) but also access to education and longevity, which is related to health [15]. It includes a measure of standard of living (measured as GDP), education (measured as adult literacy and percentage school enrollment), and health (measured as life expectancy). The inclusion of these additional factors is an attempt to circumvent the problem that wealth does not necessarily correlate well with health or self-reported happiness [16]. In Figure 2.8 we see that a per capita energy use of less than 100 million Btu is sufficient to raise the HDI to a level above 85%. It is also obvious from the graph that some countries, such as Norway and the United States, have

FIGURE 2.8

Human Development Index (HDI) versus per capita energy consumption for select countries.

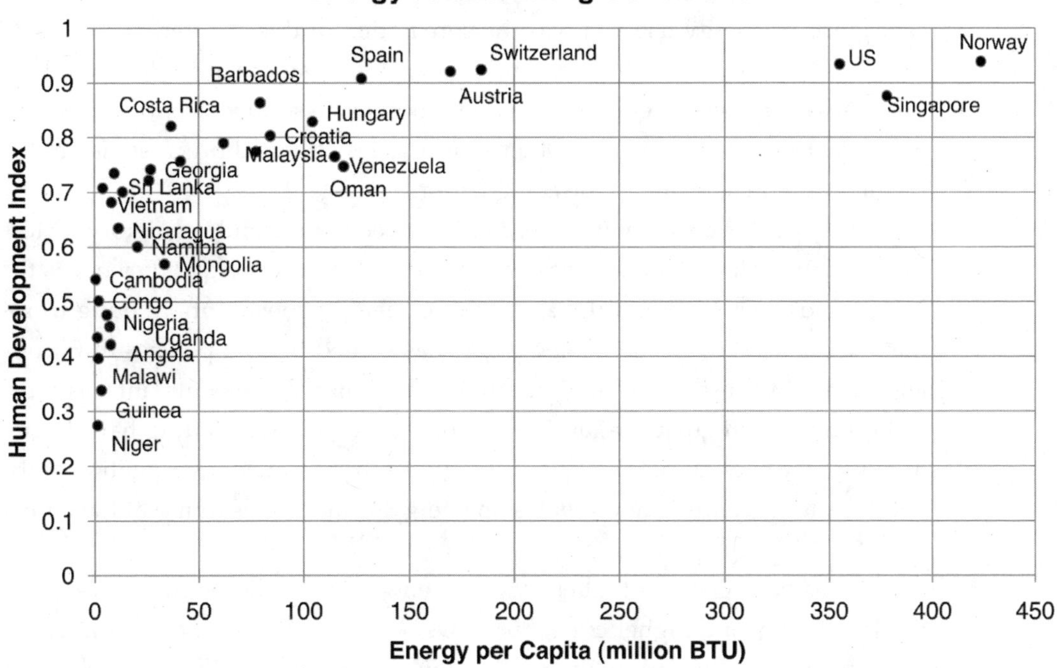

approximately the same HDI as Spain and other countries that use one third the energy per person. One could argue that the United States needs more energy because the transportation distances are longer than in many other parts of the world; however, an almost identical graph can be made by looking only at electricity use per capita [3, 17]. This clearly shows that it is not the longer driving distances in the United States that cause the difference in energy per capita but rather other aspects involving lifestyle. Clearly the quality of life as measured by the HDI does not depend strongly on energy use per capita after some minimum amount, which appears to be much less than that used in the United States.

2.8 ■ Summary

In this chapter we investigated the first law of thermodynamics and energy transfer mechanisms in addition to reviewing some basic concepts of thermodynamics. As we saw, the first law limits energy output to be at best equal to the energy input for that process; in other words, you cannot get something for nothing. The second law of thermodynamics, which in effect says you cannot break even in energy conversion processes, will be examined in more detail in the next chapter. Efficiency, defined as the ratio of the energy benefit to the energy cost, is the most important physical concept when considering practical aspects of energy use and was the core notion of this chapter. The first law prohibits a cyclic process from having efficiencies greater than 100%, but linking processes so that waste energy from one process is used in a second process can improve the overall efficiency. Otherwise, a series of linked processes is limited by the least efficient step in the combination.

The chapter closed with a discussion of economic and regulatory factors involved in achieving greater energy efficiency. Greater energy efficiency is the easiest and most economical way to reduce energy consumption, but government energy policy in the United States has lagged behind policies in Europe and Japan. In those places tax incentives, labeling requirements, international coordination of regulations, and greenhouse gas emission trading have begun to reduce the ever-increasing use of energy. A more detailed investigation of the effects of tax incentives and rebates on energy economics is found at the end of Chapter 8.

It has been claimed that energy efficiency implies discomfort, lower standards of living, and higher costs, but as we have seen this is not necessarily the case; increasing standards of living and economic growth, while decreasing costs for the consumer and using less energy are attainable goals. Achieving these goals involves knowing how energy is transmitted from one place to another and which energy conversion processes are more efficient. Other aspects of these issues are discussed in the following chapters.

Questions, Problems, and Projects

For all questions,

- Cite all sources you use (e.g., book, article, Wikipedia, Web site). Before using Wikipedia as a source, please read the policies of research and citing in Wikipedia: http://en.wikipedia.org/wiki/Wikipedia:Researching_with_Wikipedia and http://en.wikipedia.org/wiki/Wikipedia:Citing_Wikipedia. For Wikipedia you should verify the information in the article by reading the sources listed.
- For all sources, briefly indicate why you think the source you have used is a reliable source, especially Web resources.
- Write using your own words (except for direct quotes, which should be enclosed in quotation marks and properly cited).

1. A heat engine uses 2,400 J of input heat energy while exhausting 400 J of heat to the environment in one complete cycle. How much work is done in one cycle? What is the efficiency of this engine?
2. A typical power plant might output 100 MW. If a human on a stationary bicycle can generate 1 hp, how many humans would it take to replace the power plant?
3. There have been several scams over the years that claim that water can be used to improve fuel economy in cars. The general claim is that water can be broken down into hydrogen (usually using the car's battery), which can be fed back into the engine as fuel.
   a. Batteries are about 90% efficient in converting chemical energy to electricity, generators (e.g., the alternator in your car, which charges the battery) are less than 95% efficient, and the electrolysis process for making hydrogen from water and electricity is 22% efficient. What is the total efficiency of converting energy from the motor (which runs the alternator) into hydrogen? If burning hydrogen in a motor is 40% efficient in converting hydrogen into mechanical

energy, what is the total efficiency of converting energy from the motor into mechanical energy with hydrogen as an intermediary? How does this compare with the efficiency of a typical gasoline motor (20%)?

b. Electric motors can have efficiencies higher than 90% in converting electricity into mechanical energy. Find the total efficiency of converting energy from the motor into mechanical energy with an electric motor as an intermediary (instead of hydrogen). Which process is more efficient: using gasoline directly in the motor (~20%); using gasoline to charge a battery, which is then used to make hydrogen; or using gasoline to charge a battery, which is used to run an electric motor? Use reasonable estimates for unknown parameters.

c. A recent claim about a way to use water to improve gas mileage can be found at http://water4gas.com/2books.htm?hop=pesnetwork. Go to this site, summarize what the site is claiming, and critique the plan using what you know about conversion efficiencies. (If this site no longer exists, Google "water car" and pick one of the other sites to critique.)

4. Using reliable sources, summarize the early history of cogeneration in the United States and Europe. Compare this with present-day efforts at cogeneration.

5. Find a discussion from a reliable source of the wind chill factor. Summarize how the wind chill factor is calculated. Describe how the definition you found relates to the equations found in this chapter on convection. Be prepared to share the results of your investigation with the class.

6. To derive Equation (2.20) we might try to take the derivative of Equation (2.19), set it equal to zero, and solve for λ_{max}. This turns out to give a transcendental equation that cannot be solved analytically. But with a trick we can still get the result. Make the substitution $x = \frac{a}{\lambda}$, where $a = hc/k_B T$, and take the derivative and set it equal to zero. You should be able to reduce the expression to $x = 5(1 - e^{-x})$. Then plot the left and right sides of the equation as separate functions on the same graph. The two curves will intersect at zero and some other value. Set this value equal to $\frac{a}{\lambda}$, and this should give you Equation (2.20).

7. Integrate Equation (2.19) from zero to infinity to get Equation (2.18). You will probably want to consult a table of integrals and make some variable changes to make the calculation easier.
8. Find the wavelength for the peak in the light coming from the sun, assuming the sun has a surface temperature of 5,780 K. What color is this? From your answer, argue why it is advantageous to paint emergency vehicles, tennis balls, and safety vests used by road construction workers bright green.
9. (a) Use Equation (2.18) to find the energy per second (in watts) coming from the sun if the surface temperature is 5,780 K and the emissivity is 1. (b) As this energy spreads out in space it eventually reaches the earth at a distance of $r = 1.5 \times 10^{11}$ m. Divide the area of a sphere ($4\pi r^2$) with this radius into the energy coming from the sun to find the energy per meter squared at the distance of the earth from the sun. (c) Now divide by the area of a circle with the earth's radius, which is the energy arriving at the top of the earth's atmosphere. We will use this value in later chapters.
10. Find a discussion from a reliable source of the sensitivity of the human eye to various frequencies. How does this compare with other animals? Also locate information about the frequencies that easily penetrate the earth's atmosphere. How do these compare with the range of frequencies used by living creatures?
11. Find a discussion from a reliable source of electrochromic windows (you may also be able to find a video demonstration of this material on YouTube). Explain the properties, advantages, and disadvantages.
12. Using reliable sources, find out how eyeglasses that darken when exposed to bright lights work.
13. The mercury found in fluorescent light bulbs is considered a toxic substance, and there is some concern that used light bulbs will contribute to increased mercury levels in the environment. On the other hand, proponents of fluorescent lights point out that more mercury would be put into the atmosphere by burning the extra coal needed to supply electricity to incandescent lights not replaced with fluorescents. Investigate the problem of mercury found in fluorescent lighting, list pros and cons of using fluorescent lights, and report your findings.

14. A 15-W compact fluorescent bulb with a lifetime of 10,000 h produces about the same light as a 60-W incandescent, which lasts, on average, 1,000 h. Suppose electricity costs 10 cents per kilowatt-hour (1 kWh = 3.6×10^6 J).

    a. How much does the electricity for each bulb cost if it is left on for an entire year?

    b. What would be the yearly energy savings (in joules) if 100 million U.S. houses each replaced one incandescent light with a compact fluorescent bulb for 5 h a day?

    c. If a metric ton of coal contains about 8,200 kWh of energy and assuming a power plant that is 35% efficient, how many tons of coal are saved by replacing the incandescent bulbs in part (b)?

    d. Suppose the fluorescent costs $2.80 and the incandescent costs $0.45 per bulb. What is the total cost (bulb plus electricity) for 10,000 h of light in each case (this is sometimes called the life cycle economic cost)?

    e. Using reliable sources, find out how much mercury is emitted when a metric ton of coal is burned. For your answer to part (c), find out how much extra mercury would be emitted when this amount of coal is burned. How does this compare with the amount of mercury in the fluorescent lights?

15. Which of the four energy loss mechanisms is more significant for a human? Back up your answer by first making a list each of the four types of energy transfer and supplying calculations of energy loss for a range of normal situations that a human might encounter.

16. Modern construction often uses insulating material that is covered with shiny aluminum foil. Explain why that is useful, even though the material is covered over with siding or bricks.

17. Verify the conversion factor from $\mathrm{Km^2/W}$ to 1 $\mathrm{ft^2\ {}^\circ Fh/Btu}$ in Example 2.8.

18. Suppose, as in Example 2.8, the walls of a house are composed of 0.5 inch of drywall, 3.5 inches of fiberglass batting, a 1-inch thick wooden panel, and 3 inches of brick. A layer of still air clings to both the inside and outside; outside this provides insulation with an *R*-value of 0.2 and 0.7 inside (convection has more of an effect on the boundary layer

outside than the air layer inside because of wind). One wall is 2.0 m high and 4.0 m wide, and the temperature difference between inside and outside is 20.0°C. Now, however, a single pane window with area 1 m^2 takes up part of the wall. Assume the window has the same air layers clinging to it as the wall does and use *R*-values found in the text.

a. What is the rate of energy loss through the window?

b. What is the rate of energy loss through the window if it is double pane?

c. What is the total rate of energy loss of the wall and window together?

d. What is the energy savings per year of the double-pane window as compared to the single-pane window if the average temperature difference between inside and outside remains at 20.0°C?

19. Suppose 100 million single-pane windows in the United States, identical to the window in the previous problem (1 m^2) are replaced by double-pane windows.

    a. What is the energy savings per year, assuming the same temperature conditions?

    b. If 1,000 ft^3 of natural gas contains about 293 kWh of energy and assuming home furnaces with 90% efficiencies, how many cubic feet of gas are saved by replacing the single-pane windows with double-pane windows?

20. Compare convection, conduction, and radiation energy losses from a typical house. You will have to estimate the size of the house and decide of what materials it is made. (*Hint*: Keep it simple! Think box house here.) Do the convection calculation for laminar flow and for a (turbulent) wind flow of 15 m/s. Be sure to list all your assumptions.

21. In 2009, U.S. secretary of energy Steven Chu suggested that painting all rooftops white in the United States would result in a significant energy savings. Using reliable sources, track down this suggestion and report on how much energy would be saved. Attempt to verify the calculation, making reasonable assumptions about the area of rooftops involved and the savings per building due to lower use of air conditioning.

22. Find information about three of the following super–energy efficient buildings. List the methods applied to reduce energy losses, the physical principle involved, and the percentage savings over traditional

construction. Use reliable sources and be prepared to share your findings with the class. The ABN-AMRO world headquarters, Amsterdam, the Netherlands; the Szencorp Building in Melbourne, Australia; the Genzyme Corporation headquarters, Cambridge, Massachusetts; the Swiss Re Tower, London; the Menara Mesiniaga building, Subang Jaya, Malaysia; the Edificio Malecon, Buenos Aires, Argentina; and the Masdar Development in Abu Dhabi, United Arab Emirates. If you find other examples, bring those as well.

23. Using reliable sources, investigate the Leadership in Energy and Environmental Design (LEED) certification program sponsored by the U.S. Green Building Council. What are the goals of the program, how does it work, and what is involved in the certification process?
24. Go to the MIT Design Advisor(http://designadvisor.mit.edu/design/) and follow the directions to design a building. Report on two scenarios: a cheap scenario in which insulation is minimal and an energy-saving scenario in which more efficient construction materials are used. Compare the two scenarios using the menu items on the left side of the Web page. What conclusions can you draw from your results?
25. Energy loss of a building scales with (i.e., is proportional to) area, as can be seen from Equations (2.10), (2.11), and (2.22). Energy sources in a building, such as people, computers, and electronic gear, scale as volume, however. Suppose, as in our examples, the energy loss through 10-m^2 area of wall on a day when the temperature difference between inside and outside is 20°C is approximately 100 W.
    a. For a building with walls 2.5 m high and floor area 12 m^2 (say, 4 m by 3 m), find the energy loss from assuming the roof loses energy at the same rate as the walls and the floor loses no energy.
    b. Assume an office worker occupies this volume and radiates about 75 W. He uses light that gives off 100 W of energy and a computer that gives off 50 W of energy, so the total input energy in the volume is 225 W. Find the additional energy needed to heat the building assuming the loss from part (a) and the energy input from the worker, computer, and light.
    c. A larger building of area 121 m^2 (11 m by 11 m) can accommodate 10 times the workers if each occupies a floor space of 12 m^2, as did the single worker in the smaller building. For this case, what additional

energy is needed to heat the building? (Careful; the wall area is not 10 times larger!)

d. Discuss the energy savings possible by using larger buildings.

26. Using reliable sources, write a short explanation of aerogels and their use in insulating windows.

27. Using reliable sources, investigate the following lighting sources. In addition to listing various physical properties, be sure to list their efficacies and the predominant mechanism by which they emit light (blackbody, discrete energy levels, bands, or some combination): Compact fluorescent, xenon arc, high-pressure sodium, sulfur lamp, low-pressure sodium, LEDs of various colors.

28. Use Equation (2.3) to find the average kinetic energy of electrons in silicon at room temperature. Convert this energy to electron volts and compare with the energy gap in silicon, which is about 1.1 eV.

29. Repeat the arguments of Example 2.10 for transmitting current at 100,000 V and compare it to 10,000 V.

30. Estimate how much electrical energy your house uses each year (or use figures from the Energy Information Administration). Using the same assumptions as in Example 2.10, estimate how much energy is saved every year by transmitting current at 10,000 V to your house instead of 110 V.

31. From a reliable source (e.g., the Energy Information Administration) find the total residential electrical use in the United States. Use the arguments of Example 2.10 to estimate the annual energy saved in the United States by transmitting electrical power at 10,000 V instead of 110 V.

32. Using reliable sources, summarize the current CAFE standards. (a) There are about 251 million vehicles in the United States, and the average number of miles driven per year is about 15,000 miles. Verify these figures using reliable sources and calculate the amount of gasoline used in 1 year assuming the 2007 CAFE standard of 27.5 mpg. (b) How much gasoline would be saved if all cars could instantly be made to have a gas mileage 50% higher than this standard?

33. It is sometimes argued that mandating better fuel economy in cars would force automobile companies to make lighter cars that are not as

safe as heavier vehicles. Counter this argument using data from reliable sources. Two articles to start with (which you may be able to find online) are

a. Thomas P. Wenzel and Marc Ross, "Vehicles for People and the Planet," *American Scientist* 96(2) (2008):122.

b. Leonard Evans, "Traffic Crashes," *American Scientist* 90(3) (2002): 244.

34. Using reliable sources, write a short summary of the Energy Star program run jointly by the U.S. Department of Energy and the Environmental Protection Administration.

35. Investigate the historical improvements in efficiency for a common appliance (e.g., refrigerators, TVs, computers). How much has the efficiency of these appliances improved in the past 40 years? What changes were made to these appliances to cause these improvements?

36. Find three or four reliable Web sites that give hints for home energy conservation. Make a list of what actions they agree on as providing the greatest savings. Make a second list of things they disagree on.

37. Using reliable sources, either validate or refute the following statement: "The energy savings provided by increased efficiencies of refrigerators and other home appliances has largely been offset by the increase in energy used by home electronics, mostly operating in standby mode." (*Hint*: See reference [14]).

38. Using reliable sources, find out how much energy is used in supporting the Internet and cloud computing (i.e., the servers that run the Internet).

39. Using reliable sources, write an explanation of how the Human Development Index is calculated. See whether you can locate a reliable discussion of similar indices that attempt to evaluate other factors besides just wealth (e.g., the rate of use of natural resources, happiness indices) for the purpose of comparing the relative success of different countries.

40. Investigate the high energy use of Singapore and Norway shown in Figure 2.8. Why do these places use as much energy per capita as the United States?

41. Using reliable sources, download data and create a graph like Figure 2.7 but for the United States instead of the world. (*Hint*: The World Bank

[http://www.worldbank.org/] and the CIA World Factbook [https://www.cia.gov/library/publications/the-world-factbook/] have these statistics.) What differences are there between your graph and Figure 2.7?

42. Go to the World Bank Web site (http://www.worldbank.org/), pick your favorite two countries (not including the United States), download data, and make a graph similar to Figure 2.7 for these countries. What differences are there between your graph and Figure 2.7?

43. Using reliable sources, compare the average horsepower, weight, and gas mileage of cars in the United States with the averages in Europe or Asia.

44. Using reliable sources, find out how gasoline and diesel tax in Europe and Asia compare with these taxes in the United States. Be prepared to share your results with the class.

45. Find a few examples in which it has been a particular businesses advantage to go green—in which serious, real investment in new energy technology has paid off for a commercial entity. Also report on examples in which a company has simply relabeled products as green without making significant changes in operation or products.

REFERENCES AND SUGGESTED READING

1. D. V. Schroeder, *Thermal Physics* (2009, Addison Wesley Longman, New York).

2. T. Casten and P. Schewe, "Getting the Most from Energy," *American Scientist* 97(1) (Jan.–Feb. 2009):26.

3. Special Issue, "The Energy Challenge," *Physics Today* 55(4) (April 2002).

4. V. Smil, *Energies: An Illustrated Guide to the Biosphere and Civilization* (1999, MIT Press, Boston).

5. X. Zhu, S. Long, and D. Ort, U.S. Department of Agriculture, Agricultural Research Services, "What Is the Maximum Efficiency with Which Photosynthesis Can Convert Solar Energy into Biomass?" ScienceDirect, April 2008 (http://www.sciencedirect.com/science?_ob=ArticleURL&_udi=B6VRV-4S9R4K5-2&_user=10&_coverDate=04%2F30%2F2008&_rdoc=1&_fmt=high&_orig=search&_sort=d&_docanchor=&view=c&_searchStrId=1185975643&_rerunOrigin=google&_acct=C000050221&_version=1&_urlVersion=0&_userid=10&md5=9f95bb6940604152bb74e0924e2031e1).

6. J. L. Monteith and M. H. Unsworth, *Principles of Environmental Physics*, 3rd ed. (2008, Elsevier, New York).

7. A. H. Rosenfeld, T. M. Kaarsberg, and J. Romm, "Technologies to Reduce Carbon Dioxide Emissions in the Next Decade," *Physics Today* 53(11) (Nov. 2000):29.

8. Efficient Windows Collaborative (http://www.efficientwindows.org/index.cfm).

9. Lawrence Berkeley Laboratory Environmental Technologies Division (http://windows.lbl.gov/).

10. A. Bergh, G. Craford, A. Duggal, and R. Haitz, "The Promise and Challenge of Solid-State Lighting," *Physics Today* 54(12) (Dec. 2001):42.

11. J. Ouellette, "White LEDs Poised for Global Impact," *Physics Today* 60(12) (Dec. 2007):25.

12. World Bank database (http://www.worldbank.org/).

13. J. M. Deutch and R. K. Lester, *Making Technology Work* (2004, Cambridge University Press, Cambridge).

14. L. R. Glicksman, "Energy Efficiency in the Built Environment," *Physics Today* 61(7) (July 2008):35.

15. United Nations Development Programme, Human Development Indices Update 2008 (http://hdr.undp.org/en/statistics/).

16. C. Hertzman, "Health and Human Society," *American Scientist* 89(6) (Nov.–Dec. 2001):538.

17. D. Hafemeister, *Physics of Societal Issues* (2007, Springer Science, New York).

Notes

[1] The gap for most semiconductors is actually quite a bit larger (more than 1 eV) than the average electron thermal energy at room temperature (about 0.025 eV), but because there are so many electrons the distribution of thermal energies is broad enough that a significant number can span the gap. Only about 1 in 5×10^{12} electrons is promoted to the conduction band in silicon at room temperature, but this is sufficient for significant current flow.

[2] However, the advent of personal computers, standby mode on many electronic devices, and videogame consoles has nearly negated the gains obtained by higher efficiencies for newer models of appliances used in the 1970s [14].

CHAPTER 3

Efficiency and the Second Law of Thermodynamics

In Chapter 2 we saw that there is an important limitation for any process that involves energy transfer or conversion: Energy is conserved. The first law of thermodynamics says that it isn't possible to have a cyclic (or any other) process that ends up with more energy than was present initially. The second law of thermodynamics, which we discuss in this chapter, further limits energy conversion processes to be less than 100% efficient. In particular, any process that involves the conversion of heat into mechanical work is limited by a particularly stringent form of the second law called the Carnot efficiency. The next section introduces several versions of the second law as it applies to heat flow. Subsequent sections look at other energy conversions occurring in devices such as motors, engines, and generators. The discussion is somewhat detailed and theoretical at first, but it is important to understand that the second law is a limit imposed by nature, not a limit of human technological creativity. A more detailed review of the second law can be found in [1].

3.1 ■ The Second Law of Thermodynamics

The first law of thermodynamics tells us energy is conserved but does not provide any restrictions about how energy may be converted from one type to another. The second law of thermodynamics restricts the kinds of energy conversions that are possible. For example, it says that in a cyclic process we cannot only convert heat to work; some other changes have to occur at the same time.

There are several different ways of stating the second law. Each refers to a closed thermodynamic system (we track the energy flowing into and out of the system) and cyclic processes (the internal energy returns to the same value at the end). Although it might not be obvious at first glance, each different statement of the second law can be used to prove the others.

- Version 1 of the second law: Entropy remains constant or increases in a closed system.
- Version 2 of the second law: In a closed system heat flows from hot to cold; it takes input energy to make it flow the other way.
- Version 3 of the second law: The Carnot cycle is the most efficient cycle possible for a reversible heat engine operating in a cyclic process:

$$\eta = \left(1 - \frac{T_C}{T_H}\right) \text{ (Carnot cycle, } T \text{ in kelvin)} \quad (3.1)$$

Each of these statements of the second law and the connections between them will be examined in this section.

3.1.1 Entropy Remains Constant or Increases in a Closed System

Suppose we have four coins and want to know how many different results we could get from tossing them. We assume the coins are fair, meaning there is a 50% chance each will come up as heads or as tails. We also assume we can track which coin is which at all times, or, in thermodynamic terms, they are distinguishable. All 16 of the possible outcomes are shown in Table 3.1, where *H* indicates heads and *T* indicates tails.

From the results we can see that, out of 16 possible outcomes, there is only one way to have all heads but six ways to have two heads and two tails. Our conclusion is that there is a 6 out of 16 chance of getting a result of two heads and two tails but only a 1 out of 16 chance of getting all heads.

What does coin tossing have to do with thermodynamics and the second law? Suppose we have four molecules that can move randomly between two regions of a container. Let's call the left half of the container T and the right half H. Using the same reasoning, we can see that it will be 6 times as likely to find two molecules in the H side and two in the T side than it is to find all four in the

TABLE 3.1

All possible outcomes of tossing four coins.

Coins	Toss 1	Toss 2	Toss 3	Toss 4	Toss 5	Toss 6	Toss 7	Toss 8	Toss 9	Toss 10	Toss 11	Toss 12	Toss 13	Toss 14	Toss 15	Toss 16
1	H	H	H	H	T	H	H	H	T	T	T	H	T	T	T	T
2	H	H	H	T	H	H	T	T	H	H	T	T	H	T	T	T
3	H	H	T	H	H	T	H	T	T	H	H	T	T	H	T	T
4	H	T	H	H	H	T	T	H	H	T	H	T	T	T	H	T
		Three heads				Two heads						One head				

H side. In other words, molecules tend to spread out in roughly equal numbers between the two halves of the container simply because there are more ways for that to happen.

The number of different ways the same event can happen is called the number of microstates or the multiplicity, Ω, which has no units. So for the case of four coins, $\Omega = 1$ for all heads (or all tails); there is only one way to have this occur, so only one microstate is available. For two heads and two tails we have $\Omega = 6$ because there are six distinct ways for this to occur, so the multiplicity is six. We are more likely to see two heads and two tails in a random coin toss because there are more microstates available (the multiplicity is higher) for this to happen.

Entropy is defined to be

$$S = k_B \ln \Omega \,, \qquad (3.2)$$

where k_B is Boltzmann's constant ($k_B = 1.38 \times 10^{-23}$ J/K), Ω is the number of available microstates (the multiplicity), and ln is the natural logarithm. Notice that the units for entropy are joules per kelvin. The entropy for having all heads ($S_{\text{all heads}} = k_B \ln 6 = 2.47 \times 10^{23}$ J/K) is larger than the entropy of having one head ($S_{\text{one head}} = k_B \ln 1 = 0$ J/K). So from an entropy standpoint, final states with higher entropy (two atoms in one half and the two in the other half) are more likely. From this we see that the first statement of the second law, that entropy remains constant or increases in a closed system, is the most likely outcome.[1] High-entropy states are more probable because there are more possible ways for those states to occur. If we continue to toss the coins, we will see the highest-entropy state more often.

But how much more likely is a higher-entropy state? This turns out to depend on the number of objects (atoms or coins). For larger numbers of coins (or atoms) it is convenient to use equations from probability rather than trying to count individual states. The number of possible outcomes for N coins when there are two states (heads and tails) is given by 2^N, and the probability of the number of heads for a given throw is $p_H = N/2$ (the most likely outcome is half heads and half tails). Probability theory also tells us that the number of ways of arranging N molecules (or coins) between two possible states (H and T) with N_H molecules in state H is given by the binomial distribution

$$(3.3) \quad \Omega_{N_H}^{N} = \frac{N!}{N_H!(N-N_H)!},$$

where $x! = x(x-1)(x-2)\ldots 1$ is called the factorial (e.g., $4! = (4)(3)(2)(1) = 24$) For the case of four atoms divided between the two states with two on one side, we have $\Omega_2^4 = 6$, a result we got earlier by simple counting. We also get $\Omega_0^4 = 1$ for no molecules in the H state (here we have used the fact that $0! = 1$). For larger numbers of molecules, counting by hand becomes difficult if not impossible, but Equation (3.3) allows us to calculate the multiplicity.[2] Suppose we want to look at multiplicities for 40 molecules. We have $\Omega_0^{40} = 1$, or one way to put them all in one state, but now we have $\Omega_{20}^{40} = 1.38 \times 10^{11}$ different ways to split the molecules so that half are on one side. The total number of possible outcomes is $2^{40} = 1.10 \times 10^{12}$, so there is a 1 in 1.10×10^{12} chance of finding all 40 molecules in one state. However, there is a $1.38 \times 10^{11}/1.10 \times 10^{12} = 0.125$ or 12.5% chance of finding the molecules equally split between the two states, compared with a $1/1.10 \times 10^{12} = 9.09 \times 10^{-13}$ or 9.09×10^{-11}% chance of finding them all on one side.

Figure 3.1 shows the distribution of multiplicities for two possible outcomes (H or T) for two cases. In Figure 3.1a is the case of 4 coins, and in Figure 3.1b is the case of 40 coins. Notice that as the number of coins increases, the distribution of multiplicities gets very narrow compared with its height. This is another way of saying that the likelihood of not finding the coins equally distributed between the two states becomes very, *very* small for large numbers of coins or molecules.

In probability theory, a measure of the spread of a particular data set is given by the standard deviation, σ. The width of the curves shown in Figure 3.1 at half the maximum value is approximately 2.35 σ (the reader is referred to texts on probability theory for more detail [2]). For a large sample of independent, unbiased measurements of an outcome (a normal distribution), approximately 68% of the values should fall within $\pm\sigma$ of the average (or mean) value. About

FIGURE 3.1

For a small number of objects (4) the distribution of multiplicities is broad **(a)**, but for a larger number of objects (40) the distribution is very narrow **(b)** relative to the height of the curve.

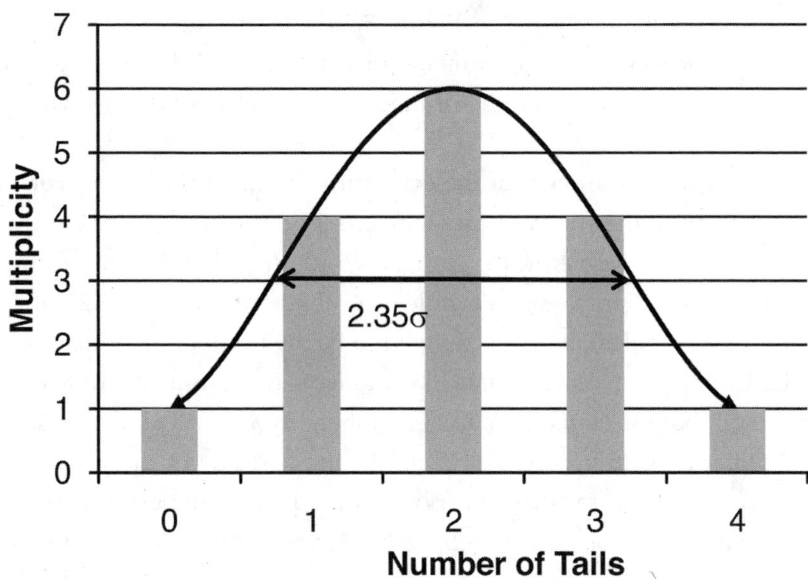

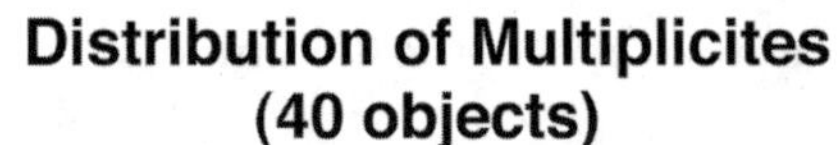

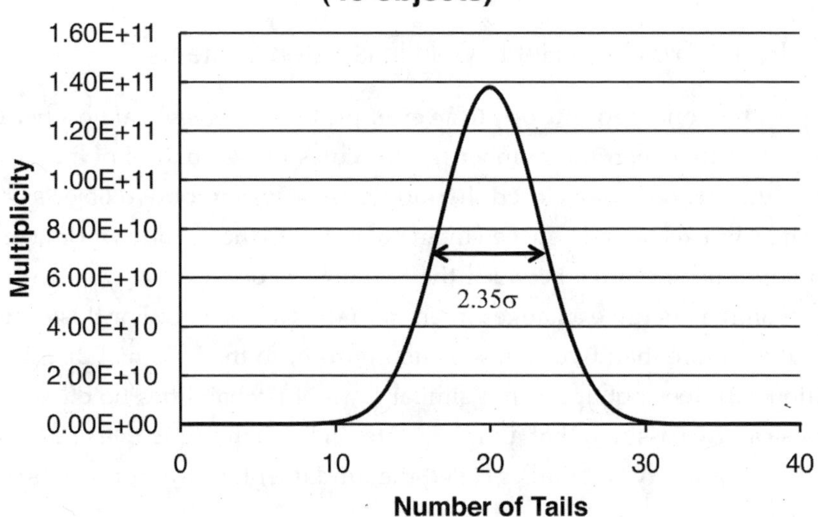

95% of the values will be within ±2 σ of the mean, and 99.7% of the values will be within ±3 σ of the mean. For the binomial distribution (Equation (3.3)), $\sigma = \sqrt{Np(1-p)}$, where p is the probability of the event. In the case of coin tosses, $p = 0.5$ or 50%, assuming the coins are equally likely to come up heads as tails. Notice that for $N = 4$ we have $\sigma = 1$, and the height of the plot is $\Omega^N_{N/2} = 6$, as shown in Figure 3.1a. For $N = 40$ we have $\sigma = 2.24$, and the height of the plot is $\Omega^N_{N/2} = 1.38 \times 10^{11}$, as shown in Figure 3.1b. Clearly the distribution curve becomes very narrow for large numbers of objects, and it becomes extremely unlikely that the objects are not distributed equally between the two states (see Problem 3.4).

Given the large numbers of molecules involved, any time we work with realistic problems in thermodynamics we can expect the system to move quickly toward a state of highest entropy with very little likelihood of returning to a lower-entropy state. For example, a mole of a substance contains 6.02×10^{23} entities (atoms or molecules), which is defined to be the number of atoms in 12 g of carbon-12. Even a gram of a substance would have huge numbers of molecules, with the result that the statistics of large numbers comes into play. This severely limits what happens in nature. For example, the first law of thermodynamics, conservation of energy, says there is no reason why all the perfume molecules escaping from a gram located in an open bottle can't collide in such a way that they exactly reverse their path at some point and migrate back into the bottle. But the second law says that because of the small multiplicity compared with a broad distribution, this is so improbable that it will never occur (or possibly occur once in a time larger than the age of the universe).

3.1.2 Heat Flows from Hot to Cold in Isolated Systems

Let's apply the concept of entropy to an example that involves heat flow between two objects rather than the number of molecules in two halves of a container [1]. Here the meaning of a closed thermodynamic system of two objects is that energy may be exchanged between them but not with the outside world. We will assume that the exchange between the two objects occurs slowly, much more slowly than the energy exchanges internal to the two objects. For this example there will be more than two states to choose from, so the binomial distribution (Equation 3.3) does not apply, but similar laws of probabilities hold. As in the previous case we assume that all microstates are equally probable and equally accessible, a property sometimes called the fundamental assumption of statistical mechanics.

For real solids the energies available to the atoms are quantized; in other words, only certain vibrational energies are allowed, just as only certain energy levels are allowed for electrons bound to an atom. To make things simple, suppose we have a solid with identical quantized energy levels, so you can add (or subtract) exactly one unit of energy or two or three units to an atom but never, say, 1.5 units of energy (this simplified model is called an Einstein solid). Let's start with a solid that has only three atoms ($N = 3$) and count the number of microstates, Ω, available for a given quantity of added energy to this solid. If no energy is added, there is only one available microstate; all three atoms have zero energy, and the multiplicity is one. For one unit of added energy the multiplicity is three; we can give the one unit to one atom, and there are three possible atoms that might have it. For two units of energy added there are six microstates. Table 3.2 summarizes these possibilities.

The general formula for finding the multiplicity for a given number of oscillators, N with a given number of energy units, q, is given by probability theory and is

$$\Omega(N,q) = \frac{(q+N-1)!}{q!(N-1)!}. \tag{3.4}$$

From this formula we see that $\Omega(3,0) = 1$, $\Omega(3,1) = 3$, $\Omega(3,2) = 6$, as in Table 3.2, $\Omega(3,3) = 10$, and so on.

Now let's look at what happens when we have two Einstein solids that have different temperatures [1]. Let's call the solids A and B and suppose there

TABLE 3.2

Multiplicity chart for distributing energy units among three atoms in an Einstein solid.

Energy and Multiplicity	Atom 1	Atom 2	Atom 3
No energy, $\Omega = 1$	0	0	0
One energy unit, $\Omega = 3$	1	0	0
	0	1	0
	0	0	1
Two energy units, $\Omega = 6$	2	0	0
	0	2	0
	0	0	2
	1	1	0
	1	0	1
	0	1	1

are N_A atoms in solid A with q_A units of energy and N_B atoms in solid B with q_B units of energy. As a simple example, let's suppose each solid has only three atoms, or $N_A = N_B = 3$, and there are six energy units to be divided between them: $q_A + q_B = 6$. Initially the solids will be isolated; for now we are just considering the possible ways to split the energy. If A gets no energy, B has all six units; if A gets one energy unit, B gets five, and so forth. Using Equation (3.4) for multiplicity, we can fill in Table 3.3 for the microstates of each solid. For example, in the second row we see that solid A has no energy and only one microstate but solid B has all six energy units and 28 different ways (microstates) to spread those six units among the three atoms in solid B.

Now put the two solids in thermal contact with each other and assume the energy will move around between them, with equal probability for each microstate. The total number of microstates for the two solids in contact is the product of the microstates for each solid. For the seven cases in Table 3.3 we see that the combined number of microstates ($\Omega_{\text{total}} = \Omega_A \times \Omega_B$) for the two solids in contact are $1 \times 28 = 28$, $3 \times 21 = 63$, $6 \times 15 = 90$, $10 \times 10 = 100$, $15 \times 6 = 90$, $21 \times 3 = 63$, and $28 \times 1 = 28$, respectively. The multiplicity for having half the energy in each solid is much larger ($\Omega_{\text{total}} = 100$) than having all the energy in one solid ($\Omega_{\text{total}} = 28$). In other words, assuming the energy will move randomly between the two solids, it is a lot more likely that energy will tend to equalize between them because there are more microstates available for this outcome. Because entropy is related to the number of microstates (Equation (3.2)), we may also say that the entropy is higher if the energy is distributed equally between the two solids.

It should be noted that we could have used the energy that flows from the hot to the cool object to do useful work (and in fact this is how heat engines work, as we will see later). However, once equilibrium has been reached the

TABLE 3.3

Multiplicity of two Einstein solids sharing six units of energy.

Energy Units for q_A	Ω_A	Energy Units for q_B	Ω_B	$\Omega_A \times \Omega_B$
0	1	6	28	28
1	3	5	21	63
2	6	4	15	90
3	10	3	10	100
4	15	2	6	90
5	21	1	3	63
6	28	0	1	28

energy is no longer available to do work unless a third object at even lower temperature is available. It is in this sense that an increase in entropy represents a change from useful energy to a random or degraded form of energy.

For our Einstein solid the only available energy states are vibrational, so two samples with the same average energy per molecule will also have the same average kinetic energy per molecule. From Equation (2.3) we see that two objects having molecules with the same average kinetic energy also have the same temperature. So equalization in energy also means equalization in temperature (at least for Einstein solids).[3] From this discussion we conclude that energy moving from hot objects to cold objects is more probable than other redistributions of energy. We have thus shown that the first two statements of the second law given at the beginning of this chapter are equivalent: Heat flow from a hot object to a cool object increases the overall entropy of the combined system. This is also the most probable outcome. Although we have used a specific example (two Einstein solids in thermal contact), it is true for any two objects (solids, liquids, or gas) that can exchange thermal energy with each other by any of the heat transfer mechanisms discussed in Chapter 2 but are otherwise isolated from their surroundings.

These concepts can be extended to larger numbers of atoms in each solid and to the case in which the number of atoms is different in each solid. When we do this we see something similar to the example of coin tosses for very large numbers of coins. Equal division of energy between two solids with equal number of atoms becomes extremely likely compared with the probability of all or even most of the energy in only one solid. For example, for two solids with 300 atoms each and 100 units of energy split between them, the calculation described earlier gives the number of microstates with all the energy in one solid as 1.7×10^{96}, a large number, but the number of microstates for an equal distribution of energy is 1.3×10^{122}, which is 10^{26} times larger.[4] For practical applications the mechanisms of heat flow discussed in Chapter 2 tell us under what circumstances and how quickly this redistribution of energy will occur. The laws of probability for large numbers of objects tell us it is extremely likely for the system to move toward equally distributed energy, compared with some other configuration of energy sharing. Thus, the second law of thermodynamics is a statement of probability for large numbers of atoms.

Although there are examples in which an association between entropy and disorder does not hold, often this connection is a useful way to think about entropy. In our example of molecules located in two halves of a container, if all the molecules are in one half, we have more information about where they are than if they are spread equally between the two halves. They are less disordered if they have a specific arrangement (all on one side) than if half is on each side. So in this case the second law says closed systems tend toward an increase in disorder. In contrast, systems that are not isolated, such as growing organisms, can increase in order because they use energy (they are not closed systems). There is an overall increase in entropy of the organism plus environment; the organism by itself lowers its entropy as it grows at the expense of the environment, which experiences an increase in entropy. If there is no energy flow into the system, the organism will die and decay into disorder.

3.1.3 Heat Engine Efficiency Is Limited by the Carnot Efficiency

As we saw in Chapter 2, a heat engine is a cyclic device that converts heat into mechanical work. Gasoline and diesel engines are important examples. Typical car engines under normal conditions have less than 25% efficiency (only 10% to 15% of the energy is actually used to make the car move; 74% is exhausted as heat). This means that 75 cents of every dollar spent on gasoline for your car is wasted. Why can't we make a more efficient gasoline engine? The short answer is that we can make engines more efficient than they are currently, but the second law limits the maximum efficiency of all heat engines. As we will see in Chapter 6, it is also possible to combine several different processes with a heat engine to create a car with overall higher efficiency.

First let's look at a schematic diagram for an ideal heat engine (Figure 3.2). Energy, Q_H, leaves the hot reservoir and does work, W (the useful work output), and waste energy in the form of heat, Q_C, is expelled to the cold reservoir. For a gasoline engine Q_H is supplied by the exploding gasoline, which pushes a piston (see Figure 2.1) connected to a rotating drive shaft. The work done is the mechanical rotation of the drive shaft. The cold reservoir is the atmosphere, with the radiator and muffler as intermediaries. Because the piston returns to its original location, the process is cyclic, and the change in internal energy for the engine, ΔU is zero for the cycle. Thus for a heat engine, conservation of energy (the first law of thermodynamics, Equation (2.1)) gives $Q_H = Q_C + W_{cycle}$.

Recall from Equation (2.9) that efficiency is defined to be $\eta = \frac{\text{benefit}}{\text{cost}}$, which becomes $\eta = W_{cycle}/Q_H$ for our heat engine. Using $Q_H = Q_C + W_{cycle}$ from the first law, efficiency can be written as

$$\eta = \frac{W_{cycle}}{Q_H} = \frac{Q_H - Q_C}{Q_H} = \left(1 - \frac{Q_C}{Q_H}\right). \tag{3.5}$$

FIGURE 3.2

Schematic operation of an ideal heat engine.

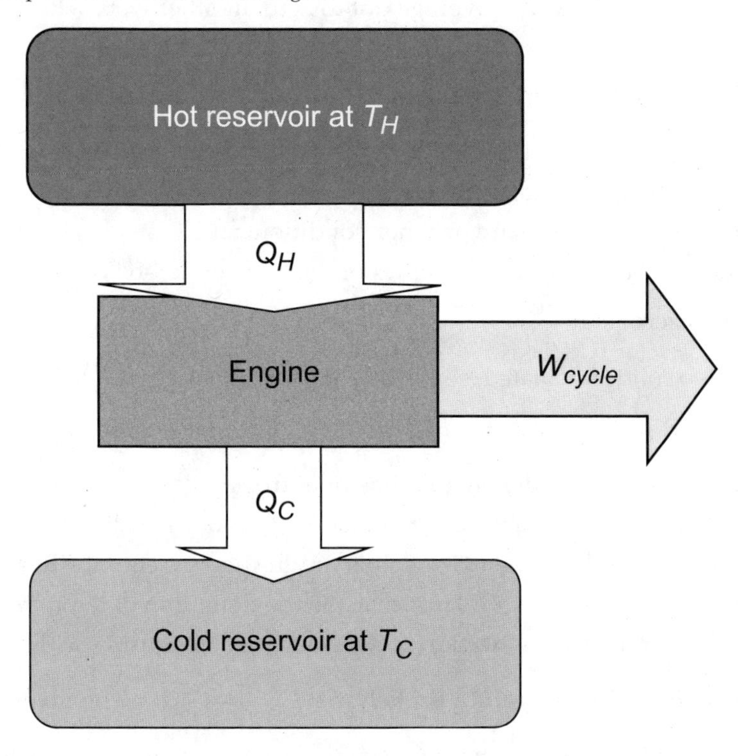

Notice that from this equation efficiency can never be 100%, with the possible exceptions that either the hot reservoir expels infinite heat or the cold reservoir takes in no heat. We have accounted for all the energy in this perfect engine; no energy is lost to friction in this hypothetical case. An infinite heat transfer from the hot reservoir is impossible, but why couldn't the heat expelled to the cold reservoir, Q_C, be zero so that all the input energy is converted to useful work and efficiency is 100%? As we show later, this is exactly what the second law says cannot happen because a change in entropy in the wrong direction would be needed, and as we have seen, decreases in entropy for isolated systems of large numbers of molecules are impossible for all practical purposes. Intuitively we should suspect this already; heat energy has no reason to flow out of the hot reservoir to do work unless it is moving to a cool reservoir.

To apply the second law to our heat engine, first notice in the example of two Einstein solids at different temperatures that energy will flow from the hot to the cold solid until the entropy is at a maximum. The solids may then exchange equal amounts of energy, but the net energy flow has ceased, and total entropy is constant.[5] Fluctuations of entropy as the result of fluctuations in either heat flow, Q, or internal energy, U, average out to zero. In other words, the change in entropy per change in heat flow or change in internal energy equals zero. So for solid A at equilibrium we may write $\frac{\partial S_{\text{total}}}{\partial Q_A} = 0 = \frac{\partial S_{\text{total}}}{\partial U_A}$, where U_A is the internal energy. Here we have used partial derivatives to indicate that we are interested only in how entropy changes with energy or internal energy (we aren't going to change the volume or pressure or other conditions; the system is isolated). The total entropy is the sum of the entropies of the two solids, so fluctuations of the total are also zero, and $\frac{\partial S_A}{\partial U_A} + \frac{\partial S_B}{\partial U_B} = 0$ at equilibrium. We have assumed the system of two solids is isolated, so the internal energy lost by solid A was gained by solid B, and we have $dU_A = -dU_B$, which allows us to write $\frac{\partial S_A}{\partial U_A} = \frac{\partial S_B}{\partial U_A}$ at equilibrium. In other words, fluctuations of entropy due to changes in internal energy in solid A are compensated by equal fluctuations in solid B.

We also know that when two solids are in thermal equilibrium the temperatures are the same: $T_A = T_B$. This in fact is the definition of temperature and allows us to formulate a connection between change in entropy with respect to internal energy and temperature (in kelvin): $T_A = \left(\frac{\partial S_A}{\partial U_A}\right)^{-1}$. The inverse of the entropy change is chosen so that increasing entropy results in lower temperatures if solid A loses heat to solid B. This choice also makes the units for temperature correct. In fact, this is a more fundamental definition of temperature than the one relating temperature to average kinetic energy; temperature is defined to be the inverse of the change in entropy as a result of heat flow. The proper definition of temperature (in kelvin) therefore is

$$(3.6) \quad T = \left(\frac{\partial U}{\partial S}\right)_{N,V},$$

where the N and V indicate that the number of particles and the volume of the system should be held constant (the system is closed).

One very important thing to notice from this discussion is that any time there is an energy flow, Q, there is also a change in entropy (assuming the number of particles and the volume stay constant). Equation (2.1), the first law of thermodynamics, says that the change in internal energy equals the heat flow plus the work done; $\Delta U = Q + W$. To see how much entropy change there is for a given heat flow when no work is done, we can use the first law and rearrange Equation (3.6) as

$$(3.7)\quad dS = \frac{dU}{T} = \frac{Q}{T}.$$

Now we are in a position to show why the heat flow to the cold reservoir for our heat engine cannot be zero, preventing us from making a heat engine with 100% efficiency. When heat flows out of the hot reservoir, the entropy of the hot reservoir decreases by $dS_H = \frac{Q_H}{T_H}$. There is no entropy change of the engine because it returns to exactly the same state from which it started; the process is cyclic. So the second law requires that the entropy of the cold reservoir, $dS_C = \frac{Q_C}{T_C}$, has to increase by at least the same amount as that of the hot reservoir. In other words, the heat flow into the cold reservoir, Q_C, cannot be zero in order to maintain constant (or increasing) entropy for the closed system. The second law requires the total entropy change to be equal to or greater than zero: $dS_C - dS_H \geq 0$. The best possible case would be an equal exchange of entropy from the hot to cold reservoir, which leads to the maximum efficiency possible.

Substituting the heat flow expressions (Equation (3.7)) into the inequality $dS_C - dS_H \geq 0$ gives $\frac{Q_C}{T_C} \geq \frac{Q_H}{T_H}$ or $\frac{Q_C}{Q_H} \geq \frac{T_C}{T_H}$. Equal signs would apply to cases in which the process is gradual enough to be reversible; at any step of the way we could return to the previous step. For irreversible processes (e.g., if there is friction) the prior state cannot be returned to, and the inequality applies. Using the previous definition of efficiency in Equation (3.5) gives the efficiency of any heat engine undergoing a cyclic process as

$$(3.8)\quad \eta \leq \left(1 - \frac{Q_C}{Q_H}\right) = \left(1 - \frac{T_C}{T_H}\right),$$

where T is in kelvin, and it is understood that the inequality applies to irreversible processes.

Notice that we arrived at these expressions without referring to any particular kind of heat engine; they are applicable to all heat engines. These expressions can never yield 100% efficiency unless we assume infinite temperatures or a reservoir at absolute zero. It is very important to notice that the second law expresses a fundamental limitation of nature that cannot be thwarted by technological innovation. Even if we eliminate all friction and all other energy losses, we cannot have any better efficiency than $\eta_{max} = \left(1 - \frac{T_C}{T_H}\right)$ for a heat engine. A physical process that has this efficiency is the Carnot cycle. Most real engines have less than the theoretical maximum efficiency because they are not Carnot cycles, and they lose at least some energy to irreversible processes such as friction.

3.2 ■ Heat Engines and Applications

In this section we start with a consideration of the most efficient heat engine possible, the Carnot cycle. Carnot engines do not exist in the real world, but the concepts introduced regarding this hypothetical case are applicable to other, real engines. In the middle of this section we compare the four most common heat engine cycles in use today. A detailed discussion of these engines is not given; the reader is referred to the technical literature, but the detailed discussion of the Carnot cycle can be applied to each of these four real processes. In the final part of this section we consider refrigerators and heat pumps, which are basically heat engines run in reverse. Mechanical work is input, and heat is moved from a cold reservoir to a hot reservoir. Refrigerators and heat pumps are subject to similar second law efficiency limitations as heat engines.

3.2.1 The Carnot Cycle: The Most Efficient Heat Engine

Engineer Nicolas Léonard Sadi Carnot showed in the early nineteenth century that there was one particular cyclic process, now known as the Carnot cycle, that achieves the maximum efficiency given in Equations (3.1) and (3.8). He started by assuming the processes would have to be perfectly reversible so that there would be no energy loss. Such a reversible process is defined to be one that is quasistatic (the process is slow enough that it can be reversed at any stage) and frictionless. Slowly compressing a perfectly smooth piston by adding a few grains of sand at a time onto the top of it would be a reversible, quasistatic process. A balloon exploding and a box sliding around on a table losing energy due to friction are examples of irreversible processes.

Figure 3.3 shows the pressure versus volume diagram for the maximum efficiency developed by Carnot. In the cycle a piston is taken through a series of steps to return to its original volume and pressure. From our discussion of the first law of thermodynamics we know that the net change in internal energy is

FIGURE 3.3

Schematic pressure versus volume and temperature versus entropy diagrams for the Carnot cycle.

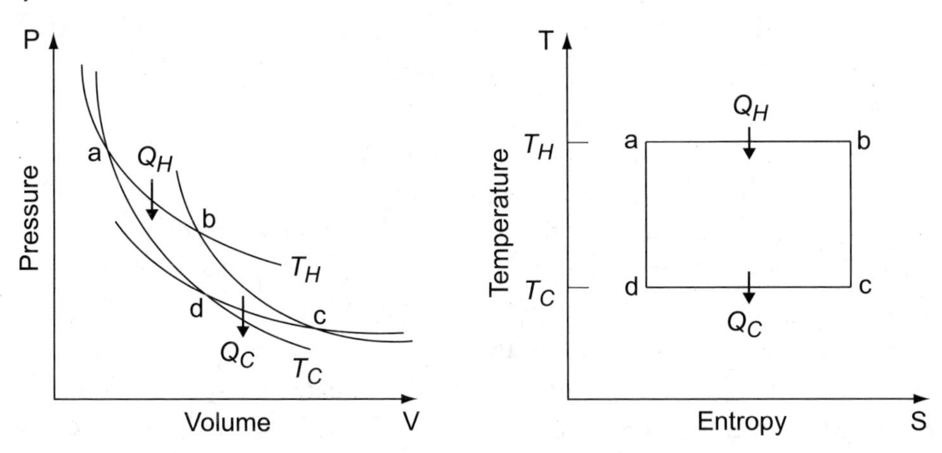

zero when the piston is returned to its starting point. As we saw in Chapter 2, the area inside the curves is the work done in the process. The following are the steps in the process:

1. Isothermal heating from $a \rightarrow b$. The system absorbs heat, Q_H, at constant temperature, T_H, by letting the piston expand from P_a, V_a to P_b, V_b while in contact with a large fixed-temperature hot reservoir.
2. Adiabatic[6] expansion from $b \rightarrow c$. The piston is thermally isolated while it continues to expand from P_b, V_b to P_c, V_c. *Adiabatic* means there is no heat exchange.
3. Isothermal cooling from $c \rightarrow d$. From P_c, V_c to P_d, V_d, heat, Q_C, at constant temperature T_C is expelled as the piston expands while in contact with a large fixed-temperature cool reservoir.
4. Adiabatic contraction from $d \rightarrow a$. Thermally isolate the piston and let it continue to contract back to P_a, V_a. No heat is exchanged.

The equation of the line on the P–V diagram for the two isothermal processes comes from the ideal gas law: $P = nRT/V$, where n is the number of moles of gas, $R = 8.315$ J/mol K is the ideal gas constant, and T is constant. The equation for the two adiabatic processes is $P = kV^{-\gamma}$, where k is a constant and γ is the ratio of specific heat at constant pressure to specific heat at constant volume (see reference [1] for a derivation of this equation).

The only entropy changes occur in the first and third steps when there is a heat flow. The entropy flow into the system at the higher temperature has to equal the entropy flow out of the system at the lower temperature, so $dS_{a\to b} = \frac{Q_H}{T_H} = dS_{c\to d} = \frac{Q_C}{T_C}$. Following the same reasoning leading to Equation (3.8) leads us to the same efficiency for the Carnot cycle:

$$\eta = \left(1 - \frac{T_C}{T_H}\right) \quad \text{(Carnot cycle).} \qquad (3.1)$$

It is also possible to verify that Equation (3.1) comes directly from definition of efficiency given in Equation (2.9), $\eta = W_{\text{cycle}}/Q_H$, by calculating the work done in each step and combining them to find the total work. Because the work is the area under the curve, each of the four curves in Figure 3.3 must be integrated from initial to final points and added (or subtracted for work done by the system), which gives the area inside the curves. This calculation is rather complicated, but this is how efficiencies for the four engines listed in the next section are carried out. In the case of the Carnot cycle the efficiency can be written in terms of temperature. For other cycles the efficiencies may be expressed as different equations that have temperature, volume, or pressure as variables.

EXAMPLE 3.1

Work Done by an Adiabatic Process

Calculate the work done by an adiabatic process. This will be the area under either the curve $b \to c$ or $d \to a$ in Figure 3.3.

As mentioned earlier, the pressure versus volume relation for an adiabatic process is $P = kV^{-\gamma}$, where k is a constant and γ is the ratio of specific heat at constant pressure to specific heat at constant volume. Thermodynamic work is $W_{1\to 2} = \int_{V_1}^{V_2} PdV$, so we have $W_{b\to c} = \int_{V_b}^{V_c} PdV = \int_{V_c}^{V_c} kV^{-\gamma}dV = k\frac{1}{1-\gamma}V^{1-\gamma}\Big|_{V_b}^{V_c} = \frac{k}{1-\gamma}(V_c^{1-\gamma} - V_b^{1-\gamma})$. We can rewrite $P = kV^{-\gamma}$ as $k = PV^{\gamma} = P_b V_b^{\gamma}$ to eliminate the constant k from our result to get $W = \frac{P_b V_b}{1-\gamma}\left(\left(\frac{V_b}{V_c}\right)^{\gamma-1} - 1\right)$ as a final result. This will be the area under the curve $b \to c$ in Figure 3.3. The work done from $d \to a$ is negative because $V_a > V_d$ and is subtracted from the work from $b \to c$ to find the total work done in the two adiabatic processes. A similar calculation (see Problem 3.8) can be done for the two isothermal processes. The work done in the cycle $a \to b \to c \to d \to a$ is the area inside this path.

In some cases it is more convenient to work with an alternative expression for work in terms of temperature and entropy. According to the first law of thermodynamics, $W = \Delta U - Q$. For a heat engine undergoing a closed, cyclic process the total change in internal energy is zero because the system returns to its original state and the heat flow is positive. Replacing heat flow with its entropy equivalent, $dS = \frac{Q}{T}$ (Equation (3.7)) yields the following alterative expression for work as the area inside a closed T–S diagram:

$$W_{\text{cycle}} = \oint T dS, \tag{3.9}$$

where the circle on the integral sign reminds us that we must add the entropy changes over an entire cycle. From Figure 3.3 it is clear that calculating the work done (the area inside the curve) in the Carnot cycle from a T–S diagram is much simpler than from a P–V diagram.

3.2.2 Comparison of Various Real Heat Engines

As mentioned in the introduction to this chapter, because of engineering constraints it is not practical to build a working Carnot engine, with the result that no engine in use today operates on the Carnot cycle. However, a number of other cycles are in use, and the Carnot cycle serves as an upper theoretical limit for these real processes. A comparison of four commonly used heat engines is given in Table 3.4, and a brief discussion of each process follows. The steps given for each cycle are for the ideal case; real engines only approximate the steps shown. All these processes are subject to the limitations of the second law of thermodynamics and are less efficient than a Carnot cycle operating at the same temperatures and pressures. It should be emphasized that in an analysis of all of these cycles, including P–V and T–S diagrams, the work output and expressions for the theoretical efficiency can be calculated following the same steps used earlier for the Carnot cycle. In some cases the calculated efficiency is found as a function of operating temperature (as in the Carnot case); in other cases the efficiency is a function of volume or pressure. All these engines have been under development for a significant amount of time, and laboratory demonstration models have reached a level of sophistication that has yielded engines very close to their ultimate, maximum efficiency. For example, steam engine efficiency has not improved significantly since the mid-1950s.

Steam engines use expanding steam to drive a piston that does mechanical work. A mixture of air and water vapor is first compressed, with no change in

TABLE 3.4

Comparison of various common heat engine cycles.

Engine (Process)	Steps	Typical Efficiency (%)
Steam (Rankine cycle)	1. Adiabatic compression (pump) 2. Isobaric heating (boiler) 3. Adiabatic expansion (turbine) 4. Isobaric condensation (condenser)	<10% for simple engines; 25–40% for superheated, triple expansion
Internal combustion (four-stroke Otto cycle)	1. Isobaric fuel intake stroke 2. Adiabatic compression of fuel 3. Isochoric fuel ignition 4. Adiabatic expansion of burning fuel 5. Isochoric cooling 6. Isobaric exhaust stroke	25% with an upper limit of about 35% for engines made of metal
Diesel (diesel cycle)	1. Adiabatic compression 2. Isobaric fuel ignition 3. Adiabatic expansion of burning fuel 4. Isochoric cooling	40% with an upper limit of about 55%
Gas turbine (Brayton cycle)	1. Adiabatic compression 2. Isobaric expansion 3. Adiabatic expansion 4. Isobaric cooling	48% for simple engine; up to 78% for complex designs

entropy (isentropic or adiabatic). The mixture is then heated at constant pressure (isobaric heating) in a boiler. The heated steam is next allowed to expand adiabatically as it pushes a piston. In other designs the expanding steam pushes a set of turbine blades. The final stage is to condense the vapor to a liquid at constant pressure. This process is called the Rankine cycle. A chief advantage of steam engines is that the steam can be formed from any source of thermal energy. Coal, oil, solar energy, natural gas, and nuclear power can all be used to generate steam for use in a steam engine.

Using 100°C (373 K) as the hot temperature and 0°C (273 K) as the exhaust temperature, the theoretical maximum (Carnot) efficiency of a steam engine would be $\eta = \left(1 - \frac{T_C}{T_H}\right) \times 100\% = 27\%$. Because the exhausted steam temperature is actually much higher than the surrounding atmospheric temperature, a simple, one-stage steam engine has efficiencies of less than 10%. However, it is possible to use superheated steam under high pressure and to capture the exhausted steam and use it in a second or even third piston, each operating at a lower temperature. A modern triple expansion steam engine using superheated steam can achieve efficiencies between 25% and 40%. It is also possible to combine the Rankine steam cycle with a gas turbine running the Brayton cycle. In a combination similar to examples of cogeneration, exhaust heat from the gas turbine is used to make steam for a steam engine, giving a much higher combined efficiency than either process operating alone. A steam power plant using cogeneration of heat for other purposes (see Example 2.4) can capture 90% of the thermal energy provided by the fuel used in the plant.

The modern internal combustion (IC) engine uses burning gasoline as the heat source. The gasoline is ignited by a spark from a sparkplug, and the burning is very rapid, theoretically occurring at a constant volume (isochoric) before the piston starts to move. The process is called a four-stroke cycle because in addition to the compression and power strokes the piston travels back and forth two extra times, first to pull the gasoline–air mixture into the cylinder and again after the power stroke to push the exhaust fumes out of the cylinder. The entire cycle is known as the Otto cycle. With the temperature of burning gasoline, 257°C, as the hot temperature and 273 K as the exhaust temperature, the theoretical maximum (Carnot) efficiency of an internal combustion engine would be $\eta = \left(1 - \frac{T_C}{T_H}\right) \times 100\% = 48\%$. In practical applications the heat transfer is slowed by the physical properties of the engine, resulting in temperatures higher than 257°C. The melting temperature of the material making up the engine and the rate at which heat can be removed place an upper limit on the efficiency of the IC engine. The cooling process needed to keep the engine from melting in effect constitutes a two-stage heat engine, which can be shown [3] to have a theoretical maximum efficiency of

$$(3.10)\quad \eta = \left(1 - \sqrt{\frac{T_C}{T_H}}\right).$$

This two-step process reduces the efficiency because the intermediate heat reservoir temperature is higher than the atmosphere's temperature, and the second stage of the process does no work. Although various ceramic materials have been investigated that have higher melting temperatures than the metals used in normal engine construction, a ceramic engine has yet to be made commercially available. Most gains in efficiency are likely to come about by reducing energy lost due to friction (rolling resistance and aerodynamic drag), ignition processes (valves, cams, variable timing), and internal friction (transmission, alternator, accessories).

The maximum efficiency of the IC engine can also be written in terms of the compression ratio [4], which is the ratio of the volumes during the two isobaric steps of the process, V_1 and V_2, where V_2 is the larger volume:

$$(3.11)\quad \eta = \left(1 - \frac{Q_C}{Q_H}\right) = \left(1 - \left(\frac{V_1}{V_2}\right)^{\gamma-1}\right).$$

Here γ is the specific heat ratio, $\gamma = C_p/C_V$, where C_p is the specific heat of the gasoline–air mixture at constant pressure and C_V is the specific heat at constant volume. This equation shows that higher compression ratios give higher efficiencies. The physical strength of the piston and cylinder and the fact that gasoline will self-ignite at high pressure then become limiting factors for efficiency in an IC engine burning gasoline. It should be noted that this efficiency is consistent with the Carnot efficiency; the equation is derived by starting with the definition of maximum thermal efficiency (Equation (3.8)) and finding the heat flow as a function of volume change for each step of the process.

The diesel cycle is actually less efficient than the Otto cycle if operating at the same pressures and temperatures. However, because diesel fuel–air mixture can be compressed to a higher pressure without prematurely igniting, the diesel engine can operate at higher compression ratios. In fact, most diesel engine designs do not require sparkplugs to ignite the air–fuel mixture; rather, the increased pressure causes the fuel mixture to ignite spontaneously. The higher compression means that the cycle encompasses more area on a P–V diagram and thus does more work per cycle than the gasoline engine. Although diesel fuel has less energy per kilogram than gasoline, diesel fuel is denser, with the result that the energy per volume is about 15% higher for diesel fuel than for gasoline. This coupled with a higher engine efficiency means that a car powered by a diesel engine typically gets as much as 40% better fuel economy than a gasoline engine. The lack of a need for an electronic ignition system and slower cycle rates of a diesel engine also make it generally more reliable and give it a longer life than a gasoline engine. Other details of car fuel economy for various combinations of engines will be considered in Chapter 6.

The Brayton or Joule cycle describes the thermodynamic process occurring in a gas turbine or jet engine. Heat is added, either as burning fuel or from an external heat source, to a continuous supply of compressed gas, which then expands against a set of turbine blades, causing them to rotate. Because the process is continuous (the heating and cooling stages occur at the same time), the temperature of the burning fuel mixture can be much higher (more than 1,000°C) than for an IC engine, which results in higher maximum efficiencies. For a simple once-through cycle, efficiencies in real gas turbines may be as high as 48% (which is still lower than the Carnot efficiency). As mentioned in the description of the steam engine, the exhaust gas from a gas turbine still contains a lot of thermal energy. This heat can be recaptured for use for heating the intake fuel–air mixture, used with additional fuel in a second turbine (afterburner), or used as a cogeneration heat source. Multistage Brayton cycles can achieve efficiencies that approach the maximum Carnot efficiency [5]. The melting temperature of the turbine blades limits the upper temperatures possible in a gas turbine or jet engine, and this in turn limits the maximum efficiency.

3.2.3 Refrigerators

Refrigerators and heat pumps are heat engines running in reverse, as shown in Figure 3.4. A fluid with a low boiling point (e.g., one of the chlorofluorocarbons, hydrochlorofluorocarbons, or bromofluorocarbons mentioned in Chapter 1) is allowed to expand in a process that absorbs heat from the cold reservoir. The fluid is then condensed and compressed in a process that expels heat to the hot reservoir. For a home refrigerator the cold reservoir is the inside of the unit, and the hot reservoir is the set of cooling coils on the back of the unit, which expel heat to the surrounding air. Air conditioner units operate in a similar fashion; heat is carried from inside the house and exhausted to the environment from the outside unit.

FIGURE 3.4

Schematic of a refrigerator.

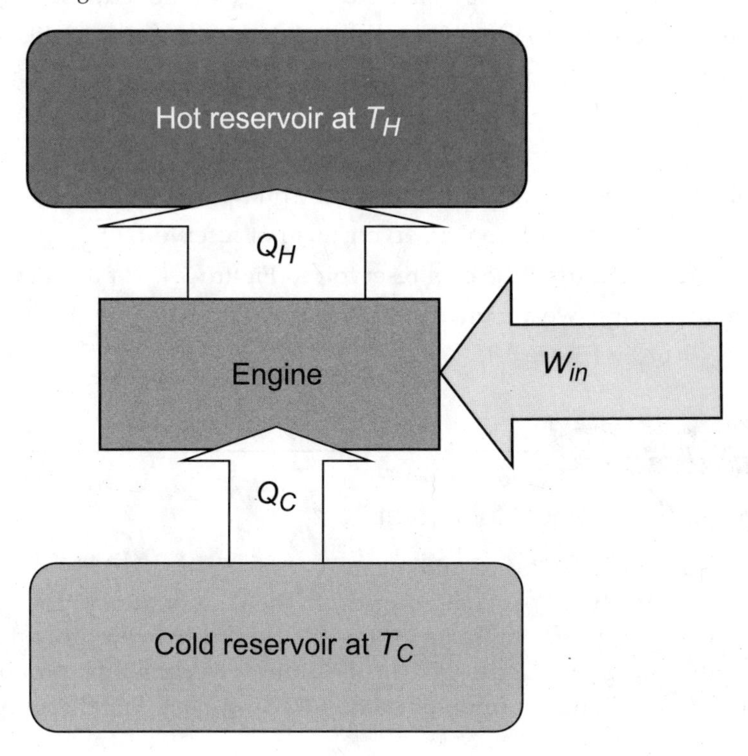

Heat pumps used for home heating and cooling work similarly to a refrigerator and often use underground water as a temperature reservoir. Because the temperature of the earth at depths deeper than about 10 m stays nearly constant in most locations, heat can be extracted from the ground during the winter to warm a building and expelled to the ground during summer to cool the building.

The efficiencies of refrigerators and heat pumps are limited by the second law, just as those of heat engines are. In this case the efficiency, defined in Equation (2.9) as $\eta = \frac{\text{benefit}}{\text{cost}}$, has a different significance. The benefit is heat removed from a cold reservoir or heat expelled to a warm reservoir, and the cost is work done by a motor. To avoid confusion with heat engines, where the efficiency is never larger than one, the thermodynamic efficiency is renamed as the coefficient of performance (C.O.P.) for these devices. The C.O.P. is defined as the useful energy transferred, $Q_{\text{delivered}}$, divided by the work input, W_{in}:

$$\text{(3.12)} \quad \text{C.O.P.} = \frac{\text{benefit}}{\text{cost}} = \frac{Q_{\text{delivered}}}{W_{\text{in}}}.$$

A heat pump or refrigerator with a C.O.P. equal to two will deliver two units of energy (e.g., remove two units from inside a refrigerator or add two units of heat to a house) for every one unit of energy used. Notice also that, unlike efficiency, which is always less than one (100%), the C.O.P. is usually greater than one for a refrigerator and always greater than one for any useful heat pump. A C.O.P. less than one for a heat pump would mean you would get more heat by just burning the fuel for heat instead of using it to run the pump. In the case of a refrigerator, which moves heat from a cool reservoir to a hot reservoir, $Q_{\text{delivered}} = Q_C$, where Q_C is heat removed from the cool reservoir in Figure 3.4. For a heat pump used to warm a building, we are interested in the heat flowing into the hot reservoir, so $Q_{\text{delivered}} = Q_H$ in Figure 3.4.

EXAMPLE 3.2

Power Requirement for a Heat Pump

Calculate the power requirement for a heat pump that is to supply a home with 10 kW of energy for heat. The temperature of the working fluid of the pump must be higher than room temperature when it is in the house in order to give off heat but lower than the ground temperature when it is underground extracting heat. For the purpose of this example, assume the working temperatures of the fluid in the heat pump are –25°C underground and +45°C in the house.

If we assume a perfect, Carnot engine, the work done will be $W_{in} = Q_H - Q_C = T_H - T_C$. Here $Q_{delivered} = Q_H$, so the C.O.P. will be

$$\text{(3.13)} \quad \text{C.O.P.} = \frac{Q_{delivered}}{Q_H - Q_C} = \frac{Q_H}{Q_H - Q_C} = \frac{T_H}{T_H - T_C} = 4.5.$$

The power needed, then, will be $P = 10\ \text{kW}/4.5 = 2.2\ \text{kW}$.

It should be noted that real heat pumps do not achieve Carnot cycle efficiency, so the C.O.P. is likely to be less than the one calculated in Example 3.2. Commercially available heat pumps currently have C.O.P. values of about two for the temperatures used in the example. Also from this example it should be clear that climates where the temperatures are more extreme will benefit less from the use of a heat pump. If the temperature extremes in Equation (3.13) are greater, the C.O.P. will be smaller.

An absorption refrigerator uses a heat source to drive a heat transfer process in a manner similar to a heat pump (see Problem 3.31). The process is inefficient, with the result that these refrigerators have been used mainly in special applications where other energy sources are not available (e.g., diesel-powered refrigerators in remote areas where electricity is not available). Because these devices could be run on waste heat (e.g., waste heat from IC engines), they have the potential for much wider application.

3.3 ■ Fuel Cells and Batteries

In the earlier discussion of heat engines and refrigerators, all the processes discussed were cyclic. By definition, in a cyclic process the internal energy returns to the same value at the end of the process, so $\Delta U = 0$.[7] In batteries and fuel cells, however, this is not the case; at the end of the process there have been chemical changes that leave the internal energy changed. These processes may also occur at constant volume and pressure, as compared to heat engines, which do work by changing pressure or volume. Nevertheless, there will necessarily be heat exchanges in order to compensate for changes in entropy. Although the second law still applies to these devices, the amount of expelled (or waste) energy is generally less than that of heat engines, so they often have higher efficiencies.

To begin a calculation of the entropy change in a fuel cell, we first define a quantity know as the enthalpy,

$$H \equiv U + PV, \tag{3.14}$$

which can be thought of as the energy needed to create a system out of nothing plus the work needed to make room for the new system [1]. Or if you destroy the system, it is the energy you get out: the energy of the system plus the work the atmosphere does in filling the space where the system was. Of course, in the real world we never create something from nothing, so usually we are interested in changes in enthalpy. Changes in enthalpy involve changes in internal energy (e.g., chemical reactions) and work done on or by the system. The change in enthalpy can be measured experimentally for a given chemical reaction, and standard reference tables exist for enthalpy changes in most important chemical reactions [6].

Because values for enthalpy include the possibility of heat flow from or to the environment, it is sometimes useful to subtract this amount and define the energy needed to create a system out of nothing, not including energy provided by the environment from heat flow, $Q = TS$. This quantity is called the Gibbs free energy and is defined as

$$G \equiv U - TS + PV = H - TS \tag{3.15}$$

Once again, we are really interested in changes in energy, so for processes occurring at constant pressure and temperature we may write

$$\Delta G \equiv \Delta U - T\Delta S + P\Delta V = \Delta H - T\Delta S, \tag{3.16}$$

or using the first law of thermodynamics this can be expressed as

$$\Delta G \equiv Q + W - T\Delta S + P\Delta V. \tag{3.17}$$

Here it should be remembered that work could be either positive (work done on the system) or negative (work done by the system). The Gibbs free energy, then, is the energy available for useful work after heat flow, entropy changes, and work done by expansion or compression have been accounted for. Values of ΔG can also be measured experimentally, and tables exist for a wide range of chemical reactions [6].

In a fuel cell, chemicals react at constant pressure and temperature to provide electrons, which can be used in an external circuit, as shown in Figure 3.4. In this sense it acts like a motor or engine; the energy from input fuel is converted into work, electrical work in this case. The most common fuel cell fuel used today is hydrogen (Figure 3.5), although similar reactions can be made to occur with many other chemicals (see Problem 3.34). In the hydrogen fuel cell the proton exchange membrane permits the movement of hydrogen ions through the cell without allowing the premature recombination of hydrogen and oxygen. If the reaction takes place without a proton exchange membrane, the device is usually called a flow battery, but the concept is the same: Incoming material at the anode undergoes a chemical reaction, which gives off electrons, while a reaction absorbing electrons occurs at the cathode. Examples of flow batteries in use today are zinc bromide and vanadium redox batteries.

EXAMPLE 3.3

Theoretical Efficiency of a Fuel Cell

For hydrogen fuel cells, the overall chemical reaction that occurs is

$$H_2 + \frac{1}{2}O_2 \rightarrow H_2O. \quad (3.18)$$

Based on available reference tables [6], the enthalpy change for this reaction is $\Delta H = -286$ kJ, which is the energy that would come from burning a mole of hydrogen at standard temperature and pressure. The entropy changes for each component of Equation (3.18), also from standard tables, are $\Delta S_{H_2O} = 70$ J/K, $\Delta S_{H_2} = -131$ J/K, and $\Delta S_{O_2} = -205$ J/K, so the total entropy change is $-131\,\text{J/K} - \frac{1}{2}\,205\,\text{J/K} + 70\,\text{J/K} = -163\,\text{J/K}$. Using these numbers and Equation (3.17) we can calculate the Gibbs free energy, which is the maximum energy available to do work:

$$\Delta G = \Delta H - T\Delta S = -286\text{ kJ} + 300\text{ K} \times 163\text{ J/K} = -237\text{ kJ}. \quad (3.19)$$

This means +237 kJ of energy is available from the combination of a mole of hydrogen with a mole of oxygen. Dividing this by the number of atoms involved gives the energy per reaction as $237\text{ kJ}/2 \times 6.02 \times 10^{23} = 1.97 \times 10^{-19}\text{ J} = 1.23$ eV. This is also the energy per electron produced by the reaction, so the basic cell of a fuel cell using this reaction will have a voltage of 1.23 volts. Higher voltages can be obtained by adding cells together in series.

The heat expelled during the process in Equation (3.18) is $T\Delta S = -300\text{ K} \times 163\text{ J/K} = 49\text{ kJ}$. The thermal efficiency, $\eta = \frac{Q_H - Q_C}{Q_H}$, can be defined as the energy input available if we were to burn the hydrogen minus the equivalent heat expelled due to entropy loss divided by the energy input. For this fuel cell we have (286 kJ – 49 kJ)/286 kJ = 83%. Hydrogen can also be burned directly as a fuel in a heat engine, but as we saw previously, the theoretical maximum efficiency of a heat engine is limited by the temperature at which it can operate and is much less than 83%.

The heat expelled by the hydrogen fuel cell results in a high operating temperature, close to that of an internal combustion engine [7]. Other chemical reactions can be used in a manner similar to the hydrogen fuel cell but occur at different temperatures. For example, the fuel cells used in NASA space missions operate at much lower temperatures and use potassium hydroxide as an electrolyte, in place of the proton exchange membrane. Other fuel cells operate at temperatures as high as 1,000°C. The fuel for many fuel cell designs is pure

FIGURE 3.5

Chemical reactions of a hydrogen fuel cell.

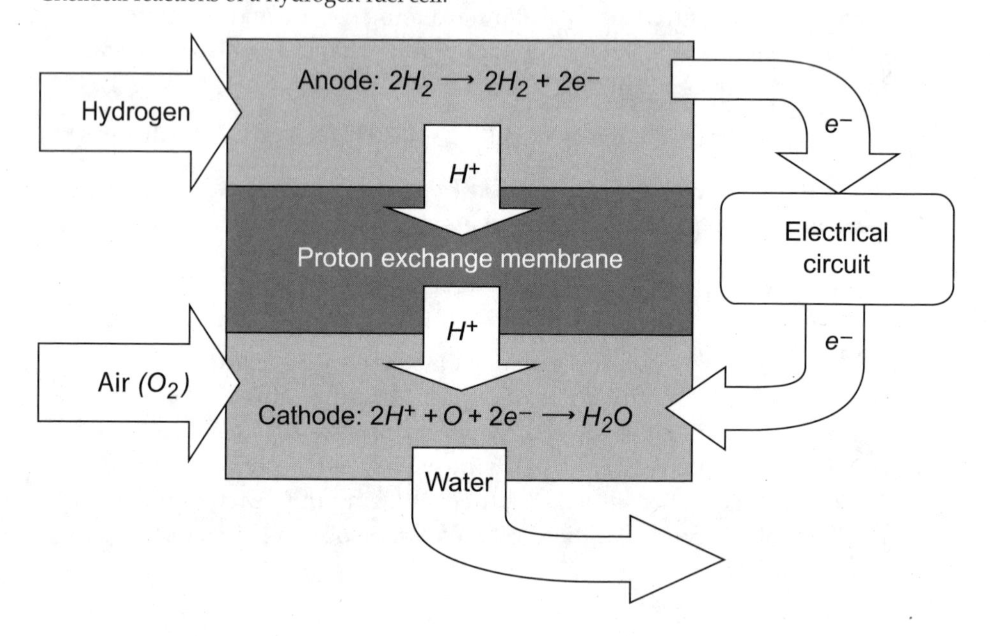

hydrogen, which is combined with oxygen, yielding water as a waste product. However, it is possible to manufacture fuel cells that will run on methane, ethanol, or other hydrogen-carrying compounds. For these fuel cells other waste products besides water are produced; for example, fuel cells using methane as a fuel generally give off carbon dioxide as a byproduct. Fuel cells have been proposed for many different applications, including energy sources for transportation, homes, and portable electronic devices such as calculators and cell phones.

The entropy changes occurring in a battery can be calculated in a similar way as the fuel cell in Example 3.3. In fact, a fuel cell can be thought of as a battery where the chemicals are continually resupplied, in contrast to a battery where the chemicals are limited by the initial quantity in the battery. Batteries will be treated in more detail in Chapter 6, but here we will mention that the entropy change when a battery is being discharged turns up as expelled heat. One engineering problem involved in making a battery-powered car is ensuring the heat from the battery does not build up to a dangerous level as the car accelerates. The process of charging a battery requires absorption of entropy or heat from the environment; batteries cool as they are charged.

3.4 ■ Electric Motors and Generators

You may recall from introductory physics that a current-carrying wire in a magnetic field will experience a force acting on it:

$$d\vec{F} = I\,\vec{dl} \times \vec{B}, \tag{3.20}$$

where $d\vec{F}$ is the force in newtons on a small length of wire, $\vec{dl}$, in meters carrying current, I, in amperes. The magnetic field, $\vec{B}$, is in tesla, and the cross product of the two vectors (sometimes called the right hand rule) tells us that the force acts perpendicular to both the magnetic field and the wire. Because of this force, a loop of wire will experience forces so that the loop will turn, as shown in Figure 3.6.

If the directions of current and magnetic field remained fixed, the loop would rotate to a position where the forces were opposite, and the loop would oscillate back and forth instead of turning continuously in the same direction. By a clever use of connectors, called brushes and commutators, we can reverse the direction of current flow every half turn so the force changes direction and keeps the loop rotating. An alternative is to reverse the direction of the magnetic

FIGURE 3.6

A loop of wire carrying current I in a magnetic field, $\vec{B}$. In **(a)** the current goes up the wire on the right, producing a force into the page, and down the wire on the left, producing a force out of the page. The top is parallel to the field, and so no force is produced on the loop. In **(b)**, a top view, the current comes out of the page on the right, producing a force upward, and into the page on the left, producing a force downward.

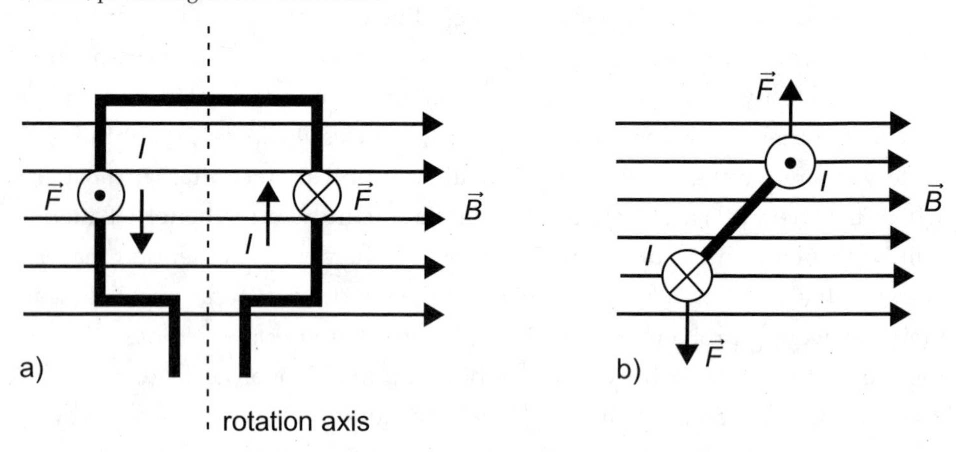

field each half turn. Overlaying many loops increases the force on the configuration, so motors typically have large coils of rotating wires. Commercial motors use three coils oriented at 60° to each other, so that there is no spot at which the force is zero as the loop turns.

Generators are essentially motors run in reverse; magnetic field and mechanical force are the input and current is the output. Recall from Chapter 2 that Faraday's law (Equation 2.34) says a voltage appears in a circuit or wire as the result of the change in magnetic flux. As an application of Faraday's law, if a wire is forced to move through a magnetic field so that it sweeps out an area, a voltage, ε, will appear across the wire:

$$\varepsilon = Blv, \quad (3.21)$$

where l is the length of wire moving at velocity v perpendicularly through a magnetic field B. The arrangement in Figure 3.6 can therefore be used in reverse; if the loop, with no current initially flowing, is forced to turn in the magnetic field, an emf (in volts) is generated. If the loop is connected to an external circuit, a current will flow. This is the basis behind an electric generator, and in many early applications of electrical circuits, generators were simply motors being run in reverse.

It should be emphasized that a force is needed to move the loop in a generator, and the amount of force needed to keep the loop turning is proportional to the amount of current generated. This is an application of the first law of thermodynamics; the energy coming out of the generator cannot be greater than the mechanical work applied to make the loop turn. The gas mileage of an automobile decreases slightly when electrical apparatus such as lights and radio are run because the motor has to work harder to create the current flow needed to recharge the battery. Any scheme attempting to use an electric motor to turn an electric generator that then runs the motor plus other devices will not work; the output current is proportional to the applied force. Even without friction losses the scheme would at best produce no more energy than was originally input.

The limits to efficiency of motors and generators due to the second law of thermodynamics are exceedingly small. An ideal motor with no friction or other loss can have a theoretical efficiency of more than 99%, and real electric motors have been built with efficiencies close to this limit. The second law limit to efficiency has to do with the power needed to overcome random thermal energy of the electrons already in the wire so that current will flow. It can be shown that this power is about

$$(3.22) \quad P = \frac{V^2}{R} = 4k_B T\, b,$$

where k_B is Boltzmann's constant and T is temperature [8]. Here b is the bandwidth, the range or variation of frequencies in hertz of the alternating current being used in the circuit. Because Boltzmann's constant is very small, the power needed to overcome thermal resistance is practically negligible, with the result that an ideal electric motor can have near perfect second law efficiencies.

For real electric motors there are mechanical friction losses and resistance losses proportional to $P = I^2R$, as discussed in Chapter 2. These losses are not limitations due to the second law of thermodynamics. Decreasing the resistance by making the wires thicker will increase the efficiency but at a higher economic cost and greater weight. For stationary motors where weight is not a factor, efficiencies can be very high. The fluctuating magnetic fields involved in the motor are also not perfectly confined to the motor, with the result that electromagnetic waves radiate into the surroundings and thus also constitute energy loss. Well-designed low-horsepower electric motors (<1,000 W) typically have efficiencies of about 80%, and larger motors (>95 kW) have efficiencies as high as 95%. It should be pointed out that these efficiencies are significantly higher than the efficiencies of fuel cells and any heat engine we have discussed. The theoretical, second law limitation to efficiency is extremely small for electrical processes. As we will see in Chapter 6, this has important implications for transportation choices in the future.

3.5 ■ Summary

As we have seen, as a consequence of the second law it is not possible, even in theory, to make a heat engine that is 100% efficient. The entropy balance for any process that involves heat flow mandates that some heat will be given off to the environment. This is the reason IC engines must have a radiator or cooling fins where heat is expelled. Current gasoline engines have average efficiencies less than 25% under ideal conditions, which means more than 75% of the energy contained in the gasoline is wasted. As mentioned in Chapter 2, only about a third of the energy contained in the fuel consumed in electrical generation plants is converted into electricity; the rest is exhausted as waste heat. The second law says we may be able to do a little better (the Carnot efficiency in Equation (3.1)) but never 100%, even if all friction and other losses are eliminated. Figure 3.7 shows the energy losses in generating electricity in the United States. Although some of the conversion losses shown in the figure can be reduced or captured by cogeneration, the loss of primary energy sources is substantial and is a direct result of the second law of thermodynamics.

It is possible to capture some of the so-called waste heat from a heat engine for other uses. For example, in the winter the waste heat from car engines is used to warm the people in the car. As we saw in this chapter and the last, factories and electric generating plants that operate heat engines can sometimes sell or otherwise use the waste heat (called cogeneration), making the overall efficiency of the combined processes much higher. Waste heat can also be used as an energy source for refrigeration using absorption refrigerators. In this case the process is called tri-generation.

Other devices such as fuel cells, batteries, and electric motors are also limited in efficiency by the second law, but because they are not heat engines (they do not convert heat energy into mechanical work) they can have greater efficiencies than heat engines. As noted in Table 2.1, some energy conversion processes have significantly higher efficiencies than others. As we have learned in this chapter, this often has to do with the second law of thermodynamics and is a limitation of nature, not technological achievement. Because fuel cells and

FIGURE 3.7

Electricity generated by primary energy sources and electricity used. Notes: [1]Blast furnace gas, propane gas, and other manufactured and waste gases derived from fossil fuels. [2]Batteries, chemicals, hydrogen, pitch, purchased steam, sulfur, miscellaneous technologies, and nonrenewable waste (municipal solid waste from nonbiogenic sources, and tire-derived fuels). [3]Data collection frame differences and nonsampling error. Derived for the diagram by subtracting the "T & D Losses" estimate from "T & D Losses and Unaccounted for." [4]Electric energy used in the operation of power plants. [5]Transmission and distribution losses (electricity losses that occur between the point of generation and delivery to the customer) are estimated as 7% of gross generation. [6]Use of electricity that is self-generated, produced by either the same entity that consumes the power or an affiliate, and used in direct support of a service or industrial process located in the same facility or group of facilities that house the generating equipment. Direct use is exclusive of station use [9].

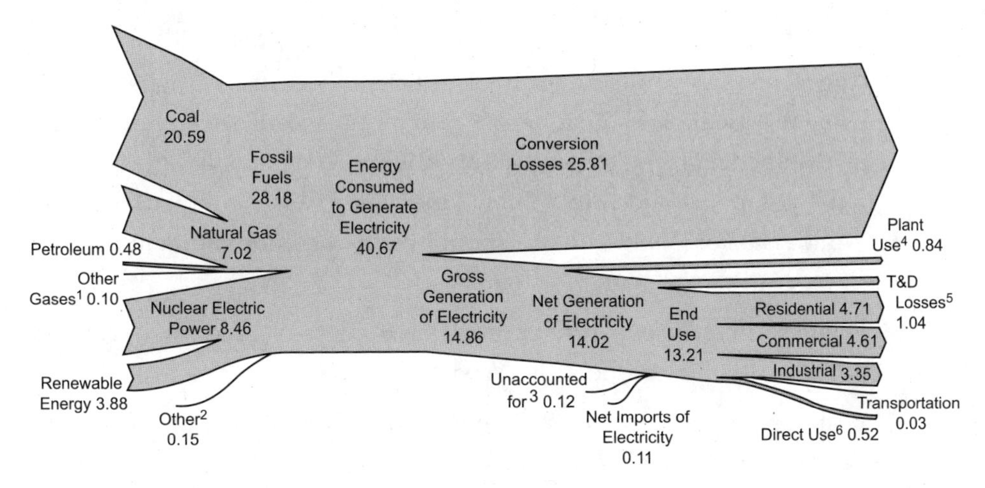

electric motors don't use a heat flow to do work, the entropy changes are calculated differently, with the result that, for example, electric motors can be more than 95% efficient and fuel cells can have theoretical efficiencies greater than 80%. In general, replacement of heat engines with electrical devices is desirable to reduce energy losses required by the second law of thermodynamics. Even though the electricity used would be generated from heat engines, a modern power plant with cogeneration can achieve much higher second law efficiencies than isolated heat engines operating on a smaller scale.

Heat engines can be run forward or backward, and this is the basis of refrigerators and heat pumps. For a refrigerator application, work is input, and this causes heat to flow from the cold reservoir (inside your refrigerator) to outside (the coils on the back of the refrigerator). Refrigerators are subject to the same limits in efficiency as heat engines, but the efficiency for refrigerators and heat pumps is given as the coefficient of performance (C.O.P.), which is the ratio of useful heat transferred to work input. The useful transfer may involve heat removed from a cool reservoir (refrigerator) or heat added to a warm reservoir (heat pump). Refrigerators and refrigeration units have experienced the greatest advances in efficiency, improving by more than 50% in the past 10 years and 75% in the past 40 years. This has been the result largely of government incentives and regulations.

Questions, Problems, and Projects

For all questions,

- Cite all sources you use (e.g., book, article, Wikipedia, Web site). Before using Wikipedia as a source, please read the policies of research and citing in Wikipedia: http://en.wikipedia.org/wiki/Wikipedia:Researching_with_Wikipedia and http://en.wikipedia.org/wiki/Wikipedia:Citing_Wikipedia. For Wikipedia you should verify the information in the article by reading the sources listed.
- For all sources, briefly indicate why you think the source you have used is a reliable source, especially Web resources.
- Write using your own words (except for direct quotes, which should have quotation marks around them and a reference).

1. Work out a probability table similar to Table 3.1 for six coins.
2. Find the entropy for the multiplicities Ω_0^{40}, Ω_{10}^{40}, and Ω_{20}^{40} in the 40-coin case. Which state has the higher entropy?
3. Explain in your own words why the graph in Figure 3.1b is narrower than the one in Figure 3.1a.
4. Using a spreadsheet or other software (see Appendix B), construct a graph similar to Figure 3.1 for 100 coins. Find the standard deviation, σ, and compare the results for 100 coins with the 40-coin case.
5. Using a scientific calculator with a factorial function, find the largest value of N for which you can calculate $N!$ (most calculators will return a value of ∞ once N is too large). Look up Stirling's approximation and find the value $N!$ for this N using Stirling's formula.
6. Work out a multiplicity table similar to Table 3.3 for $N_A = N_B = 4$ with eight energy units to be divided between them ($q_A + q_B = 8$).

7. For the table you found in Problem 3.6, find the multiplicity for having the energy equally divided between the two solids, Ω_{total}. How do these compare with the case in Table 3.3? How do they compare with the multiplicity of having all the energy in one of the solids?

8. Use the definition $W = \int_{Va}^{V_b} PdV$ to find the work done by an ideal gas ($PV = nRT$, where n is the number of moles of gas and R is the ideal gas constant) undergoing the isothermal stages of the Carnot cycle. *Hint*: The answer $= nRT \ln \frac{V_b}{V_a}$.

9. Watch the drinking bird video on YouTube: http://www.youtube.com/watch?v=WOEwcemjt6I. The bird will continue to dip into the cup of water as long as there is water in the cup. Explain how this does not violate the second law of thermodynamics.

10. Salt crystals will form from saltwater as the water evaporates. A salt crystal is much more ordered than salt dissolved in water. Explain how the process of evaporation does not violate the second law of thermodynamics.

11. A single cell in a chicken egg eventually becomes a complex, multi-celled organism. Explain how this does not violate the second law of thermodynamics.

12. What is the Carnot efficiency of an engine that burns gasoline at a temperature of 500°C if the cold reservoir is outside air at 20°C?

13. Repeat the efficiency calculation of Problem 3.12 for a more realistic engine that obeys Equation (3.10). Compare the two results.

14. For a simple once-through cycle, efficiencies in real gas turbines operating at 1,000°C may be as high as 48%. How does this compare with the theoretical maximum?

15. Suppose an inventor wants you to invest in a new type of engine. He claims the engine operates between 520°C and 30°C, uses 10,000 J of heat input, expels 1,000 J of heat to the environment, and does 9,000 J of work. (a) Does this violate the first law of thermodynamics? Explain. (b) Does this violate the second law of thermodynamics? Explain. (c) Would you invest in this engine? Why or why not?

16. Compression ratios for gasoline engines are between 8 and 12 to one ($V_2 = 12\ V_1$), whereas for diesel engines the ratios are between 14 and 25 to one. Use Equation (3.11) to approximate the maximum theoretical

efficiency of a gasoline versus a diesel engine. Use a specific heat ratio of $\gamma = C_p/C_V = 1.3$ for both engines.

17. The percentage of diesel engines used in cars in Europe is much higher than in the United States. Using reliable resources, discuss the pros and cons of replacing gasoline engines with diesel engines. Explain why diesels are more popular in Europe.

18. Using reliable sources, find out more details about how steam, gasoline, and diesel engines work and write a summary your findings for each. Include a P–V and S–T diagram for each. Find a Web link for a simulation of a working engine for each. What are some advantages and disadvantages for each?

19. Using reliable sources, for one of the engines listed in Table 3.4, compare the P–V or T–S diagram for the theoretical engine to the diagrams for a real engine. Discuss the differences and the reasons for those differences.

20. Using reliable sources, find out more details about two-stroke internal combustion (IC) engines, four-stroke IC engines, Wankel engines, Stirling engines, gas turbine engines, diesel engines, and jet engines. For each case state the cycle the engine uses, its theoretical Carnot efficiency, where the engine is or has been used, and any other interesting details.

21. Using reliable sources, compare the pollution produced by each of the following engines per unit of work generated: gasoline, diesel, gas turbine, and jet engine.

22. The Atkinson cycle is a gasoline internal combustion cycle that is more efficient than the Otto cycle, although it provides less power. It is used in the Toyota Prius, coupled with an electric motor that is available for acceleration. Find P–V and T–S diagrams for the Atkinson cycle and describe the stages involved. How is the Atkinson cycle different from the Otto cycle?

23. Make a plot of Equation 3.1 for a cold reservoir of 300 K and hot reservoir temperatures from 300 K up to 2,500 K. At what temperature is the Carnot efficiency 50%? 75%? 100%?

24. In Example 3.2 the temperature difference used was 70°C. What temperature difference would give a C.O.P. equal to one?

25. Repeat Example 3.2 but for the case of a refrigerator (so $Q_{\text{delivered}} = Q_C$. What is the C.O.P.? (Heat pumps can be used to cool as well as heat a building.)
26. Using reliable sources, find the most efficient heat pump available on the market today. Describe how it works and any interesting features.
27. A salesperson wants to sell you a heat pump with a maximum C.O.P. of one. Explain why you would want to consider other means of heating your home.
28. Using a reliable source, make two lists, one of regions where it would be advantageous to use a heat pump (high C.O.P.) and one of regions where it would not be advantageous to use a heat pump. Explain your choices.
29. Explain why you cannot cool your kitchen by leaving the refrigerator door open. However, if you were willing to do some remodeling, you could use a refrigerator as an air conditioner. How would you do this?
30. Using reliable sources, find out more details about the following and write a summary of your findings: gas-powered (absorption) refrigerator, thermoacoustic cooling, and thermoelectric cooling.
31. Albert Einstein and Leó Szilárd designed and received several patents for a type of absorption refrigerator that has no moving parts. (a) Explain the difference between a compression type refrigerator and an absorption refrigerator. (b) Describe the Einstein–Szilárd refrigerator and how it works. (c) Explain how an absorption refrigerator could be more energy efficient than a compressor type refrigerator if all energy inputs are included.
32. Cogeneration is the use of heat created during electricity generation for other purposes. Explain what tri-generation is. (*Hint*: What is needed for an absorption refrigerator?)
33. Repeat the calculation of efficiency in Example 3.3 for a fuel cell using methane given the following information. The reaction is $CH_4 + 2O_2 \rightarrow 2H_2O + CO_2$. $S_{CH_4} = 186.26$ J/molK, $S_{O_2} = 205.14$ J/molK, $S_{H_2O} = 69.91$ J/molK, $S_{CO_2} = 213.74$ J/molK, $\Delta G = 630.64$ kJ, and the temperature is 300 K.
34. Using reliable sources, write a summary of the advantages and disadvantages of various types of fuel cells. Include information about

efficiencies, fuel sources, operating temperatures, and technical problems. In Chapter 7 we will consider carbon dioxide (CO_2) emissions. Based on Example 3.3 and the previous problem, what additional considerations besides efficiency might be appropriate for choosing between hydrogen and other hydrocarbons as a fuel for fuel cells?

35. From reliable resources, write a summary of how a flow battery works. Give some examples of the chemical reactions involved and applications of these batteries.

36. Make a list of the various types of fuel cells now under investigation, using reliable sources. List the chemicals used, the differences in applications, the operating temperatures, the pollution produced, and the efficiencies.

37. Currently three main types of batteries are in use: lead storage, lithium ion (Li-ion), and metal hydride (NiCd and NiMH). Using reliable sources, compare the performance and properties of these batteries. What applications are more suitable to each? Which ones are used in cell phones, laptops, automobiles, and other devices? Which ones should be charged more frequently for the longest life? Which ones should be fully discharged periodically for a longer life?

38. Find and summarize a discussion from reliable sources of the heat problem experienced by the lithium ion batteries used in the Tesla all-electric vehicle.

39. Using reliable sources, describe, with diagrams, how the brushes and commutators of an electric motor work to reverse the current flow each time the motor coil goes around half a turn.

40. Use Equation (3.22) to find the power loss in a circuit due to the second law. Use room temperature and a bandwidth of a few hertz. How does this compare to the power loss due to resistive heating, $P = I^2R$ for a 1-ohm wire with 2 amps flowing through it?

41. Engines or motors that supposedly produce more energy than they use are called perpetual motion machines. Using a reliable source, find definitions of perpetual motion machines of the first and second kinds. Also find an example of each of the two types. For the examples you find, explain the flaw in the design of each.

42. Flaws in the design for motors that supposedly break the laws of physics are often difficult to detect (http://web.mit.edu/newsoffice/1999/

perpetual-0519.html). Pick a perpetual motion machine for which the design flaw is difficult to detect from one of the following Web sites and report on it. Explain the how the device is supposed to work and which laws of physics are being violated. Speculate on what detail is being overlooked that would make the flaw hard to detect.

a. The Museum of Unworkable Devices (http://www.lhup.edu/~dsimanek/museum/unwork.htm).
b. Magnetic motors (http://magnetmotor.go-here.nl/).
c. Wikipedia on the history of perpetual motion machines (http://en.wikipedia.org/wiki/Perpetual_motion_machine).

43. Read the description of the Perepiteia engine and scan through some of the blog comments (http://www.physorg.com/news121610315.html). Based on the description, the comments and your knowledge of physics, what do you think is going on here? If this link no longer works, see whether you can find some other perpetual motion device that is currently receiving attention in the news and report on it.

44. The U.S. Patent Office has a policy that rejects any device claiming to be a perpetual motion machine; however, several devices that appear to be perpetual motion machines have been patented. Find a description of one of these and report on your findings.

45. Using reliable resources, write a summary of the electrocaloric effect.

46. (a) Calculate the overall efficiency of the U.S. electrical supply system, shown in Figure 3.7. (b) Assume the coal plants involved are 35% efficient, and the natural gas power plants are 60% efficient. What is the overall efficiency of the remaining sources of electricity? (c) What could be done to improve the efficiency of the present system?

REFERENCES AND SUGGESTED READING

1. D. V. Schroeder, *Thermal Physics* (2000, Addison Wesley Longman, New York).

2. M. Glazer and J. Wark, *Statistical Mechanics: A Survival Guide* (2001, Oxford University Press, Oxford).

3. E. Boeker and R. van Grondelle, *Environmental Physics* (1999, Wiley, New York).

4. A detailed calculation of the Otto cycle can be found at http://web.mit.edu/16.unified/www/SPRING/propulsion/notes/node25.html.

5. Methods for improving the efficiency of a hypothetical Brayton cycle can be found at http://www.qrg.northwestern.edu/thermo/design-library/airstd/brayton.html.

6. D. R. Lide, *CRC Handbook of Chemistry and Physics*, 86th ed. (2005, CRC Press, Boca Raton, FL).

7. H. Petroski, "Fuel Cells," *American Scientist* 91(5) (Sept.–Oct. 2003):398.

8. H. Nyquist, "Thermal Agitation of Electric Charge in Conductors," *Physical Review* 32 (July 1928):110.

9. Tables 8.1, 8.4a, 8.9, A6 (column 4), and Energy Information Administration, Form EIA-923, "Power Plant Operations Report," *Energy Information Administration Annual Energy Review 2008* (http://www.eia.doe.gov/emeu/aer/elect.html).

Notes

[1] Fluctuations from the highest-entropy state will obviously occur, but, as shown in the following, deviations from this maximum entropy state become smaller as the number of atoms increases. The topic of fluctuations is beyond the scope of this brief presentation.

[2] Most calculators and computers are limited to calculating factorials of numbers less than 500 because the values become very large. For higher numbers of entities, the Stirling approximation and other mathematical tricks must be used to evaluate Equation (3.3) [1].

[3] For real solids there are other available energy states (e.g., rotational, bending), and these introduce additional multiplicities. In the end the same result holds: Equalization in energy between microstates of the two solids (the most probable outcome) is equivalent to equalization of temperature. In classic thermodynamics this result is known as the equipartition theorem [1]. A calculation including quantum mechanical effects changes this picture somewhat, but the second law still holds.

[4] It should be clear now why it useful to use the natural logarithm in the definition of entropy. For the first case (all energy in one solid) we have $\ln(1.7 \times 10^{96}) = 221.6$, and for evenly split energy we have $\ln(1.3 \times 10^{122}) = 281.6$. For multiplicities involving moles of atoms, the entropy will be a manageable number after we take natural logarithms and multiply by Boltzmann's constant.

[5] This is sometimes known as the principle of detailed balance. Basically it says that small exchanges of identical molecules or energy cannot be detected. Although it cannot be proved in a strict sense, it seems reasonable for large numbers of molecules.

[6] Some references use the term *isentropic* for this process, meaning that the entropy remains constant. As we have seen, a heat flow is equivalent to an entropy change, so an isentropic process is also an adiabatic process (one with no heat flow).

[7] Entropy changes of the burning fuel were not included in calculating the efficiency of the heat engine itself.

CHAPTER 4

Nonrenewable Energy

Despite concerns about dwindling oil supplies and the climate impacts of fossil fuels, today more than 86% of energy used comes from nonrenewable sources. In this chapter we review sources of nonrenewable energy, starting with a look at how much energy is used currently, how it is used, and projected increases in demand. The next two sections discuss the estimated reserves of fossil fuels and estimates of how long they will last at current rates of usage. Carbon emissions from these fossil fuel sources and their associated effects are treated in Chapter 7. We finish this chapter with a brief discussion of nuclear energy and reactor design. The risks associated with nuclear reactors and nuclear waste are treated in Chapter 8.

4.1 ■ World Energy Consumption

In Figure 4.1 the world's proportional energy use and population are given by region for the year 2005. Notice that North America has only 5% of the world's population but uses 28% of the world's energy. In comparison, Asia and Oceania (which includes Australia, New Zealand, and the South Pacific islands) use about the same amount of energy but have 61% of the world's population. Clearly, for the per capita energy use in Asia to rise to the level of North America, which it appears to be doing, a significant amount of new

FIGURE 4.1

Energy use and population (in parentheses) by region, 2005 [1–3].

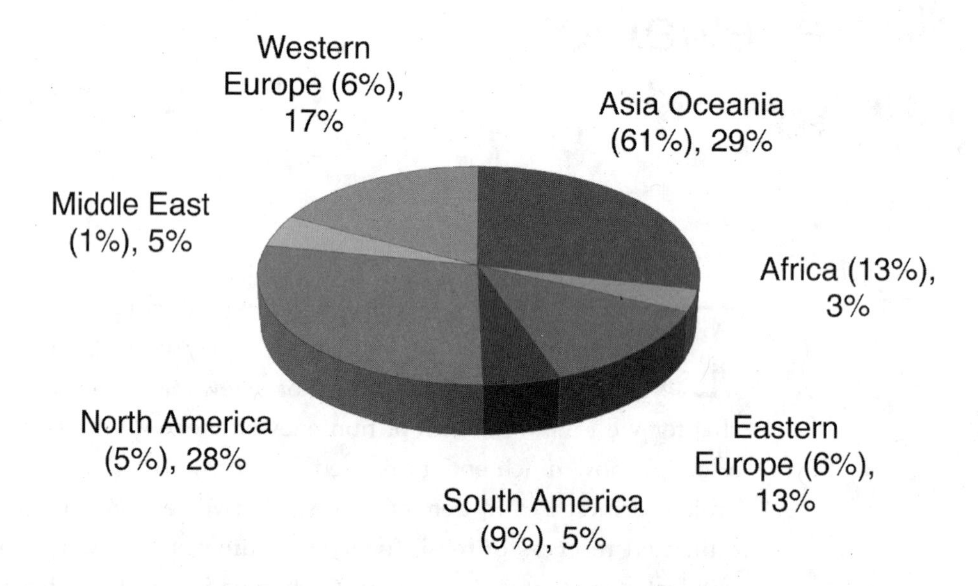

energy will have to be found. It should also be noted that Western Europe, which has a standard of living approximately the same as North America, has about the same population but uses significantly less energy.

A more detailed look at energy use for a few select regions is shown in Table 4.1. Of the 446.7 quadrillion Btus (1 quad = 1.055×10^{18} J; approximately an exajoule) of energy used globally in 2004, the United States used 100.7 quad, or about a fifth of the world total. This figure dropped slightly to 99.9 quad in 2007 but has remained near 100 quad since 2000. The percentage of energy used for industry is appreciably lower in the United States, reflecting the departure of energy-intensive heavy industries to markets where labor is cheaper. In contrast, the commercial use of energy is higher in the United States, reflecting the growth of service industries there. It is often argued that the enormous economy of the United States merits the larger percentage of world energy used, and in general it is true that larger economies need greater amounts of energy. But as Figure 2.7 illustrated, gross domestic product (GDP) and energy use are not tightly linked, and from Figure 2.8 we also saw that large disparities in energy

TABLE 4.1

Percentage energy use by sector for select countries and regions, 2004 [1]. OECD is the Organisation for Economic Co-operation and Development [4]. South and Central America figures do not include Brazil. The five sectors do not add up to 100% because electrical use is calculated separately and includes uses in other categories.

Entity	Residential (%)	Commercial (%)	Industrial (%)	Transportation (%)	Electrical (%)	Total Quad
World	10.7	5.5	36.6	19.6	39.2	446.7
OECD	11.8	7.6	30.2	24.1	38.7	239.8
Non-OECD	9.4	3.0	44.1	14.4	39.8	206.9
United States	11.3	8.3	26.0	27.7	38.6	100.7
China	7.4	3.0	47.1	7.4	46.0	59.6
India	11.7	3.2	40.9	9.7	48.1	15.4
Africa	9.5	2.2	40.1	21.9	38.0	13.7
South and Central America	8.9	3.7	40.7	25.2	33.3	13.5

use per capita do not always indicate higher GDP and other indicators of human well-being.

It is interesting to note that the percentage of electrical use (which includes use in other sectors) is roughly the same in OECD[1] countries and non-OECD countries. The rapidly growing economies of India and China have slightly higher than average percentages of electrical energy use.

Another interesting point in Table 4.1 is that the percentage of national energy use for residential purposes is similar for developed nations and less developed nations. This is not to say that populations in developing countries enjoy lifestyles similar to those of people living in developed nations. Per capita energy use in the United States is about 350 GJ per year, nearly 8 times as large as China's at 47 GJ and 25 times as large as India's at 14 GJ per person per year.

Table 4.2 lists the energy consumption of select appliances in the United States, which is typical of developed countries in this respect. Refrigerators consume the largest percentage of annual electrical energy of any appliance except air conditioning (heating, hot water, lights, cooling, and air conditioning make up 32%, 13%, 12%, 10%, and 9%, respectively, of residential energy used in the United States). The total number of households in the United States in 2001 (the last year for which these data are available) was 107.0 million, so this table also includes the interesting data that a very large percentage of homes had refrigerators, televisions, and lights but only about 70% had a computer. It might be assumed that energy consumption will level off as markets become saturated

TABLE 4.2

U.S. residential electricity use in 2001 for select appliances [5]. The total number of households was 107.0 million.

Appliance	Households (millions)	Annual per Household (× 10^9 J)	Total (× 10^{14} J)	Percentage of Total
Refrigerator	106.8	5.4	5.8	13.7
Air conditioning				
Central	57.5	10.4	3.9	14.1
Room	23.3	3.5	20.8	1.9
Space heating	43.8	14.9	4.3	10.1
Water heating	40.8	9.5	3.9	9.1
Lighting	107.0	3.5	3.7	8.8
Clothes dryer	61.1	4.0	2.4	5.8
Dishwasher	56.7	1.9	1.1	2.5
Electric stove	59.7		1.2	2.8
Color TV	105.8	1.2	1.2	2.9
Personal computer	68.4	1.5	0.7	1.6

and appliances become more efficient. For example, once everyone has a sufficient number of televisions, refrigerators, and computers, the demand for energy for these appliances should roughly reflect any population increase. However, as mentioned in Chapter 2, lower energy consumption due to increased efficiency in refrigerators has been nearly offset by increased energy demand by the use of personal electronics such as video games and cell phones.

Energy used for transportation in the United States, including transportation of goods, has increased from 18.6% of total energy use in 1973 to nearly 30% in 2004. Petroleum currently supplies more than 95% of this energy, and slightly less than 65% of the petroleum used in the United States is used for transportation. The percentage of total energy used for transportation in the United States is somewhat higher than the world average of about 20%; however, it should be noted that South and Central America are not too far below the United States and other OECD countries. The percentage of energy used for transportation in India and China, on the other hand, is less than half the global average. Again, if the populations of China and India begin to drive and move goods to the same extent as North America or Europe, the world will need significantly more energy.

A breakdown of transportation energy use in the United States is shown in Table 4.3. Walking uses about 0.2 MJ in the form of food per kilometer, and cycling

TABLE 4.3

Energy use figures for transportation in the United States, 2006 [6]. "Personal trucks" include SUVs. "Demand response" includes taxis, limousines, and other types of on-demand transportation. "Other" includes agricultural, construction, and other industrial transportation.

Transportation Method	Number of Vehicles (thousands)	Passenger km ($\times 10^9$)	Load Factor (persons/vehicle)	Energy Intensity (MJ per passenger km or MJ per metric ton km)	Energy Use (quad)	Energy (%)
Cars	136,430.6	4,295.1	1.57	2.2	9.331	30.9
Personal trucks	80,817.7	2,380.3	1.72	2.7	6.403	27.4
Motorcycles	5,767.9	17.9	1.1	1.4	0.025	<0.1
Demand response	37.1	1.5	1	8.7	0.013	<0.1
Vanpool	5.9	0.9	6.4	0.78	0.0007	<0.1
Buses					0.194	
Transit	65.8	34.2	8.7	2.7	0.091	0.6
Intercity				0.5	0.030	0.1
School	677.2				0.073	0.5
Air, commercial						8.4
Passenger	219.0	882.9	90.4	2.7	2.414	
Freight				15.8		
Recreational boats	12,770.0				0.247	
Rail						2.2
Passenger	19.1	50.1	23.7	1.9	0.093	
Freight				0.25	0.566	
Heavy trucks	8,482.0			2.80	4.535	15.5
Pipeline				0.20	0.822	2.9
Water				0.37	1.300	4.6
Walking		42.2	1	0.20		
Bicycle		10.0	1	0.15		
Other					2.203	6.5

cuts this figure to 0.15 MJ/km. Most transportation methods use approximately 10 times the energy per kilometer compared to walking. Vanpools, motorcycles, and rail are next in energy efficiency. Rail and bus figures are highly dependent on the number of passengers. Most buses can hold up to 30 passengers; a fully loaded bus would use a third of the energy per passenger kilometer of a personal automobile. Air transportation averages out to be about the same as cars or light trucks because most of the energy is involved in gaining altitude; once airborne, aircraft can travel a very long distance with little additional fuel. For freight, however, air transportation is the least efficient at 15.8 MJ per metric ton kilometer. Therefore, in order to achieve the greatest energy efficiency, air travel should be reserved for passengers who are traveling long distances.

4.2 ■ Future Demand for Energy

In order to predict future energy demands, especially for countries with newly emerging economies, it is instructive to look at changes in past energy use in developed nations. Rising economies will be able to avoid some of the past mistakes made by industrial economies and will have access to fully developed technology rather than having to invest in new knowledge. But we can still expect their energy history to be parallel to that of mature economies. Figure 4.2 shows U.S. energy consumption over the past 60 years by source. The major source of energy in the United States 250 years ago was wood. Coal gradually replaced wood, and petroleum use increased sharply in the 1930s, basically replacing further increases in the use of coal, which did not reach the same rate of consumption until the late 1970s. Fifty years ago the United States used about one third of the energy it uses today, and petroleum and coal contributed equally to the energy used. Petroleum now provides almost twice as much energy as coal in the United States. The use of natural gas has also increased significantly over the past 60 years. The energy used per capita in the United States increased 55% from 1949 to the late 1980s and has leveled off somewhat since then.

Two dips in petroleum use occurred, as can be seen in Figure 4.2; one around 1973 and the second beginning in 1979. The first dip was the result of an oil embargo by the major Middle East oil producers against countries that supported Israel in conflicts against Syria and Egypt in 1973. The second, much more significant drop in oil consumption was the result of several interrelated events including the fall of the Shah of Iran and the invasion of Iran by Iraq in 1980, both of which interrupted oil production in the Middle East, resulting in higher oil and natural gas prices. Both of these dips appear in world energy use as well. These events give important clues about the degree to which energy markets react to political and economic factors.

In the late 1970s the United States administration of President Carter instituted several programs including economic incentives for energy efficiency and development of new energy technology such as windmills, which resulted

FIGURE 4.2

U.S. energy use by source since 1950 [1]. Biomass before 1950 was mostly wood.

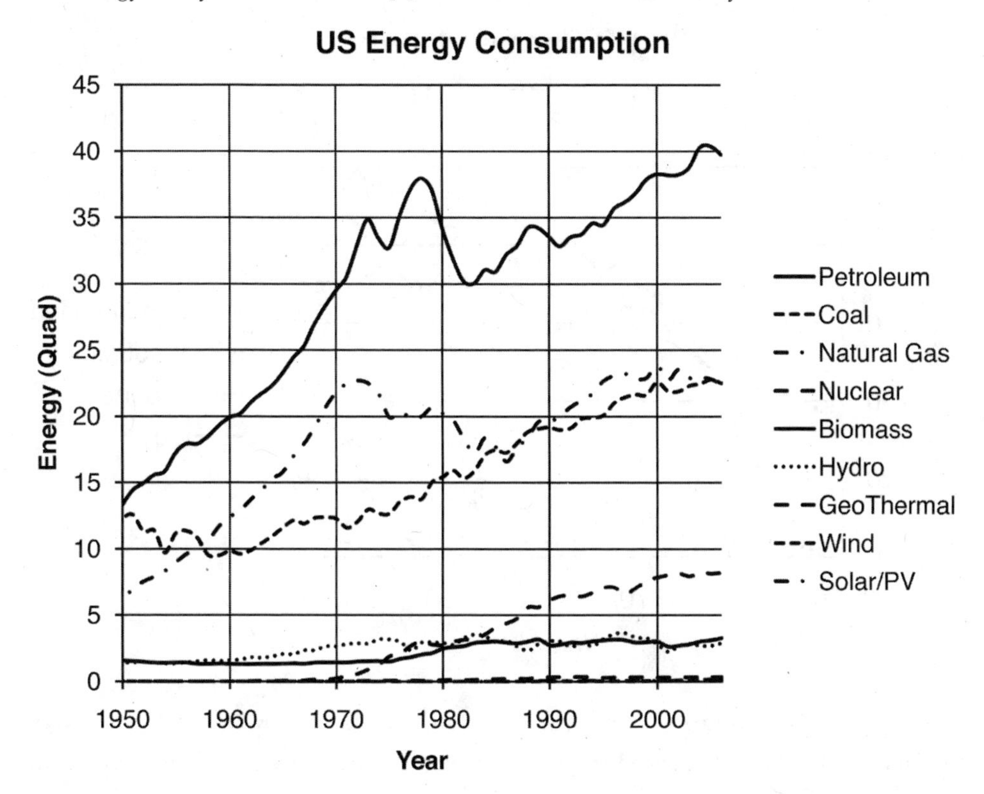

in increases in national energy efficiency. The increased energy efficiency and higher energy prices resulted in a 10-year period from the mid-1970s to the mid-1980s when energy use in the United States remained approximately level, as can be seen in Figure 4.2. Both the increased energy efficiency and the drop in energy consumption were mirrored in other nations globally. The U.S. GDP continued to rise during the period, in part because of new economic opportunities in the energy sector. However, as world oil prices fell in the late 1980s, most government-sponsored energy conservation programs were canceled and oil consumption began to rise again. Identical trends were seen globally.

A slightly different view of the history of energy use in the United States is seen in Figure 4.3. Energy production remained nearly constant after the early 1980s, but total energy consumption continued increasing. This necessitated a much greater dependence on imported energy, which is clearly seen in the graph. Energy imports now make up about a third of the energy the United States uses

FIGURE 4.3

Historical energy consumption, production, exports, and imports for the United States [1].

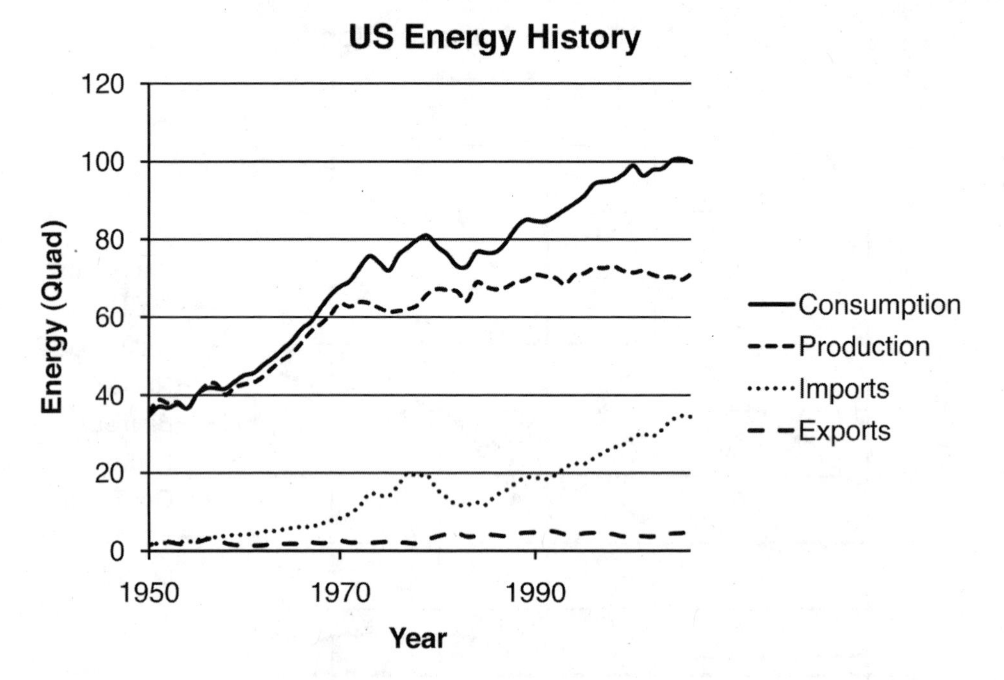

($700 billion in foreign oil in 2008), compared to less than 10% 50 years ago. Similar trends have occurred in other developed countries and are expected to occur in other non–oil-producing countries as they reach comparable levels of development. For example, in 1950 the United States had 76% of the world's automobiles. This percentage fell to 30% by 1990 and is currently about 21% of the world total despite a more than tripling of the number of automobiles in the United States since 1950 [6]. This is only one indication that as countries become more developed they use more energy. World energy consumption increased 6%, from 447.2 Quad in 2004 to 472.3 Quad in 2006. Nearly all this increase in energy use occurred in China and India. During this same time period energy consumption in the United States actually declined by about 0.5 Quad.

In 2004 the top 10 suppliers of crude oil to the United States were Canada (1,616,000 barrels per day), Mexico (1,598,000), Saudi Arabia (1,495,000), Venezuela (1,297,000), Nigeria (1,078,000), Iraq (655,000), Angola (306,000), Kuwait (241,000), United Kingdom (238,000), and Ecuador (232,000). It is sometimes mistakenly thought that the United States could simply buy more oil from Canada or Mexico if there were interruptions of supply from other countries. However, as seen in the dips in the graphs of Figures 4.2 and 4.3 (which were mirrored in the global market), the disruption of one supplier in the global market causes prices to go up for all buyers, regardless of which source is providing oil to any given country. The global market for petroleum also explains why the United States exports oil at the same time it imports oil. It makes sense to export oil from Alaska to Asia and import to the lower 48 states from South and Central America. Because the price is fixed globally, the distance the oil travels is the only important factor determining where imports and exports occur. A more detailed analysis of political factors affecting the market for petroleum can be found in Chapter 3 of reference [7]. Clearly, if increases in energy use by rapidly developing countries such as China and India follow the historical trends of the United States, there will be additional geopolitical pressure on the global energy market.

Figure 4.4 shows the global price of crude oil since 1860 in nominal dollars (the price paid at the time) and real dollars (what a barrel then was worth in today's dollars). Initially petroleum was a scarce commodity, so real prices remained high. The spikes in prices in 1973 and 1979 are clearly evident and were mirrored by fluctuations in gasoline prices at the pump. It is not clear

FIGURE 4.4

Nominal and real price of a barrel of crude oil, in 2006 dollars, through 2009 [1].

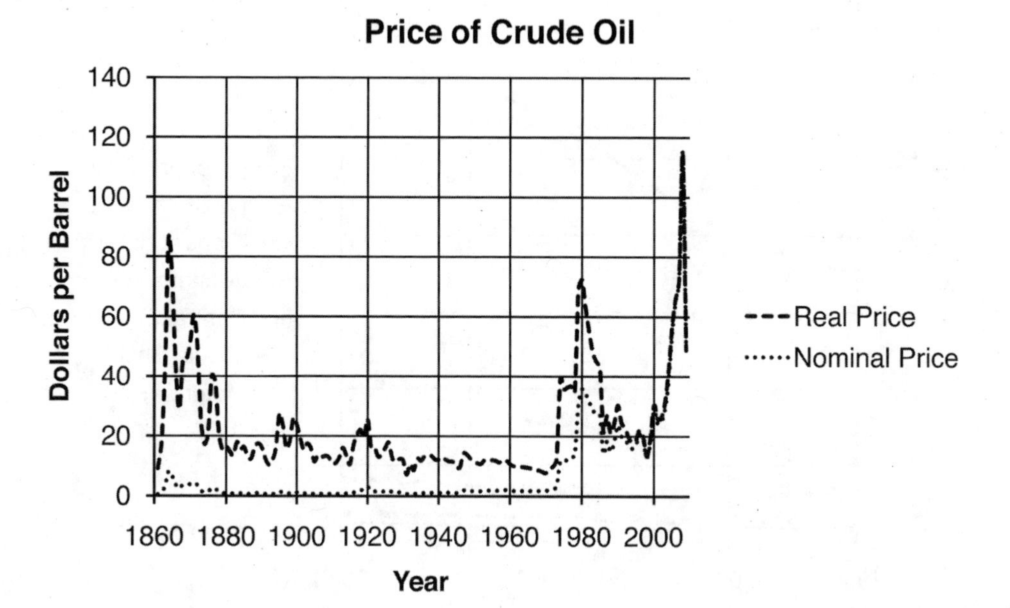

whether the sharp increase in oil prices in 2008 was the result of real shortages in supply or a temporary fluctuation. At least part of the spike in gasoline prices that occurred at the same time had to do with a sharp reduction in the number of refineries available to convert crude oil into gasoline over the past 25 years. A reduction in refinery capacity causes prices at the pump to go up even if there is plenty of crude oil. However, it should be noted that the consumers in the United States enjoy some of the lowest gasoline prices in the world (a handful of countries in the Middle East being the exception). Pump prices for gasoline in Europe and Japan are approximately three times higher than in the United States because of taxes and other government policies designed to encourage higher efficiency.

As shown in Figure 4.5, the U.S. Energy Information Administration (EIA) predicts that world energy demand will grow by about 60% in the next 25 years,

FIGURE 4.5

Projected (as of 2009) energy demand by world region [1]. "Emerging Asia" includes China, Korea, and India. "Transitional" includes former Soviet Union territories and Eastern European nations.

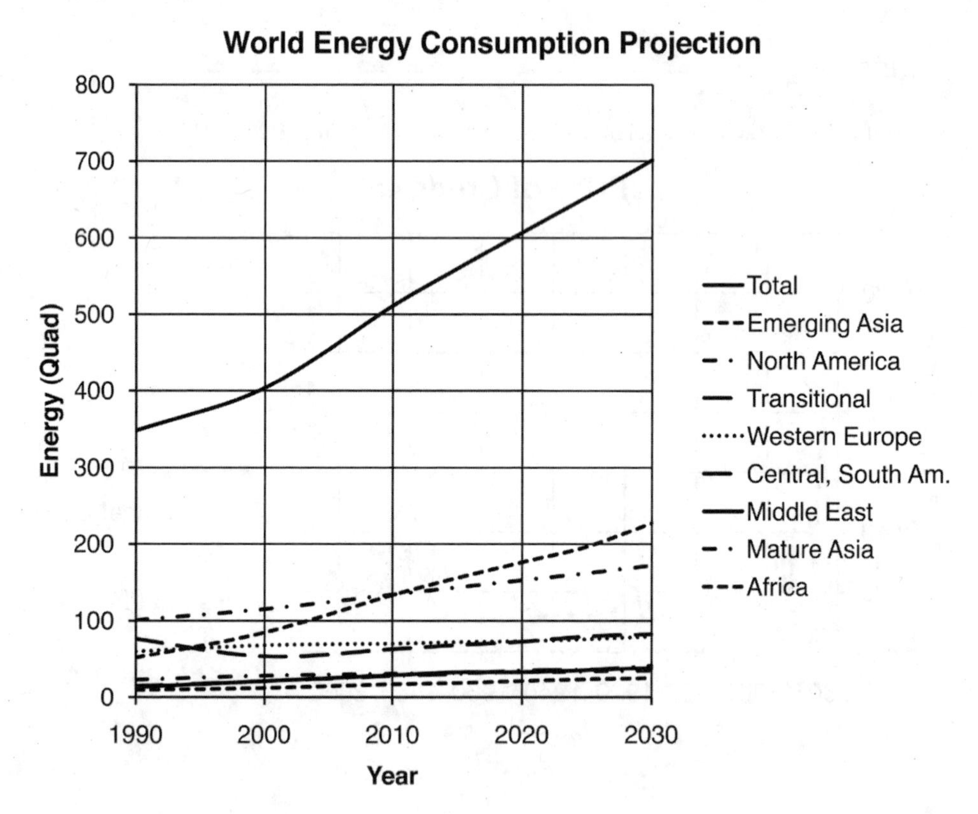

despite the economic downturn of 2008–2009. This is consistent with the 70% increase in global energy use since 1971. The use of liquid fuels (petroleum) is projected to increase by 1.4% per year, natural gas by 1.9%, coal by 2.2%, nuclear by 1.4%, and other fuels (hydro, renewable) by 1.9% each year over the next 25 years. Electrical demand is expected to double in the next 25 years. Most of this demand will come from nations experiencing rapid economic growth, such as China, India, and Russia, whose economies grew at a rate of 11.4%, 8.5%, and 7.4%, respectively, in 2007, whereas the U.S. economy grew at a rate of 2.2% [1, 3]. The recession of 2008–2009 reduced growth in the United States to near zero, but China experienced more than 6% growth and India more than 5% in 2009. China is expected to increase its liquid fuel use by 3.7%, natural gas by 6.5%, coal by 3.0% per year, and electricity use by 3.5% per year over the next 25 years. The figures are similar for India. For the United States these percentage increases are 1.0% or less for fossil fuels and 1.2% for electricity use per year for the next 25 years.

At this point it should be mentioned that long-term energy projections are often wrong, sometimes significantly so [7]. Clearly factors such as the rate of economic growth, population increases, the prices of energy, and proposed limitations on carbon emissions will affect energy demand. The oil embargo of 1973 and the political turmoil that resulted in the oil crisis of 1979 were not predictable events, nor was the economic downturn of 2008–2009. The numbers in Figure 4.5, calculated by the EIA [1], reflect what is known today about the world's economy, population, and political situation. If the world economy grows faster or slower, the figures in the graph could be different, and a practical strategy is to calculate a range of predictions. The EIA predicts an increase in global energy use as high as 2.1% per year for a world with high economic growth but an increase of only 1.4% per year for a world with stagnant economic growth over the next 25 years. If oil prices continue to rise, this will restrain energy use, and the EIA projects an increase of 1.6% per year for the next 25 years. But if oil prices level off and remain steady, the EIA projection is a 1.9% per year increase in global energy use.

There is a gathering global political momentum to cap carbon emissions because of the threat of global warming. All fossil fuel technology in use today emits carbon dioxide. The 100 Quad energy use of the United States in 2006 represents the emission of some 5.9×10^{12} metric tons of carbon dioxide into the air (19.8 metric tons per person, compared to a global per capita average of 4.5 metric tons). A significant cap on CO_2 could drastically affect the energy market because about 86% of the world's energy currently comes from fossil fuels. Improved energy efficiency could also have an effect on demand. As we saw in Figure 2.8, the energy needed to provide a given increase in the GDP has declined over time. This is also likely to continue as heavy manufacturing becomes more energy efficient, new energy technologies come online, and the world economy shifts toward a less energy intensive service economy. Nevertheless, given the expected increase in population discussed in Chapter 1 and the current economic growth in places such as China and India, we can expect significant increases in energy use in the short term.

4.3 ■ Global Sources of Energy

Considering the energy used today and projections for the future, you might wonder, "Where does the world get more than 450 quad of energy per year?" As we have learned, roughly 86% of the world's energy currently comes from fossil fuels. As shown in Figure 4.6, only about 1% of the world's energy currently comes from renewable resources, or about 7% if hydroelectric is included. Although renewable energy is the fastest-growing energy sector, it has a long way to go to catch up with our current use of fossil fuels.

A more detailed look at fossil fuel use for select regions is shown in Table 4.4. As seen in the table, the percentage of fossil fuel use is slightly higher in less developed nations than in the OECD nations. South and Central America's reliance on hydroelectric and biofuels, particularly in Brazil, makes it one of the regions least dependent on fossil fuels. Even so, this region still depends on nonrenewable resources for more than 80% of its energy needs.

A few interesting fine points are obscured by the broad outline shown in Table 4.4. For example, the world gets 6.2% of the total energy used from nuclear power, which is about 17% of the electricity generated globally. France has the highest commitment to nuclear energy, supplying more than 75% of its electricity from nuclear. This figure is about 30% for Japan, which must import about 80% of its total energy use. More than 19% of the electrical energy in the United States is generated from nuclear power, 8.1% of total energy used, despite the fact that no new nuclear power plants have been ordered since 1978. Renewable and hydroelectric sources vary similarly from one local region to another. For example, Brazil relies on hydroelectric for more than 80% of its electricity, and Norway generates nearly 100% of its electricity from hydropower. Brazil also supplies about 20% of its transportation needs from ethanol derived from sugar cane. These renewable fuels will be discussed in the next chapter.

FIGURE 4.6

World energy use by source, 2009. About 86% of the world's energy comes from fossil fuels [8].

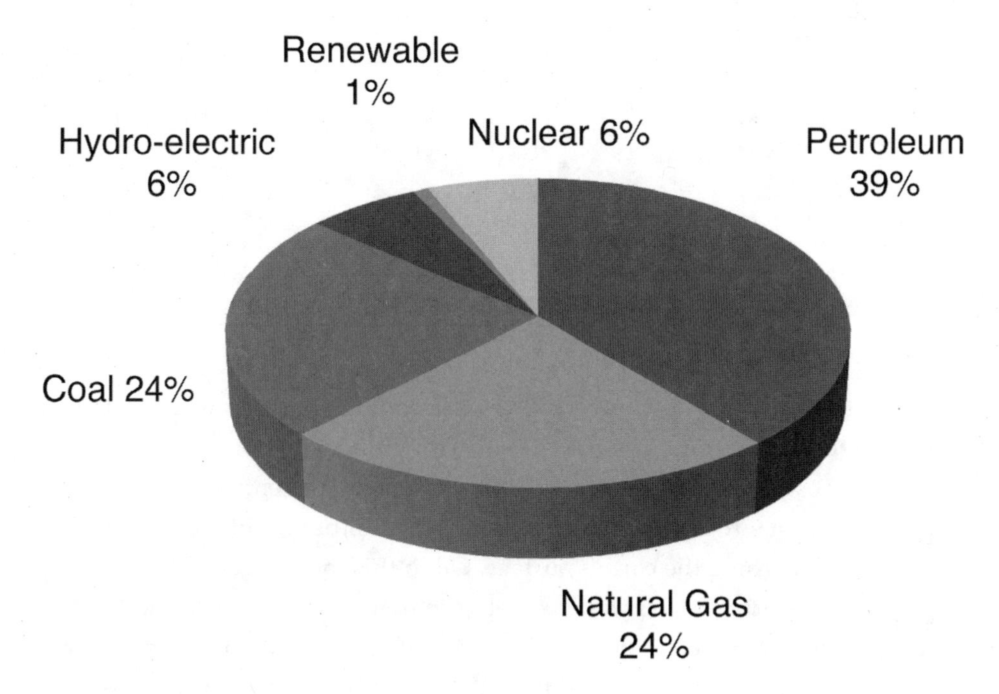

4.4 ■ Fossil Fuel Resources

World fossil fuel reserves as estimated in 2005 by the U.S. Department of Energy are given in Figure 4.7. Although these figures are only approximate, they illustrate what is currently known about global energy reserves. From the figure it can be clearly seen that the Middle East holds the majority of the known petroleum resources, whereas North America, Eastern Europe, and Asia (mostly China) have large reserves of coal. Coal clearly will play a role in the world's energy future. Shale oil and tar sands contribute only a very small fraction of the world's energy, but they may eventually be an important energy source. The discussion in this section is focused mainly on petroleum because more is known about that particular fossil fuel, given its expanded use over the last 50 years, but it should be kept in mind that many of the same points can be made about natural gas, coal, shale oil, and tar sands.

TABLE 4.4

Percentage energy use by source for select countries and regions [1]. Fossil fuels include liquids, natural gas, and coal. Nuclear, renewables, and fossil fuels add to 100% (with some rounding errors). South and Central America figures do not include Brazil. OECD = Organisation for Economic Co-operation and Development.

Entity	Liquids (%)	Natural Gas (%)	Coal (%)	Nuclear (%)	Renewables and Hydro (%)	Fossil Fuels (%)
World	37.7	23.1	25.6	6.2	7.4	86.4
OECD	41.2	22.1	19.4	9.7	7.5	82.8
Non-OECD	33.5	24.3	32.8	2.1	7.4	90.6
United States	40.5	22.9	22.4	8.1	6.0	85.9
China	22.0	2.7	69.0	0.8	5.5	93.6
India	32.5	7.8	52.6	1.3	5.8	92.9
Africa	41.6	20.4	29.9	0.7	6.6	92.0
South and Central America	51.1	27.4	3.0	0.7	18.5	81.5

The term *crude oil* or *petroleum* refers to a combination of a wide variety of organic carbon compounds found in a surprising diversity of geological formations. The most easily accessed petroleum, known as conventional oil, is found trapped in porous rock formations capped by an impermeable layer that prevents it from reaching the earth's surface. Oil in this form is generally found in large underground reservoirs layered between natural gas, which is lighter, and very saline water, which is denser. Because the rocks of the oil formation are permeable, the oil may be pumped from the ground easily once a well has been drilled through the impermeable capping layer.

Petroleum may also be found trapped in layers of shale, where it is called shale oil, or mixed with sand or clay, where it is called tar sand or bitumen. In these forms the oil adheres to the rock or sand because of surface tension and cannot be easily pumped out of the ground. Extraction techniques for shale oil and bitumen are still in the experimental stage, but two methods have been used so far. Cavities can be drilled into the clay or sand deposit and steam injected to make the oil less viscous, after which it collects in the cavities and can be pumped out. A second method involves surface mining the shale clay or sand and then vigorously mixing it with hot water and air. The petroleum particles in the sand adhere to air droplets in the mixture and rise to the top, where they can be separated from the remaining material [12]. Both methods use significant amounts of energy to extract the petroleum because they involve heating the material that contains the deposit. Another problem is that the process results in an expansion of the remaining waste by 35%, making the disposal of waste material a problem. The debris cannot just be put back into the location where the shale was removed because it has a larger volume and is in the form of a semiliquid sludge.

Oil, natural gas, and coal form over geological time spans (between a few million and a billion years) from large quantities of organic material that have been buried in the earth's crust by tectonic activity. All the fossil fuel reserves mentioned in Figure 4.7 were formed millions of years ago; any new formation of fossil fuel will not be available until millions of years in the future. Organic marine sediments are the source of oil, whereas terrestrial plant material is the source of coal, as can be verified by the fossils often found in coal. As this material sinks and is covered by other sediment or folded into the earth's crust by tectonic activity, it becomes more exposed to heat emanating from the earth's

FIGURE 4.7

Estimated fossil fuel reserves by region in 2005 [1, 9–11]. FSU = former Soviet Union. One quad is 1.055×10^{18} J, approximately the energy found in 172 million barrels of oil.

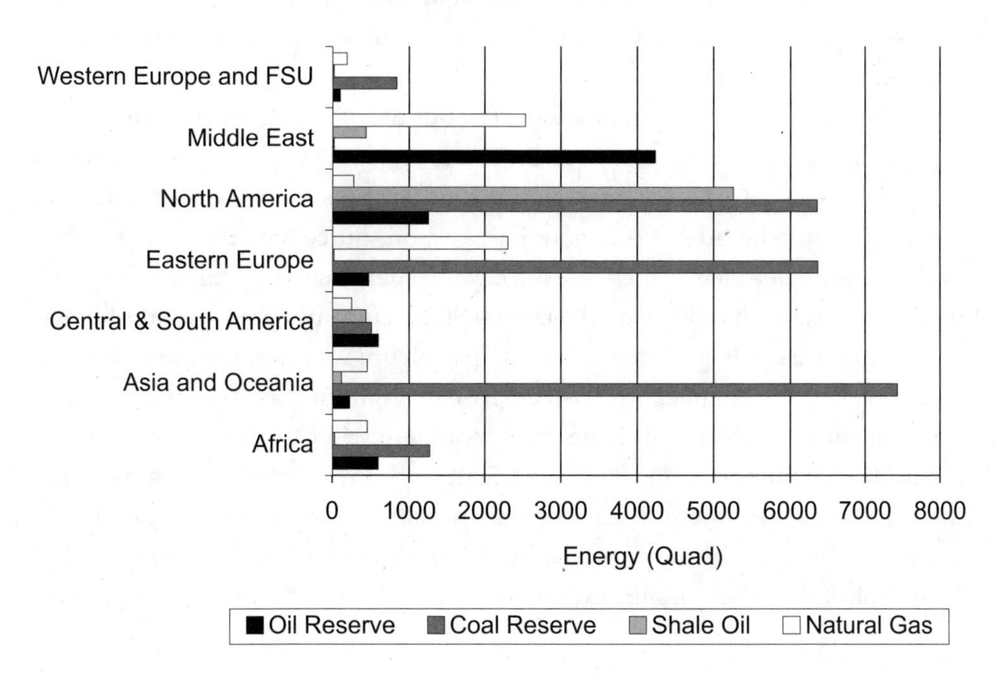

core. Organic material under anaerobic conditions undergoes a chemical process called pyrolysis when heated between 50°C and 120°C to form oil and natural gas. The longer the heating process is maintained, the more natural gas is formed. As the depth of the organic material increases, the temperature and pressure also increase so that eventually the oil and natural gas fracture the surrounding rock and migrate upward until they reach an impermeable boundary to collect as conventional oil, or the oil adheres to sand or shale, where it is found as shale oil or bitumen. Once they have escaped the heat and pressure from deeper in the earth's crust, pyrolysis stops and the oil and natural gas are stable. If petroleum reaches the surface, bacteria will begin to feed on the hydrocarbon compounds in the oil and eventually break the oil down into other organic compounds.

Coal is formed from a process very similar to oil and natural gas but starting with terrestrial plant material, typically peat. Crustal folding eventually brings the coal deposit closer to the surface, where it can be mined. The resulting resource has a variable carbon content depending on when and exactly how it formed. Anthracite has a carbon content of about 95% and is the oldest type of coal. Bituminous coal has a carbon content between 50% and 80% and formed some 300 million years ago. Lignite is the youngest coal, formed about 150 million years ago, and has a carbon content less than 50%. Burning any of these various types of coal produces carbon dioxide as a pollutant, but burning anthracite gives off the least amount of other pollutants such as sulfur because it is nearly pure carbon.

Natural gas is mostly methane (CH_4) but usually contains trace amounts (up to 5%) of other carbon compounds. The processes that create coal and petroleum also create natural gas; most natural gas deposits are associated with coal and oil reserves. Because it is a more pure carbon source than the various types of coal or even petroleum, methane is a cleaner energy source, emitting only carbon dioxide when it is burned. The United States currently produces most of the natural gas it uses. In part this is because petroleum was exploited first, leaving natural gas deposits intact. It is also the case that natural gas is more difficult to transport than petroleum. It is unlikely that there will be a significant increase of natural gas imported to the United States from resources such as the Middle East because of the difficulty of transportation. Liquid natural gas (LNG) cooled to –160°C is the most efficient form for transportation if a pipeline is not feasible, but there is a significant potential for explosion in the event of a tank rupture or leak.

There are several difficulties involved in trying to estimate the amount of oil or other fossil fuels available globally [7]. A first estimate gives how much oil is believed to be in the ground, an amount called the oil in place. This number is evaluated based on known geological features, data from seismology, gravimetric and magnetometer measurements, and sample drilling. An appraisal, generally based on test wells, of the amount of oil that can be extracted with 80% to 90% confidence using current extraction techniques is known as the proven reserve. This category can be further broken down into developed reserves (fields that already have wells) and undeveloped reserves. Unproven reserves may be divided into possible reserves (with 10% confidence that oil can be extracted using current techniques) and probable reserves (50% confidence). Neither estimate of proven or unproven reserves generally includes nonconventional oil such as oil shale or tar sands, although the tar sands in Canada have recently been included in the list of proven reserves. The estimated ultimately recoverable oil is the cumulative oil production to date (the oil that has already been extracted) plus the proven reserves. Similar terminology applies to natural gas, coal, shale oil, and tar sand deposits.

The reserves shown in Figure 4.7 are the current estimates of proven reserves for oil and natural gas. Shale oil and tar sands for North America (mostly Canada) are now considered to be proven reserves. Six of the top 10 proven oil reserves are found in the Middle East; the top 10 countries and their respective oil reserves are Saudi Arabia (1,531.1 quad), Iraq (655.5 quad), United Arab Emirates (568.5 quad), Kuwait (562.7 quad), Iran (522.1 quad), Venezuela (423.5 quad), Russia (284.3 quad), Libya (174.0 quad), Mexico (162.4 quad), and China (139.2 quad).

A second problem in knowing how much oil is ultimately available has to do with extraction techniques. Typically in a new reservoir the oil and natural gas are found under pressure and can be extracted by merely drilling a well. About 20% of the oil in a typical reservoir can be tapped by this primary extraction technique. Once the naturally occurring pressure drops to the point that the well is not producing efficiently, another 15% to 20% of the oil in the reservoir can be extracted with secondary techniques such as pumping or pressurizing the oil by injecting natural gas, water, air, carbon dioxide, or other gases. Tertiary extraction techniques include heating the oil in the reservoir to make it less viscous, either by injecting steam or burning some of the oil while it is still in the reservoir, which then heats the remaining oil. Detergents can also be injected to make the oil less viscous and easier to pump. The total amount of oil that can be extracted from a typical reservoir using today's technology varies depending on the type of oil, in particular its viscosity, but generally is less than 60% of the oil in the reservoir. Proven oil reserves are estimates based on current extraction techniques, so as new technologies are developed the estimated global reserves will tend to increase, even if no new oil is found.

The flow of liquids and gases through a porous solid is governed by an empirical law known as Darcy's law. The equation can be applied to water flow through an aquifer as well as petroleum through an oil field. The rate of flow, Q (in cubic meters per second), of the liquid is proportional to the pressure gradient, the-cross sectional area of the flow, A, and the permeability of the solid, κ (in square meters). The flow rate is inversely proportional to the dynamic viscosity, μ, of the liquid (in kg/s/m) = Pa · s) and the distance the fluid will travel, L. Darcy's law for flow in one dimension is

$$(4.1) \quad Q = -\frac{\kappa}{\mu}\frac{A\Delta P}{L},$$

where the pressure gradient, ΔP, is the difference in pressure between two locations, such as the pressure in the oil field at one location and the pressure (assumed to be close to atmospheric pressure) at the surface where a well is to be drilled.

Permeability is a measure of the ability of a solid to allow fluids to flow through it. Gravel and loose sand have permeabilities near 1×10^{-8} m^2 , whereas solid granite has a permeability of about 1×10^{-18} m^2. Oil field permeabilities typically range from 1×10^{-11} m^2 to 1×10^{-13} m^2. Dynamic viscosity is a temperature-dependent parameter indicating how easily an incompressible fluid will flow. The viscosity of water at room temperature is approximately 1×10^{-3} Pa · s and varies by a factor of 10 between freezing and boiling temperatures. The dynamic viscosity for crude oil varies between 5×10^{-3} Pa · s and 20×10^{-3} Pa · s depending on the exact chemical makeup of the oil and its temperature. A third factor is the degree of saturation of the permeable solid by the liquid. Porous solids that are not completely saturated are resistant to fluid flowing through them because of capillary action; the fluid clings to the surfaces of the channels through which it is flowing. This additional resistance to flow is sometimes called soil–liquid suction. The effect is the same as that of a decrease in the permeability for unsaturated solids.

EXAMPLE 4.1

Water Flow Rate through an Aquifer

Suppose rain falls in the hills surrounding a farm. The water flows through a porous layer of soil with permeability 1.0×10^{-10} m^2 from a height of 6 m to a height of 2 m, where it comes out of the ground as an artesian well. What will be the flow rate if the cross-sectional area of the porous layer is 0.5 m^2 and the water must travel 1,000 m from where it is absorbed to where it flows out of the ground? Assume a dynamic viscosity of 1.0×10^{-3} Pa · s for water.

The pressure difference between where the rain enters the ground and where water comes out at the artesian well is $\Delta P = h\rho g = 4 \text{ m} \times 1000 \text{ kg/m}^3 \times 9.8 \text{ m/s}^2 = 39000$ Pa. Darcy's law then gives a flow rate of $Q = -\frac{\kappa}{\mu}\frac{A\Delta P}{L} = 2.0 \times 10^{-6} \text{ m}^3\text{/s}$, which is about 0.2 m^3/day.

Primary extraction of petroleum, a process similar to an artesian well for water, uses very little energy because the liquid flows out of the well due to the higher pressure underground. About 5% of the energy recovered is used in the extraction process (see [12–15] and references therein). This does not include energy used in refining the petroleum into a usable form such as gasoline, which uses another 15% of the energy. Secondary and tertiary extraction techniques obviously use more energy for the process because they involve pumping or heating the oil to make it flow more readily. Current extraction techniques for tar sand and shale oil use as much as 50% of the energy extracted. In addition, the shale oil extraction process uses 3 gallons of water for every gallon of oil produced, putting additional stress on water resources. Although these methods will undoubtedly become more efficient over time, it should be clear that at some point the amount of energy needed to extract the remaining resource will be more than the energy contained in the extracted oil. At the point where the energy gain from the extracted resource is less than the energy needed to remove the remaining amount, it makes no sense to continue extraction, no matter how much oil remains in the reservoir. This is a physical limitation for extraction that is immune to economic considerations; an increase in the cost of a barrel of oil does not affect the point at which the amount of energy used supersedes the energy extracted.

Aside from the technical factors involved in estimating the size of oil reserves, there are political and economic uncertainties. Most technical information about oil reserves is maintained by private oil companies, which have an economic incentive to inflate their worth by overestimating the amount of oil they have access to. Likewise, the governments of countries that produce oil gain in economic and political status, both at home and abroad, by reporting inflated energy resources. Figure 4.8 shows changes in proven oil reserves over the past 25 years. Some analysts suspect that the sudden increase in proven oil reserves that occurred in the late 1980s in the Middle East was actually a reevaluation based on political concerns rather than any real increase in the amount of oil available. The increase in 2003 reflects the decision to include the Canadian tar sands as a proven reserve.

FIGURE 4.8

Estimated proven oil reserves have changed over time [1, 14].

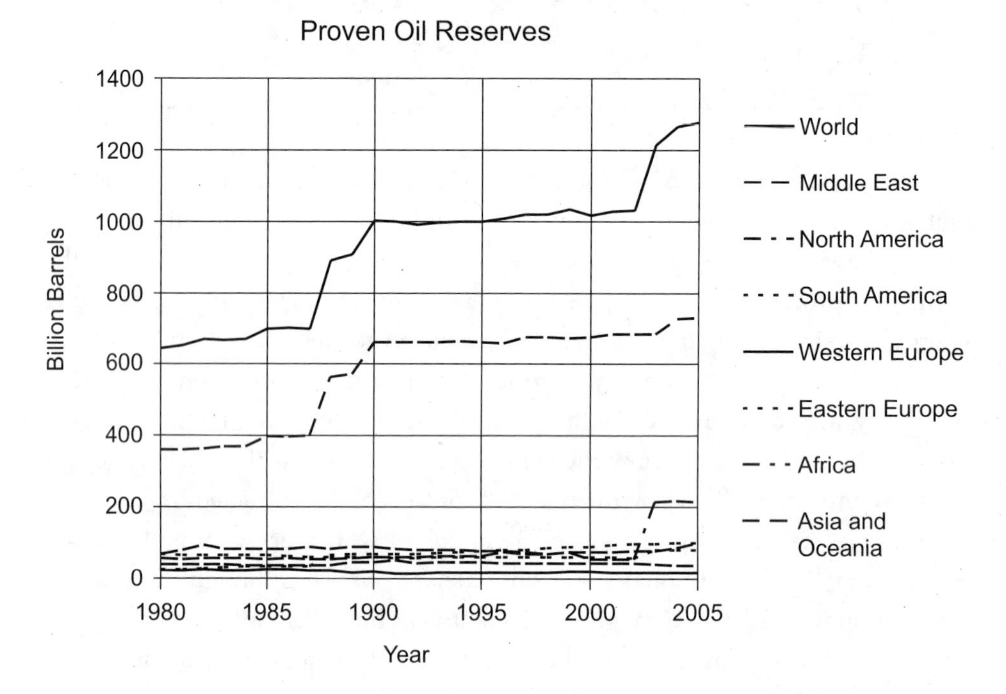

It is logical to ask how much more oil might be discovered in the world. New oil fields are found every year, but there is reason to think that no more extremely large fields, such as those found in the Middle East, will be found. One reason is that all of the terrestrial surface and most of the shallow water on the globe has already been explored thoroughly [14, 15]. Most geological formations that have porous layers capped by an impermeable layer and thus are potential oil reservoirs have been explored. As mentioned earlier, oil exposed to high temperature over long periods of time eventually turns to natural gas. This means that any oil that existed at depths greater than current wells probably has been converted into natural gas. Therefore, drilling deeper is not likely to uncover significant amounts of new oil. A graph of the total oil discovered per year over the past 40 years is shown in Figure 4.9. As can been seen, the amount of oil being discovered is declining. Eighty percent of today's oil comes from fields discovered more than 40 years ago. In the past few years there have been several new oil discoveries touted in the press as "huge" or "major." For example, in 2007 a large field was discovered in Brazil, but the total estimated

FIGURE 4.9

Global oil discovered by year [14–16].

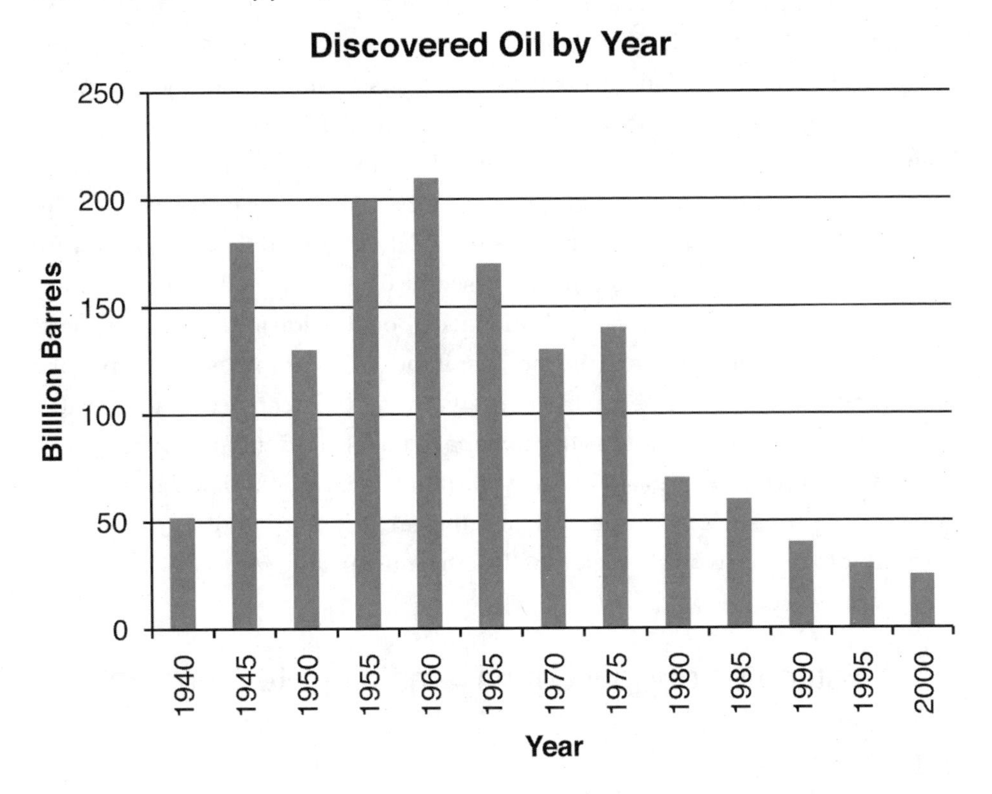

size was only 8 billion barrels, equivalent to approximately 46.4 quad, less than half the total energy used by the United States in 1 year. A large field discovered in the Gulf of Mexico in 2006 may have as much as 15 billion barrels, but again, this is less than the total energy the United States consumes each year. The high estimate for the oil reserves in the Alaskan North Slope is 11.8 billion barrels, which is about 64.9 quad, equivalent to 2 years of the petroleum used annually by the United States.

Finding other kinds of fossil fuels is possible, and in fact there has been recent discussion of clathrate hydrates, which consist of methane trapped in ice. *Clathrate* is a generic term for a crystalline structure that contains trapped gas, and generally they can exist only under high pressure and low temperature. In nature, methane clathrate hydrates are found on the deep sea floor, in arctic permafrost, and in ocean sediments. These deposits have long been known to exist and are often found during drilling for oil or natural gas in the deep sea floor or in arctic regions. Initial estimates of how much of this resource exists were as high as 3×10^{18} m^3 of methane, which would be about 1,000 times more energy than the estimated coal reserves cited in Figure 4.7. Subsequent surveys and a better understanding of how clathrates form have revised that estimate downward by a factor of about 1,000. If it does exist in these quantities, methane clathrate could be a potential energy source, equivalent to world coal deposits. However, because this resource is not found in concentrated reservoirs, as oil, coal, and natural gas are, and because it forms only in deep ocean or under frozen sediments, it is not clear how or even whether it could be captured for human use [17]. Another concern, discussed in Chapter 7, is that methane is a fossil fuel, and its use will release additional carbon dioxide into the atmosphere. Connected with this concern is the fact that if the earth continues to warm at the present rate, thawing the permafrost and warming the deep ocean, there is the potential for enormous amounts of methane to be released into the atmosphere. Methane is a greenhouse gas that acts very much like carbon dioxide and water in increasing the surface temperatures of the earth. Vast quantities of methane released into the air would accelerate the current warming trend.

4.5 ■ Estimates of How Long Remaining Resources Will Last

As mentioned in the previous section, projections of how much fossil fuel will be needed for human consumption, the amount available, and the rate at which it will be used in the future are probably not accurate for projections longer than 15 or 20 years [7]. This is because of our inability to predict political and economic changes and technological innovation, all of which strongly affect the energy landscape. It is also the case that switching from one fuel to another, for example from petroleum to natural gas, will delay the depletion of one fuel at the expense of another. Nevertheless, several mathematical models have been applied in attempts to predict how much of a resource remains and when the peak production year will occur for that fuel. Here we introduce the method of M. K. Hubbert, which at least has the merit of being one of the most popular methods of predicting peak oil production. In 1956 Hubbert predicted that oil production in the continental United States would reach a peak between the late 1960s and early 1970s and decline thereafter. A figure from his article is shown in Figure 4.10. Oil production did peak in the continental United States in 1970, and Hubbert's prediction of the amount produced that year was also fairly accurate. Oil production for the United States did not continue on the curve shown in the figure because of new oil found in Alaska; however, oil production in Alaska peaked in 1988. As we saw in Figure 4.3, the energy production in the United States leveled off after the 1970s, and the increase in consumption of petroleum after 1970 was possible only because of increasing imports.

Hubbert started with the assumption that, much like any new resource, the discovery of oil would initially increase exponentially but then level off (curve in Q_D Figure 4.11) at some finite value, Q_∞ after which all the available oil is found. The cumulative production curve, Q_P would have a similar shape but lag behind the discovery curve by a few years as markets expanded. The amount remaining in the reserve as a function of time would be $Q_R = Q_D - Q_P$ which clearly drops to zero once all the discovered oil has been produced. The curve Q_R has been modeled as a Gaussian or normal distribution curve, which gives reasonable results [13]. A slightly more sophisticated approach follows. The area under the curve Q_R is the ultimate recoverable resource, Q_∞—the total amount of oil that will ever be recovered from the reservoir.

FIGURE 4.10

Excerpt from Hubbert's article predicting the peak of oil production in the United States [18]. The area under the curve up to 1956 is the cumulative production, Q, in billions of barrels (bbls) for 1956.

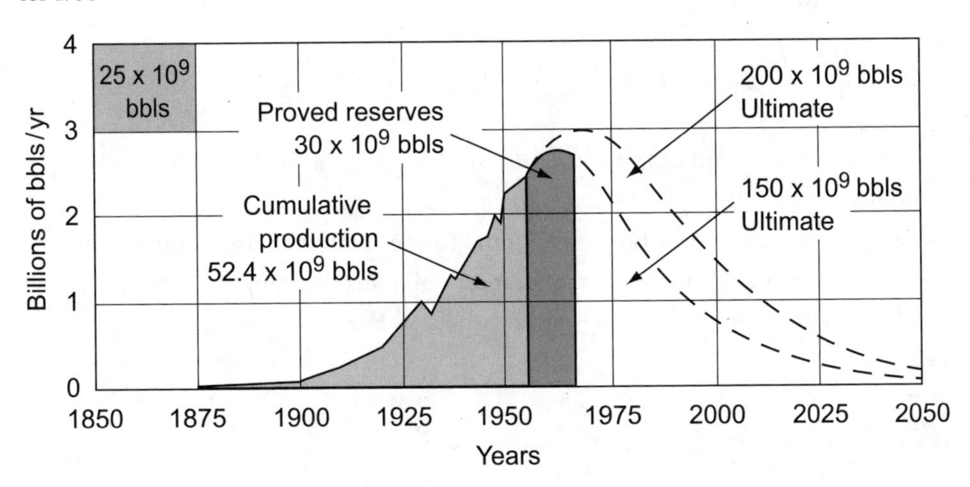

FIGURE 4.11

Theoretical resource discovery, Q_D, production, Q_P, and amount remaining, $Q_R = Q_D - Q_P$. The time of peak production, t_{peak}, is shown.

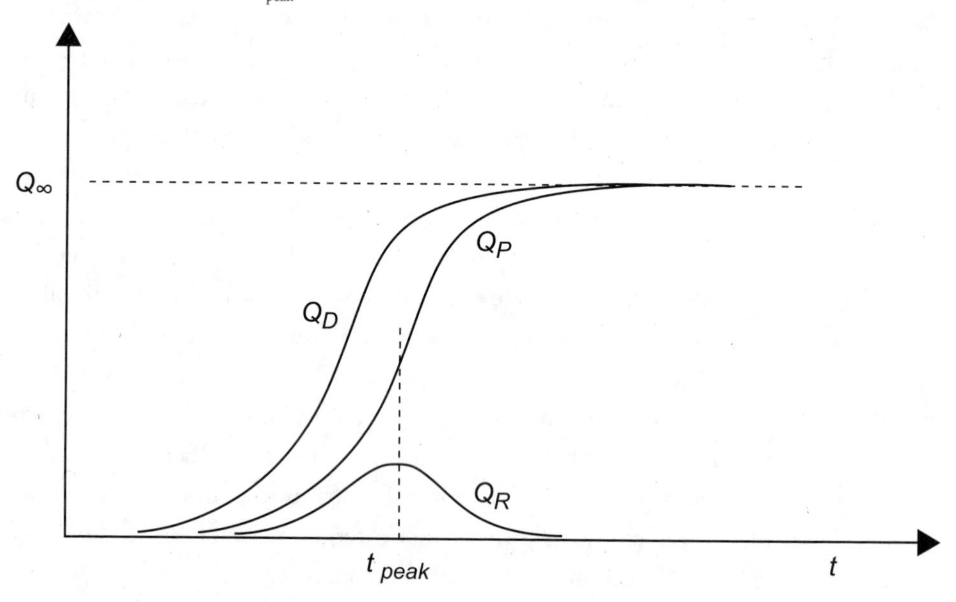

As we saw in Chapter 1, for exponential increase of a quantity, Q, we have $Q(t) = Q_o e^{\lambda t}$. Here $Q(t)$ is the cumulative production at time t, Q_o is the known cumulative production at some initial time, which we will call $t = 0$, and λ is the exponential growth rate. The rate of production, $\frac{dQ}{dt}$, for exponential increase is thus given by the equation

$$\frac{dQ}{dt} = \lambda Q_o e^{\lambda t} = \lambda Q. \tag{4.2}$$

Hubbert realized that the rate of resource production would increase exponentially only at first. Once a significant proportion of the resource has been extracted, the rate would be proportional to the fraction of remaining resource, $1 - Q(t)/Q_\infty$, where Q_∞ is the total resource available assuming the costs of extraction and extraction techniques remain constant. He therefore replaced Equation (4.2) with

$$\frac{dQ}{dt} = \lambda Q\left(1 - \frac{Q}{Q_\infty}\right), \tag{4.3}$$

which is known as the Verhulst or logistic equation. Equation (4.3) can be solved to give

$$(4.4)\quad Q(t) = \frac{Q_\infty}{1 + ae^{-\lambda t}},$$

where $a = (Q_\infty - Q_0)/Q_0$ and Q_0 is the cumulative production at some initial year when $t = 0$. This solution has the same shape as the discovery or production curves in Figure 4.11 and is called the logistic curve. The derivative of $Q(t)$ gives the rate of production,

$$(4.5)\quad P = \frac{dQ}{dt} = \frac{Q_\infty a\lambda e^{-\lambda t}}{(1 + ae^{-\lambda t})^2},$$

which has a shape similar to the curve Q_R in Figure 4.11 because $\lim_{t\to 0} \frac{Q_D - Q_P}{t} = \frac{dQ}{dt}$. The time of peak production, t_{peak}, can be found by setting the derivative of P with respect to time equal to zero and solving for t to get

$$(4.6)\quad t_{peak} = \frac{1}{\lambda}\ln a.$$

In his earliest articles Hubbert used the known cumulative production, Q_0, at some initial time and the proven reserves known at that time to hand fit Equation (4.4) or (4.5) to get an estimate of the ultimate resource available, Q_∞, the year of peak production, t_{peak}, and the production during the year of peak production, $P(t_{peak})$. It has been pointed out that trying to fit Equation (4.4) or Equation (4.5) to available data is difficult and can result in significant error [13]. A slightly more sophisticated analysis using the same equations is given next, along with a sample application.

As we saw in Chapter 1, it is generally easier to match data to a straight line, and this can be done with the Verhulst equation by replacing the rate of production in Equation (4.3) with $P = \frac{dQ}{dt}$ to get

$$(4.7)\quad P/Q = \lambda - \frac{\lambda Q}{Q_\infty}.$$

A plot of P/Q versus Q will be a straight line with (negative) slope λ/Q_∞. The x intercept will be equal to Q_∞. The only data needed to make this plot is the annual production rate, P, and the cumulative production, Q, over time. The total oil that can be extracted, Q_∞ (the estimated ultimately recoverable oil), and the production rate, λ, are found from the graph of the data. The year of peak production can be found from Equation (4.6), where a is found by inverting Equation (4.4) using data from a known year.

EXAMPLE 4.2

A Prediction of Remaining World Oil Reserves

Use the Hubbert model to predict the remaining world oil reserves and the peak year of production. The daily production from 1960 to the present is published on the EIA Web site [1]. Multiplying by 365 days gives the annual production, P. The cumulative production, Q, is found by starting with an estimate of the cumulative production up to 1960 and then adding each year's production to the sum of the previous years. A plot of P/Q versus Q using 100 bbls (bbls = 1 billion barrels) as the cumulative production in 1959 is shown in Figure 4.12.

As can be seen, the graph does not settle down into a straight line until about 1984, when the cumulative production was about 511 bbls. Variations before 1984 reflect the dip in global oil production in the 1970s. The equation for the straight part of the graph from 1984 until the present (calculated with the linear fit function in Excel) is $y = 2.434 \times 10^{-5}x + 0.0498$. The global rate of oil production for this period of time, according to the model, is approximately $\lambda = 0.0498 \approx 5\%$. Solving the equation for $y = 0$ gives the x-intercept, which is the total recoverable world oil supply; $Q_\infty = 2048$ bbls or 11,264 quad.

We can calculate the peak year by first solving Equation (4.4) for a to get

$$a = \left(\frac{Q_\infty}{Q(t)} - 1\right)e^{\lambda t}. \quad (4.8)$$

Picking 1990 as year $t = 0$ and using 637.8 bbls = Q_0 from the data for the cumulative production in that year gives $a = \left(\frac{Q_\infty}{Q_0} - 1\right)e^{\lambda 0} = 2.21$. Equation (4.6) then gives $t_{peak} = \frac{1}{\lambda}\ln a = 15.9$ years. So the predicted peak year of world oil production is $1990 + 15.9 \approx 2006$.

At this point it should be emphasized that the logistic equation is an empirical model, not a law of physics. As seen in Figure 4.12, the actual world data before 1984 are not a straight line and therefore do not fit the model. Clearly

FIGURE 4.12

Global annual crude oil production divided by cumulative production versus cumulative production in gigabarrels, (data from reference [1]).

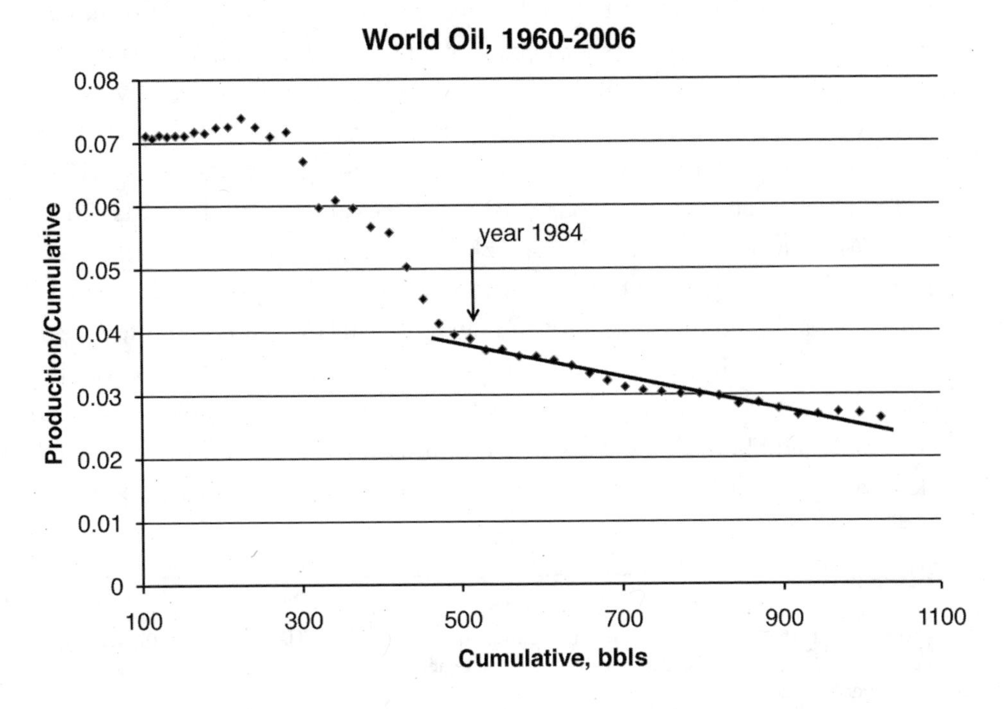

the dips in oil production seen in the 1970s do not match the smooth production curves of the logistic curve shown in Figure 4.11. As noted earlier, economic and political changes and new technology will affect the estimation of the ultimate amount of oil extracted from a given reserve, so that Q_∞ is not a constant, which is a fundamental assumption of the model. In particular, the Hubbert bell-shaped curve (Equation (4.5)) is likely not to be symmetric because after the peak, as the amount of oil left declines, economic incentives will encourage more efficient use of the resource and spawn technological advances, which will improve extraction methods, making the decline in production less steep [16]. Switching to alternative fuels will also modify the shape of the curve after the peak. Estimates of when the world peak oil will occur using other techniques range from 1998 to 2050 (an estimate of 1998 is not as ridiculous as it might seem; the top of the production curve will be flat, so that the peak year will not be evident for some years after the event). Current estimates of the ultimate recoverable amount of oil in the world made independently from a logistic equation analysis vary from 1,300 bbls to 2,400 bbls [7, 15, 19].

In support of Hubbert's model, it does seem that oil production has peaked in the continental United States and in a number of other countries, as seen in Figure 4.13. In fact, it now seems clear that nearly every major oil field except those in the Middle East has reached peak production. If the global peak has been reached, as some analysts suspect, it will take several years for the downward trend to become evident. The logistic model does work slightly better than, for example, fitting data to a Gaussian curve, and it can be modified to take into account rising oil prices, which affect the ultimate recoverable resource [16].

The Hubbert curve method has also been successfully applied to other finite resources such as copper, phosphorus, lithium, coal in some regions, and even some renewable resources where extraction now exceeds replenishing, such as ocean fisheries and aquifers (see [20] and references in [13]). Although much less is known about natural gas reserves, shale oil, and tar sands, the method can be used to estimate the total reserves for these resources as well. Table 4.5 shows the estimated Hubbert model peak year and ultimate recoverable resources for the known fossil fuel reserves and uranium. The figures are only estimates. If the

FIGURE 4.13

Oil production for select countries, not including the Middle East [1]. Each line represents the sum of the production of the countries below it. The actual country production is given by the thickness between the lines.

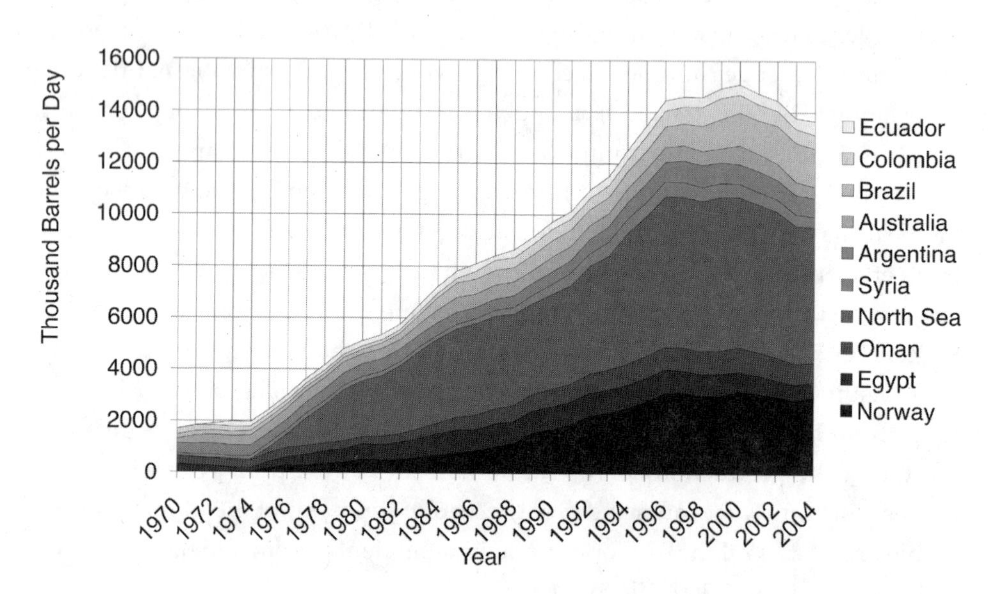

TABLE 4.5

Hubbert model estimates of the ultimate recoverable resource and year of peak production for various energy sources, calculated by the author based on Department of Energy data [1].

Resource	Peak Year	Ultimate Recoverable Resource (quad)
Petroleum	2006	11,264
Natural gas	2025	9,307
Petroleum including shale oil	2033	21,935
Coal	2060	22,696
Uranium	2113	5,080,000

world, or even just a few countries that are large consumers of energy, switches from one fuel to another, the estimate for any one of these fuels might change dramatically. There is some evidence that this is indeed occurring; China counts on coal for a much larger percentage of its energy needs than other countries, and the United States is in the process of switching from petroleum and coal to natural gas as a main fuel source. Shortages in petroleum could also be compensated by making oil from coal using the Fischer–Tropsch method (see Chapter 6). However, the energy loss for this process is about 50% of the energy, which would accelerate coal consumption greatly.

Adding the total energy available from fossil fuels (not counting what has already been used) and dividing by the approximately 450 quad per year the world currently uses gives less than 150 years until fossil fuel resources run out. Clearly economic and political changes will negate the outcome of such a crude approximation, particularly if a decision to control carbon emissions is made, but it should be equally clear that new energy resources will have to be found or energy will have to be used much more efficiently for the world economies to continue expanding.

4.6 ■ Nuclear Energy

One kilogram of uranium-235 contains about as much energy as 3,000 metric tons of coal or 14,000 barrels of petroleum. Nuclear power does not emit carbon dioxide, a known greenhouse gas. The expected decrease in lifetime due to exposure to the processes involved in operating a nuclear power plant is about

30 minutes, compared with 3.6 years for being overweight and 7 years for a one-pack-a-day smoker [21]. The number of accidental deaths per gigawatt-year of energy production is 4 deaths for coal mining, 2.6 for surface mining of coal, 0.4 for oil drilling, and 0.2 deaths for uranium oxide mining. The worst nuclear accident, Chernobyl, is estimated to have caused about 4,000 cancer deaths and fewer than 60 direct deaths, which is a factor of 10 less than the annual number of automobile deaths in the United States. For these reasons, many countries, France and Japan in particular, have opted for nuclear power over fossil fuels for electricity generation. Waste disposal and possible terrorist threats remain unsolved problems, but it seems clear that many countries, particularly China and India, are moving forward into the nuclear era. In the following we consider the processes involved in making electricity from nuclear power. An assessment of the risks involved due to radiation and waste treatment is found in Chapter 8.

4.6.1 Nuclear Physics

Atomic nuclei are specified by two numbers: the atomic number, Z, which gives the number of protons and therefore determines the element, and the neutron number, N. The sum of these two numbers, the total number of nucleons, is the mass number, A. Different isotopes of an element have the same atomic number but different numbers of neutrons and therefore different mass numbers. For example, ${}^{12}_{6}C$ and ${}^{14}_{6}C$ have the same chemical properties because they have the same number of protons (and therefore the same number of electrons, which give the element its chemical properties) but different numbers of neutrons. Electrons are designated as e^- and neutrons as ${}^{1}_{0}n$ in the following. Nuclear reactions also involve short-lived positrons, e^+, which are essentially positively charged electrons. The neutrino, v, and antineutrino, $\overline{v}$, are also created in nuclear reactions. Neutrinos are nearly massless particles that do not interact very strongly with normal matter, making them very difficult to detect.

Although the protons found in the nucleus are positively charged and repel each other, they are held together by the strong nuclear force, which can be as much as 100 times stronger than the electric force at short range. Neutrons in the nucleus also interact with each other and the protons via the strong force but do not have charge. The nucleus of different isotopes, then, can be expected to be roughly stable depending on the number of neutrons because a larger number of neutrons increases the relative amount of nuclear force without increasing the repulsive electric force. This is particularly true for heavier elements; the ratio of neutrons to protons for isotopes that are stable increases with the atomic number. The nuclear binding energy per nucleon is defined as the energy needed to separate all the protons and neutrons in a nucleus from each other divided by the number of nucleons present. A plot of binding energy per nucleon versus the number of mass number is shown in Figure 4.14. As can be seen, iron has the highest nuclear bonding energy per nucleon and thus has the most stable nucleus of any element. The fact that all elements heavier than iron have lower binding energies means there is at least the potential for these elements to lose nucleons and become more stable. This is the basis behind radioactive fission.

The nuclei of certain isotopes of some elements undergo spontaneous decay to form new, lighter elements. The fragments from the decay have less total mass than the original nuclei; the missing mass is converted to energy according to the famous equation $E = mc^2$, where $c = 2.999 \times 10^8$ m/s is the speed

FIGURE 4.14

Binding energy per nucleon for select isotopes.

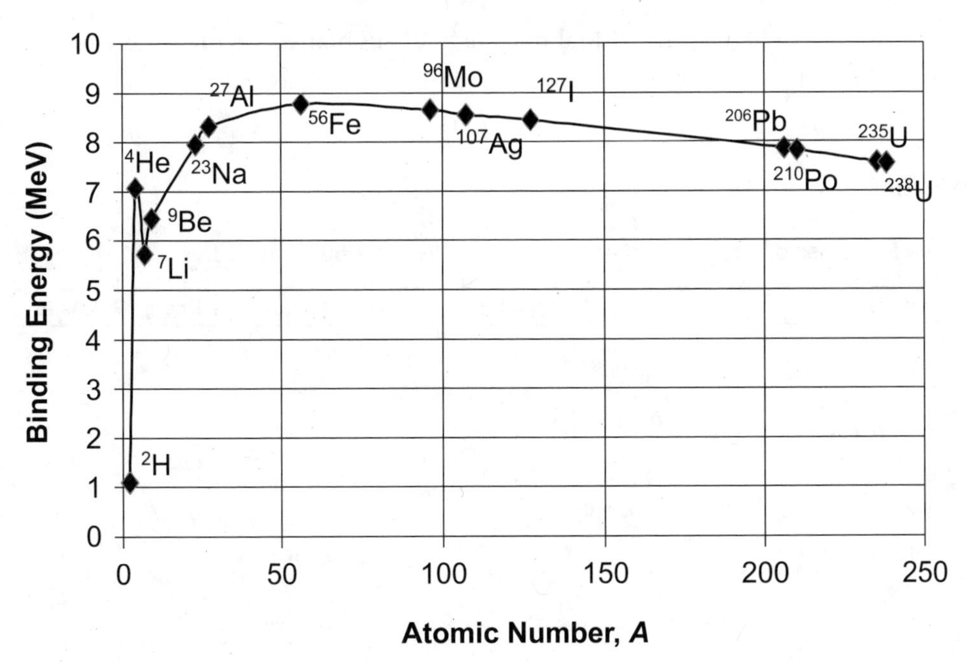

of light. From this equation it is clear that only a small amount of mass is needed to provide large amounts of energy. There are four major types of decay processes. Table 4.6 shows an example of each type of decay.

In alpha decay the nucleus emits a helium nucleus, which has two protons and two neutrons. Gamma decay can occur for an arbitrary element, X, in an excited energy state (designated by the asterisk). The emitted product is a high-energy electromagnetic photon, called a gamma photon, with a frequency above 10^{20} Hz. Beta decay emits an electron and a neutrino; the neutrino emission is needed for momentum conservation. Radioactive carbon dating is based on beta decay. Fission is the splitting of a nucleus into roughly equal parts. Two possible results of the fission of uranium-235 are shown in Table 4.6; there are dozens of other possible decay products depending on how the nucleus splits.

In Table 4.6 we see that fission typically produces more energy than the other decays and that a uranium nucleus can fission into a variety of daughter products.[2] All four of the processes may occur spontaneously, but the fission examples shown are triggered by a collision with a neutron. In each of the four cases the total mass of the products is less than the mass of the original nucleus, and the missing mass turns up as kinetic energy of the products according to $E = mc^2$. A more detailed discussion of radioactive decay can be found in [22].

Exactly when an individual nucleus will spontaneously decay is not predictable by the laws of physics. However, if there are a large number of identical

TABLE 4.6

The four types of radioactive decay processes. 1 eV = 1.602×10^{-19} J.

Process	Examples	Approximate Energy Released
Alpha decay	${}^{238}_{92}\text{U} \rightarrow {}^{234}_{90}\text{Th} + {}^{4}_{2}\text{He}$	4 MeV
Beta decay	${}^{14}_{6}\text{C} \rightarrow {}^{14}_{7}\text{N} + e^{-} + \overline{v}$ ${}^{12}_{7}\text{N} \rightarrow {}^{12}_{6}\text{C} + e^{+} + v$	9 MeV (requires energy)
Gamma decay	${}^{A}_{Z}X^{*} \rightarrow {}^{A}_{Z}X + \gamma$	4 MeV
Fission	${}^{1}_{0}n + {}^{235}_{92}\text{U} \rightarrow {}^{141}_{56}\text{Ba} + {}^{92}_{36}\text{Kr} + 3{}^{1}_{0}n$ ${}^{1}_{0}n + {}^{235}_{92}\text{U} \rightarrow {}^{140}_{54}\text{Xe} + {}^{94}_{38}\text{Sr} + 2{}^{1}_{0}n$	200 MeV

nuclei, they will collectively obey an exponential decay law according to the formula

$$(4.9)\quad R = \lambda N_0 e^{-\lambda t},$$

where R is the activity, measured in becquerels (Bq); 1 Bq = 1 emission/s. Here N_0 is the original number of radioactive atoms, and the original activity is $R_0 = \lambda N_0$, where λ is the decay constant. Rather than using the decay constant, λ, the time needed for half the material to decay, called the half life, is sometimes more useful:

$$(4.10)\quad t_{1/2} = \frac{\ln 2}{\lambda} = \frac{0.693}{\lambda}.$$

Half lives of various decay processes can vary from fractions of a second to millions of years, a property that makes nuclear waste disposal particularly difficult.

As mentioned earlier, the two isotopes of carbon are chemically identical, but ${}^{14}_{6}C$ undergoes beta decay with a half life of $t_{1/2} = 5{,}730$ yr. This isotope is created at a constant rate in the upper atmosphere from nitrogen being struck by cosmic rays from space. From measurements made of the atmosphere, there is a constant ratio of ${}^{14}_{6}C$ to ${}^{12}_{6}C$ in CO_2 of approximately 1.3×10^{-12}. This will also be the ratio of ${}^{14}_{6}C$ to ${}^{12}_{6}C$ found in living organisms because all organisms continue to take in CO_2 as they respire, maintaining the same ratio. Once an organism dies, however, the ${}^{14}_{6}C$ begins to decay and the ratio changes. Because this occurs at a known rate, the approximate date of death can be determined by comparing the ratio of carbon isotopes in the dead organism to the atmospheric constant ratio of 1.3×10^{-12}.

EXAMPLE 4.3

Radioactive Carbon Dating

Suppose a reed basket is found in an ancient burial site, and the ${}^{14}_{6}C$ activity is measured to be 6.5 Bq in a sample that contains 100 g of carbon. About how long ago was the basket buried?

From Equation (4.10), the decay constant is $\lambda = \dfrac{0.693}{t_{1/2}} = 3.84 \times 10^{-12}/s$.

When alive, the sample had 100 g $\times 1.3 \times 10^{-12} = 1.3 \times 10^{-10}$ g of ${}^{14}_{6}C$. A mole of ${}^{14}_{6}C$ weighs 14 g, so the original number of ${}^{14}_{6}C$ molecules is $N_0 = 1.3 \times 10^{-10}$ g $\times 6.02 \times 10^{23}$ atoms/14 g $= 7.8 \times 10^{13}$ atoms. Using Equation (4.9), we have 6.5 Bq $= 7.8 \times 10^{13} \times 3.84 \times 10^{-12}/s\ e^{-(3.84 \times 10^{-12}/s)t}$. Solving for t gives $t = 10.0 \times 10^{11}$s = 31630 years, so the reed used to make the basket died about 32,000 years ago.

4.6.2 Nuclear Reactors

If each of the neutrons emitted from a fission reaction strikes another uranium-235 nucleus, it will stimulate another fission decay. So, taking the first fission reaction in Table 4.6 as typical, after one reaction 200 MeV of energy is released and the three emitted neutrons hit three nuclei, releasing 600 MeV more in energy along with nine new neutrons. The process of neutrons being emitted, which cause more decays and more emitted neutrons, is called a chain reaction. A reaction in which the number of neutrons remains the same at each stage (because some of them do not collide with another nucleus or do not have the necessary energy) is said to be critical. If enough uranium-235 is present (a critical mass) so that each new neutron given off finds a nucleus to collide with (and the neutrons have the right energy), the amount of energy given off at each stage would grow as 3^n, where n is the number of decays. This is called a supercritical chain reaction and, if allowed to continue indefinitely, becomes an atomic explosion. If fewer neutrons occur at each stage, the reaction is subcritical, and the reaction eventually dies out. Notice that for a series of decays to become a chain reaction (subcritical, supercritical, or critical), a critical mass of ${}^{235}_{92}U$ must be present; other isotopes of uranium will not support a chain reaction.

These three types of chain reactions are complicated by the fact that the emitted neutrons generally travel too fast to be effective in stimulating further emissions. They bounce off the nucleus rather than cause a decay. To maintain a chain reaction the neutrons usually must be slowed down using a moderator. These slower neutrons are called thermal neutrons. In nuclear reactors the concentration and arrangement of uranium-235 are carefully designed so that the chain reaction cannot continue to be supercritical past a given point. Because the concentration of fissionable uranium is low, usually about 4%, most reactors also need a surrounding material called a reflector, which returns some of the neutrons to the uranium fuel in order to maintain the chain reaction at the critical level. Most reactor designs also include control rods, which are made from substances that absorb neutrons so that the number of available neutrons with the right speed to stimulate decay can be controlled. This allows the reactor to sustain a critical chain reaction at different levels of energy output.

The most common type of nuclear reactor is the pressurized light water reactor (PWR). In this reactor the energy emitted by the radioactive fission of uranium is captured as thermal energy by water under high pressure (>150 atmospheres) and used to drive a steam turbine. The water acts as both moderator, to slow the emitted neutrons down to the right speed for further decay, and heat transfer medium. The fuel is in the form of uranium pellets about the size of a dime, with a concentration of $^{235}_{92}U$ of about 4%. These pellets are located in a fuel rod about 1 cm in diameter and 3–4 m in length. The fuel rods are arranged in assemblies of about 250 rods, and a reactor core has about 200 fuel rod assemblies. Control rods made of a material that absorbs neutrons, such as silver–indium–cadmium alloys, are interspersed between the fuel rods; these can be raised or lowered into the core to control the rate of decay. Details of other reactor designs can be found in [23].

New reactor designs, called Generation IV reactors, are designed to be smaller, with lower fuel concentrations and fewer moving parts. They are planned to be inherently safe, meaning that they will automatically shut down under their own power and cool using natural convection in the event of operator error, mechanical failure, or failure of external and backup electric power. One such design is the gas-cooled pebble bed reactor. In this reactor the fuel is in the form of thousands of billiard ball–sized spheres made of a carbon ceramic compound that encase mustard seed–sized grains of uranium or other fissionable material. The reactor can be fueled while in operation by the addition of the fuel balls at the top and removal of spent fuel balls from the bottom. Because the balls are already coated with a tough, heat-resistant, chemically inert shell, they can be disposed of without further processing. It is also impractical to use fuel in this form in a nuclear weapon, an additional safety feature. Helium is used as the heat transfer medium, which means that the risk of steam explosion is eliminated, and the reactor can operate at higher temperatures, making the second law thermal efficiency higher. Several experimental reactors with this design are now operating or are under development around the world.

Naturally occurring uranium is only about 0.7% $^{235}_{92}U$, with the remaining ore consisting of the $^{238}_{92}U$ isotope, which is more stable and does not undergo spontaneous fission. For PWRs the concentration of uranium must be enriched to about 4%; however, other reactor designs such as the Canadian deuterium–uranium (CANDU) reactor can use unenriched uranium. Weapon-grade uranium is generally more than 85% $^{235}_{92}U$, but it is possible to make a crude atomic bomb with enrichments as low as 20%. The fact that nuclear reactors use 4% enriched uranium means it is impossible for a reactor to explode like an atomic bomb, although other kinds of problems can occur.[3]

Two other reactor fuels, ${}^{233}_{92}U$ and ${}^{239}_{94}Pu$, do not occur in nature but can be manufactured inside a reactor. The process that creates plutonium starts with the nonfissile isotope of uranium:

$$(4.11)\quad {}^{238}_{92}U + n \rightarrow {}^{239}_{92}U \rightarrow {}^{239}_{93}Np + e^{-} + \overline{v} \rightarrow {}^{239}_{94}Pu + e^{-} + \overline{v}.$$

The ${}^{239}_{94}Pu$ then undergoes fission in a manner similar to ${}^{235}_{92}U$. In this way, the nonfissile ${}^{238}_{92}U$, which makes up more than 95% of the uranium fuel rod, actually provides most of the fuel for a modern reactor. Placing a layer or blanket of ${}^{238}_{92}U$ or other nonfissile material around the core of a reactor, in front of or in place of the reflector, allows neutrons escaping from the core to be used to change this material into fissionable fuel. Such an arrangement is called a breeder reactor. All reactors are breeder reactors to some degree because fissionable ${}^{239}_{94}Pu$ is being produced via the reaction in Equation (4.11), but only those designed so that the produced fuel can be removed for use in other reactors are called breeder reactors.

Because isotopes of a particular element all have the same chemical properties, the enrichment process for changing uranium from 0.7% ${}^{235}_{92}U$ to 4% or higher is very difficult. The most common method of enrichment is to gasify the uranium by reacting it with fluoride to form gaseous UF_6 and allowing it to diffuse through a set of semipermeable membranes. The slight mass difference between the isotopes means molecules with the lighter isotope will diffuse faster than those with the heavier isotope. Specialized centrifuges and other processes (e.g., laser isotope separation, the Becker nozzle process) can also be used to take advantage of this small mass difference to separate the two isotopes.

As the ${}^{235}_{92}U$ in a fuel rod decays, the concentration of ${}^{235}_{92}U$ eventually decreases to less than 4%, and although only about 1% of the fuel has been used, a chain reaction cannot be sustained. However, it is possible to use enrichment processes to capture the ${}^{235}_{92}U$ and other fissionable material remaining in a fuel rod after it has been used in a reactor, in a process known as a closed fuel cycle. Closed fuel cycles are used in France and Japan; spent fuel rods are periodically removed from a reactor, and the fissionable uranium and plutonium are extracted for use in new fuel rods. Fuel cycles for commercial nuclear power plants in the United States are open; the spent fuel rod material is simply stored rather than recycled. This is done because it is currently more cost effective to use new fuel sources than to recycle spent fuel because of the complexity of the recycling process. In addition, the technological complications involved in the enrichment and recycling processes have so far prevented terrorists from making enough enriched uranium for an atomic bomb, and some experts fear that if this technology were widely known it would more easily become available to terrorists. This reasoning led the Carter administration to ban nuclear fuel reprocessing in 1977. Although the ban was lifted by President Regan in 1981, the lack of economic incentive has meant that reprocessing of spent nuclear reactor fuel has not been reestablished in the United States.

On a much smaller scale, the alpha decay shown in Table 4.6 can also be used to generate electricity. Deep space probes such as the *Viking* Mars lander and the *Pioneer*, *Voyager*, and *Cassini* probes to Jupiter, Saturn, Uranus, Neptune, and Pluto used plutonium-238 as an alpha source to provide heat. The heat was turned into electrical energy using thermocouples. The advantage of working with an alpha source is that, unlike in fission and the other decays, the amount of shielding needed is low, thus lowering the weight of the power source.

4.6.3 Radioactive Waste

The three main obstacles currently blocking an expansion of nuclear power in the world are the public perception of risk, potential access to radioactive material by terrorist groups, and the problem of nuclear waste. The real and perceived risks of nuclear radiation and nuclear accidents are considered in Chapter 8. Although the majority of nuclear waste found in storage today comes from military defense projects, waste from reactors is also a serious and growing problem [24]. The daughter products from the two fission processes shown in Table 4.6 are only a small sample of such fission products. Half lives for these products range from 1 year for ${}^{106}_{44}Ru$ to 2.1×10^5 years for ${}^{99}_{43}Tc$. The breeding process mentioned earlier also creates new radioactive nuclei, both in the uranium fuel, which actually becomes more radioactive than it was to begin with, and in the various metals making up the reactor structure, which also end up as different isotopes due to neutron bombardment. The half lives of these new heavy nuclei range from 13 years for ${}^{241}_{94}Pu$ to 2.1×10^6 years for ${}^{237}_{93}Np$. These elements are sometimes also chemically reactive and can be poisonous. Clearly a way must be found to keep these byproducts sequestered from humans for very long periods of time.

Radioactive waste is classified as low level, including material used in hospital treatments and research laboratories; medium level, which includes nuclear reactor parts and some material used in industry; and high level, which includes spent fuel rods. Various proposals have been made for disposal of nuclear waste, including transmuting the material into another radioactive material with a shorter half life, recycling the material in a closed fuel cycle, and disposing of it in ocean trenches or deep geological formations or even in outer space. Currently the only economically feasible disposal is in deep, stable geological structures. The waste is first modified into a chemically stable form such as synthetic rock or glass. The stabilized material is then buried in a dry rock formation such as a salt mine, in a zone where there is not likely to be much earthquake activity. The health risk factors associated with ionizing radiation such as X-rays and alpha, beta, and gamma decay are discussed in Chapter 8.

4.6.4 Nuclear Fusion

To conclude this section, we describe the process of nuclear fusion. Notice that elements lighter than iron also have lower binding energy per nucleon, just as elements heavier than iron do. However, lighter elements become more stable by gaining nucleons in a process called fusion rather than by losing nucleons in decay processes. Some of the mass is turned into energy; the combined nucleons weigh less than the individual components. The main obstacle to fusing lighter nuclei together is the electrical repulsion of the protons in the nucleus. To overcome this repulsion, the atoms must be put under enormous pressure at high temperature. Two places where this is known to occur are in the sun and in hydrogen bombs.

The nuclear process that gives stars and hydrogen bombs their energy occurs in several stages, but the proton–proton cycle can be summarized as

$$6\,{}_{1}^{1}H \rightarrow {}_{2}^{4}He + 2\,{}_{1}^{1}H + 2e^{+} + 2v + 2\gamma + 25MeV. \tag{4.12}$$

Heavier elements up to iron can be formed in a star if it is large enough so that pressures and temperatures in the center remain high. Our sun will stop shining after it finishes turning all the hydrogen in the core into helium, after which the pressures and temperatures will be too low to fuse heavier elements. This process is expected to take about another 4.5 to 5 billion years to complete. All the solid material in the universe was initially hydrogen; the heavier elements, such as carbon and iron, of which we are formed, all came from stellar fusion processes. Elements lighter than iron were formed by fusion in stars much larger than our sun that exploded at the end of a multistage fusion process due to instabilities in the balance between gravity and fusion. Elements heavier than iron were formed during these explosions by chance collisions of lighter elements.

Several different experiments are being conducted around the world in an effort to achieve controlled fusion for use as an energy source. Achieving controllable hydrogen fusion would result in a nearly perfect fuel because, unlike fission, there are no dangerous fuels or waste products, and the available energy is limitless for all practical purposes. Although fusion research has been ongoing since the 1950s, there are still substantial technical barriers to building a fusion power plant. Current estimates are that fusion may become a possible energy source by 2050.

4.7 ■ Summary

From the data presented in this chapter it should be clear that fossil fuels, which currently make up 86% of the global energy supply, are going to become more scarce in the foreseeable future. The projected 60% increase in energy use over the next 25 years will exacerbate this problem. It has taken less than 200 years to use up approximately half of the world's oil supply. If the Hubbert model is correct, it will take about another 200 years to use up the remaining oil, at an increasing cost. In this case the oil era will have lasted 400 years of the 10,000 years of human civilization, or about half the lifetime of the Roman Empire. In addition to questions about carbon emissions and global warming (which will be discussed in Chapter 7), these facts make it apparent that alternative energy sources will be needed [25, 26].

The average building has an 80-year lifetime, and power plants last about 40 years. The construction of a nuclear power plant takes as long as 10 years. It takes about 15 years from the inception of a new energy savings idea for it to be incorporated into a significant number of cars on the highway. If we are going to have energy-efficient buildings and cars in operation by the time fossil fuel supplies begin to show stress, now is the time to implement new energy saving technologies. If power plants 20 or 30 years in the future are going to use fuels other than fossil fuels, now is the time to start designing and building them.

Questions, Problems, and Projects

For all questions,

- Cite all sources you use (e.g., book, article, Wikipedia, Web site). Before using Wikipedia as a source, please read the policies of research and citing in Wikipedia: http://en.wikipedia.org/wiki/Wikipedia:Researching_with_Wikipedia and http://en.wikipedia.org/wiki/Wikipedia:Citing_Wikipedia. For Wikipedia you should verify the information in the article by reading the sources listed.
- For all sources, briefly indicate why you think the source you have used is a reliable source, especially Web resources.
- Write using your own words (except for direct quotes, which be enclosed in quotation marks and properly cited).

1. Go to references [1], [2], and [3] and update the data in Figure 4.1 and Table 4.1 to reflect the current year. Discuss any changes between the numbers from earlier years and the present.
2. Go to the EIA Web site http://tonto.eia.doe.gov/country/index.cfm, pick two countries, and compare their energy profiles (e.g., imports, exports, consumption, historical changes in consumption, types of fuel used). How do these countries compare with the same figures for the world? Write a summary of your findings.
3. Go to the EIA Web site and find the figures for the historical U.S. (or global) increase in energy provided by petroleum. Determine whether this increase is exponential, using the methods in Chapter 1. Is the rate of increase larger or smaller than for the population over the same time frame?
4. Find a discussion from a reliable source of the oil embargo of 1973 and the global oil crisis of the late 1970s. Summarize the effects of these political events on the global energy market.
5. Find a discussion from a reliable source of the various methods used to search for oil and natural gas mentioned in the text: seismology and gravimetric and magnetometer measurements. Describe each, including how they work and under what circumstances they are used.

6. Go to the International Energy Agency Web site and find the percentage of the price at the gasoline pump that is represented by taxes in the United States and five other countries of your choice (search for "energy price and taxes"). Also find the average car fuel economy in these countries. Using reliable sources, make two arguments, one against higher gasoline taxes and the other for higher taxes to improve gasoline efficiency.
7. From the list of the electrical energy used by appliances in the United States on the EIA Web site (http://www.eia.doe.gov/emeu/recs/recs2001/enduse2001/enduse2001.html), suggest where the greatest gains in energy savings could be made by improving the efficiency of a select few appliances.
8. Using the passenger kilometers traveled and the energy intensity figures in Table 4.3, estimate what the annual fuel savings would be if all cars and light trucks were replaced with (a) bicycles, (b) buses, (c) motorcycles, and (d) vanpools.
9. Find a discussion from a reliable source of the theory of abiogenic creation of petroleum. Evaluate these claims and discuss the probability of them being correct.
10. Find a discussion from a reliable source of the extraction of oil from tar sands and shale oil. Locations where this has been tried include Pechelbronn and Wietze, Germany; Yarega, Russia; and the Athabasca tar sands in Canada. Include a discussion of the methods being used to extract this oil and the energy cost.
11. Find a discussion from a reliable source of other uses for petroleum, natural gas, and coal besides directly for energy (e.g., plastic, medicine). Include some detail about the amounts of each used to create these other manufactured goods and the process to create the product. Comment on the long-term advisability of using fossil fuels as fuels compared with these other uses in the case that fossil fuels start to run out.
12. In your own words, explain the difference between proven and unproven oil reserves, oil in place, and estimated ultimately recoverable.
13. Using reliable sources, write a summary of the concept of energy return on energy invested (EROEI). Provide some estimates of the EROEI for

petroleum extraction from tar sands. There is further discussion of EROEI in the next two chapters.

14. The United States currently uses about 27 quad of energy for transportation per year, nearly all of which comes from petroleum. The Alaskan oil fields are estimated to have as much as 64 quad of oil. How many years could the Alaskan oil reserves supply U.S. transportation needs at the current rate of use? Some analysts estimate that cars and trucks could be made 60% more efficient than they are currently. How long will the Alaskan oil reserves last if the annual transportation energy demand is reduced by 60%?
15. Using reliable sources, find where coal, oil, and natural gas deposits are located in the United States (or a country of your choice). Make a list or map of where they occur and the approximate era when the deposit was formed. List some differences in the chemical makeup of different reservoirs.
16. A large natural gas deposit was discovered in 2009 in the United States. Using reliable sources, find out how much gas is expected to come from this new resource. Find the energy content of this resource and compare it with the annual total energy use of the United States of 100 quad. How many years could this new source supply the total U.S. energy needs?
17. Suppose the pressure in an oil field is 175 $lb/inch^2$. Suppose you put a well into this reservoir from the surface, where atmospheric pressure is about 15 $lb/inch^2$. Also suppose the oil, with viscosity 12.0×10^{-3} Pa · s, travels 1,000 m through rock with a permeability of $2.0 \times 10^{-12}\,m^2$. If the area of the well head is 12 m^2, at what rate will oil flow out of the well?
18. Find a discussion from a reliable source of Iceland's plans to become free of fossil fuels by 2050. Describe how they plan to do this, how much remains to be done and give an evaluation of the likelihood for success.
19. Find a discussion from a reliable source of estimations of the amount of clathrate hydrates. This topic is somewhat controversial. First, find a source that makes a reasonable argument that this resource will solve all the world's energy problems for the foreseeable future. Then find a source that argues that this resource will not be a significant energy

source. List the arguments that each source uses and evaluate which one is more likely to be correct.

20. Find two estimates of the ultimate recoverable oil resource (the cumulative production to date plus the proven reserves). Give the highest estimate you can find and the lowest and indicate why they are different and which is more likely to be correct.

21. Go to the British Petroleum Web page (http://www.bp.com/statistical review) and download the spreadsheet of historical data from 1965 to 2006. Use the production figures and the method outlined in Example 4.2 to find the ultimate recoverable resource and the year of peak resource for petroleum, natural gas, or coal. You will have to make an approximation of the cumulative resource up to 1965 to get started. The first few columns of the spreadsheet used in Example 4.2 are shown here.

TABLE 4.7

A	B	C	D	E	F
1	Year	Crude Production (million barrels/day)	Annual Crude Production, *P* (in gigabarrels)	Cumulative, *Q*	*P*/*Q*
2				Estimate of cumulative in 1959 = 100	
3	1960	20.99	C3*365/10^3 = 7.66135	E2 + D3 = 107.6614	D3/E3 = 0.071162
4	1961	22.45	8.19425	115.8556	0.070728
5	1962	24.35	8.88775	124.7434	0.071248
6	1963	26.13	9.53745	134.2808	0.071026

22. Find a discussion from a reliable source of applications of Hubbert's model to resources other than oil. Find at least a few examples in which the calculation turned out to be accurate, pick one such calculation, and explain it in detail. Pick one calculation for which the outcome is still uncertain and give reasons why the calculation may turn out to be wrong.

23. Radioactive dating can be performed with sources other than carbon-14, such as uranium–thorium dating. Find a discussion from reli-

able sources about how some of these other techniques work and write a description.

24. Repeat the calculation of Example 4.3 for a sample that is found to have 20.5 Bq of activity.

25. Because of changes in cosmic ray flux and other influences, radioactive carbon dating is accurate to only about 1% for times back to 30,000 years. (a) By how much does this error change the answer of Example 4.3? (b) Using reliable sources, write a discussion of the calibration methods used to make sure radioactive carbon dating is accurate. Include a discussion of corrections made to raw measurements.

26. Write a short explanation of each of the following methods of uranium enrichment: (a) gaseous diffusion, (b) thermal diffusion, (c) gas centrifuge, (d) separation by lasers, (e) aerodynamic (Becker) nozzle, (f) electromagnetic separation, and (g) plasma separation. Be sure to include pros and cons and indicate the extent to which the process is being used today.

27. What is the difference in velocity of UF_6 between the two different isotopes of uranium. *Hint:* $\frac{3}{2}k_BT = \left\langle \frac{1}{2}mv^2 \right\rangle$. Try a couple of reasonable temperatures.

28. Like $^{238}_{92}U$, thorium is not fissile but can be converted to $^{233}_{92}U$ in a reactor to become a reactor fuel. A thorium-based nuclear economy has been proposed because thorium cannot be used as a nuclear weapon, and the amount of radioactive material involved in using thorium is less than for uranium. Find a discussion from a reliable source of the pros and cons of using thorium as a reactor fuel. Write a discussion with explanations of how a nuclear economy based on thorium might work.

29. Write a short explanation of each of the following types of nuclear reactors: (a) PWR, (b) BWR, (c) CANDU, (d) RBMK, (e) GCR, and (f) LMFBR. Include pros and cons of each type, where they are used, and how many are being used today.

30. Reactor designs are sometimes assigned to a "generation." To date there are four generations of design (Generation IV reactors were mentioned in the text). Using a reliable source, give a definition of each of the four generations, list the reactor designs that belong to each, and explain the purpose of the four designations.

31. Using a reliable source, provide a discussion of "inherently safe" reactor designs such as the pebble bed reactor or a "grid appropriate" reactor such as the Toshiba 4S (super-safe, small, and simple) reactor.
32. Using reliable sources, write a summary of the history and current status of the gas-cooled reactor. Include a discussion of the various designs that have been used and are planned, the pros and cons of gas-cooled reactors, and current plans for commercial application.
33. Using reliable sources, write a summary of the use of alpha sources in the deep space probes mentioned in the text. Explain in detail how this process works.
34. Using reliable sources, explain how a thermocouple works. List some applications.
35. Hubbert type calculations of the long-term availability of uranium have raised questions about the idea of committing to an expanded use of nuclear energy. Find high and low estimations of ultimately recoverable uranium using reliable sources. Include an account of the idea of extracting uranium from seawater. Summarize the arguments and draw a conclusion. *Hint*: Look for the International Atomic Energy Agency's "red book" as a start.
36. Write a short explanation of the classifications *low-*, *medium-*, and *high-level radioactive waste* used by the U.S. Nuclear Regulatory Commission. What kinds of waste treatments are considered appropriate for each? What is the level of radioactivity for each?
37. Find two discussions from reliable sources, one in support of the Yucca Mountain storage facility, the other opposed to it. Summarize the arguments and draw a conclusion from your findings. Based on your research, why do you think the Obama administration decided against using the Yucca Mountain site?
38. The sun emits energy at a rate of about 3.8×10^{26} W. About how much mass per second is being turned into energy by the fusion process?
39. How much mass is converted to energy in the process shown in Equation (4.12)? What percentage of the initial mass is this? Based on this calculation, why do you think it took so long to discover the equation $E = mc^2$?

40. Find a discussion from reliable sources and write a summary of the two types of fusion reactors under investigation as potential power supplies (inertial and magnetic confinement). Describe how each works and indicate how long these projects have been under way and how close they are to building a successful power plant.

41. Most projections of future energy use have been wrong [7]. One reason this is true is that humans, acting on the assumption that the projection will be correct, change their behavior. List some possible changes in human behavior that would change the EIA's projection of an increase in energy use by 60% in the next 20 years.

REFERENCES AND SUGGESTED READING

1. U.S. Energy Information Administration, Department of Energy (http://www.eia.doe.gov/emeu/international/contents.html).

2. J. Goldemberg, *World Energy Assessment: Energy and the Challenge of Sustainability*, United Nations Development Programme Report, 2000 (http://www.undp.org/energy/).

3. *Central Intelligence Agency World Factbook* (https://www.cia.gov/library/publications/the-world-factbook/).

4. Organisation for Economic Co-operation and Development (http://www.oecd.org/).

5. U.S. Energy Information Administration (http://www.eia.doe.gov/emeu/recs/recs2001/enduse2001/enduse2001.html).

6. Oak Ridge National Laboratory (http://cta.ornl.gov/data/index.shtml).

7. V. Smil, *Energy at the Crossroads: Global Perspectives and Uncertainties* (2005, MIT Press, Boston).

8. U.S. Energy Information Administration (http://www.eia.doe.gov/emeu/international/reserves.html).

9. British Petroleum, *Statistical Review of World Energy 2007* (http://www.bp.com/statisticalreview).

10. Special Issue, "The Energy Challenge," *Physics Today* 55(4) (April 2002).

11. P. B. Weisz, "Basic Choices and Constraints on Long-Term Energy Supplies," *Physics Today* 62(3) (March 2009):31.

12. M. Gray, Z. Xu, and J. Masliyah, "Physics in the Oil Sands of Alberta," *Physics Today* 57(7) (July 2004):47.

13. "Revisiting Hafemeister's Science and Society Tests," *American Journal of Physics* 75 (Oct. 10, 2007):916.

14. C. J. Campbell and J. H. Laherrere, "The End of Cheap Oil," *Scientific American* 279(3) (March 1998):78.

15. K. S. Deffeyes, *Hubbert's Peak: The Impending World Oil Shortage* (2001, Princeton University Press, Princeton, NJ).

16. D. Hafemeister, *Physics of Societal Issues* (2007, Springer Science, New York).

17. T. E. Williams, K. Millheim, and B. Liddell, "Methane Hydrate Production from the Alaska Permafrost," Department of Energy DE-FC26-01NT41331, March 2005.

18. M. K. Hubbert, "Nuclear Energy and the Fossil Fuels," Shell Development Publication 95 (presentation to the Spring Meeting of the Southern District Division of Production, American Petroleum Institute, March 7–9, 1957).

19. "Peak Oil Forum" (collection of articles predicting when peak oil production will occur), *WorldWatch* 19(1) (2006):9–24.

20. R. Heinberg, *Peak Everything: Waking Up to the Century of Declines* (2007, New Society Publishers, Gabriola Island, BC).

21. R. Wilson and E. A. C. Crouch, *Risk–Benefit Analysis* (2001, Harvard University Press, Cambridge, MA).

22. D. Bodansky, *Nuclear Energy: Principles, Practices and Prospects*, 2nd ed. (2004, Springer, New York).

23. J. R. Lamarsh and A. J. Baratta, *Introduction to Nuclear Engineering* (2001, Prentice Hall, Upper Saddle River, NJ).

24. D. Bodansky, "The Status of Nuclear Waste Disposal," *Physics and Society* 35(1) (2006):4 (http://www.aps.org/units/fps/newsletters/2006/january/article1.html).

25. Global Energy Perspective, presentation by the Lewis Group, Division of Chemistry and Chemical Engineering, California Institute of Technology (http://nsl.caltech.edu/energy.html).

26. J. Deutch, E. Moniz, S. Ansolabehere, M. Driscoll, P. Gray, J. Holdren, P. Joskow, R. Lester, and N. Todreas, "The Future of Nuclear Power," 2005 (http://web.mit.edu/nuclearpower/).

Notes

[1] The Organisation for Economic Co-operation and Development (OECD) countries are Australia, Austria, Belgium, Canada, Czech Republic, Denmark, Finland, France, Germany, Greece, Hungary, Iceland, Ireland, Italy, Japan, Korea, Luxembourg, Mexico, the Netherlands, New Zealand, Norway, Poland, Portugal, Slovak Republic, Spain, Sweden, Switzerland, Turkey, United Kingdom, and United States [3].

[2] It is these daughter products that constitute nuclear waste. Many of them are radioactive and chemically harmful to living organisms. A further discussion of safety issues is found in Chapter 8.

[3] For example, in the case of the Chernobyl accident a steam explosion lifted the roof off the reactor building, and an intense graphite fire lofted radioactive material high into the atmosphere, allowing it to spread over large parts of Europe. This reactor design is no longer in use.

CHAPTER 5

Renewable Energy

In Chapter 4 we saw that the world's finite supply of fossil fuels is quickly being spent. Simultaneously, population growth and rising consumption continue to increase strain on these resources. To meet growing demand, societies around the globe are seeking to develop renewable energies, defined by the U.S. Energy Information Administration as "energy sources that are naturally replenishing but flow limited." In other words, renewables are "virtually inexhaustible in duration but limited in the amount of energy that is available per unit of time." These sources include hydropower, wind, biomass, solar, tidal, and geothermal.

In this chapter we review the sources of renewable energy being developed today and prospects for their increased use. It should be pointed out that, in reality, many of these sources are actually solar in origin. Waves and wind are driven by the incoming energy flux from the sun. Hydropower comes from rain, which is evaporated from oceans by sunlight. The current application of most of these renewable sources involves the generation of electricity or direct heating of water or living space. As we will see in Chapter 6, the implication for transportation is significant; few of these renewable energy sources are easily made portable.

5.1 ■ Hydropower

As late as 1940 the United States got 40% of its electrical energy from hydroelectric dams. Because of an increase in total energy use and a faster increase in the use of nuclear and coal energy, hydropower constitutes only 8% of current U.S. electrical generation, even though it has grown by 10% in the last 70 years. Globally hydropower is the most developed of the renewable resources, supplying about 6% of the world's total energy use and 20% of world's electricity. This amounts to a total of about 9.6 quad of electrical energy currently being produced globally from hydropower. The five largest producers of hydropower are Brazil, Canada, China, Russia, and the United States, with Brazil generating 84% of its electricity from hydropower [1].

Figure 5.1 shows the regional generation of hydroelectricity for the year 2005 and the potential for further development. The total projected future potential, based on currently existing technology, is estimated to be about 50 quad [2]. As we will see in this chapter, it is not possible to capture all the kinetic energy in moving water; the water has to continue to flow away from the power plant after producing electricity, so it still has some uncaptured kinetic energy. The total energy of all the rivers in the world (excluding ocean currents) is about 350 quad [3]. Exactly how much energy could possibly be extracted using advanced technology is unclear, but most experts agree that 50 quad is a reasonable estimate for the foreseeable future of hydropower [1, 3, 4].

It should be clear from these estimates that hydropower is not going to solve the projected increases in energy demand described in Chapter 4. Annual

FIGURE 5.1

Current (2005) and potential annual hydropower by region [1, 2].

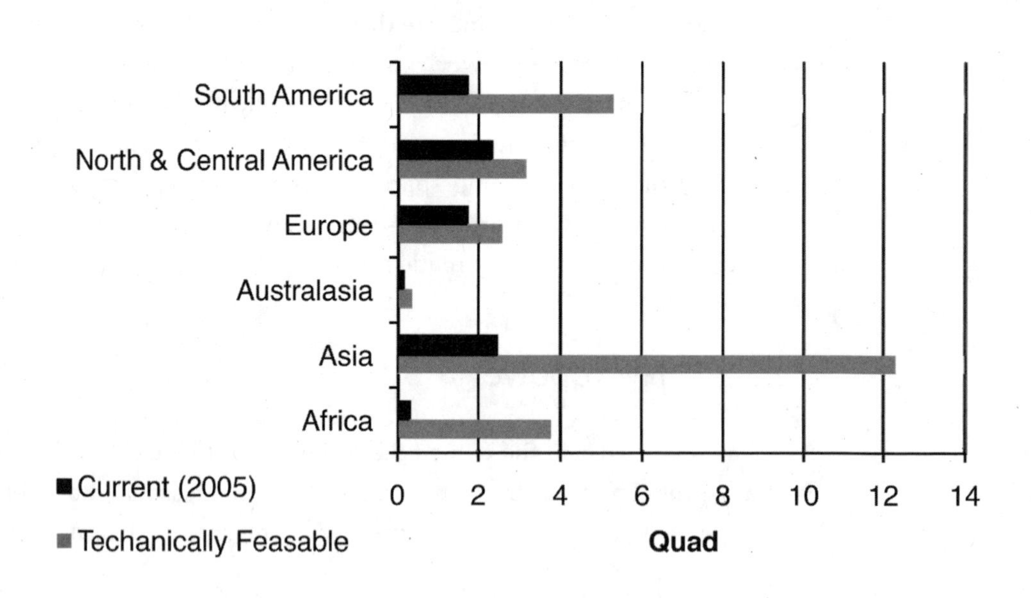

global energy use stood at 472 quad in 2006 and is increasing at a rate of 2% per year. Current projections for future hydropower using current technology would supply only 12% of this figure. Europe is estimated to already be using 75% of the available hydropower, North America close to 70%, and South America about 33% [2]. Norway generates nearly all its electricity from hydropower, Nepal and Brazil generate more than 90%, New Zealand generates 55%, Canada generates 58%, and Sweden generates about half of its electricity from hydroelectric. From Figure 5.1 it is also clear that Asia and Africa have the greatest potential for expansion of the use of hydropower.

In addition to being the least expensive form of renewable energy, hydropower is also the least carbon intensive, saving about 15% of the carbon emissions that would occur if hydroelectricity were replaced with fossil fuels [3]. Water in dams can also be used in a variety of ways in addition to generating electricity, such as for irrigation, drinking water, recreation, and pumped hydroelectric energy storage. Hydropower has long been considered the most environmentally benign of the renewable energy resources.

However, recent studies have pointed out that in the long term, decaying vegetation in the lakes behind a dam can emit significant amounts of carbon dioxide and methane. Disruption of natural habitats is now recognized as a significant outcome of dam building. It is also the case that, for large dam projects such as the Three Gorges Dam in China, millions of people may be displaced, causing severe economic and social problems. Another potential problem is that the weight of the water behind large dam projects may cause geological shifts and earthquakes. At some point the energy and environmental investment involved in capturing the energy in smaller and smaller water flows will outweigh any benefit of building new dams. Environmental and economic issues have already slowed dam development in North America; this coupled with technical issues will probably limit hydropower to a small but significant fraction of the global energy supply.

Elementary mechanics tells us that the gravitational potential energy, in joules, of a mass, m is $U = mgh$, where g is the acceleration of gravity and h is the distance of the mass above some reference point. We may imagine a "block" of water passing over a dam as having volume Ax, where A is the frontal surface area of the "block" (the cross-sectional area of the pipe of water going to a generator or the cross-sectional area of a spillway) and x is an arbitrary length. The mass of water may then be expressed using density, ρ, as $m = \rho V = \rho Ax$, and the potential energy is $U = \rho Axgh$. Dividing by time gives the power available, and using $v = x/t$ for the velocity of the water, we can write

$$(5.1)\quad P = \frac{U}{t} = \rho Agvh.$$

This expression is sometimes called the hydraulic head and is the main source of energy for a traditional, large-scale commercial dam. Equation (5.1) is sometimes also written in terms of the flow rate, $Q = Av$, as

$$P = \rho Qgh, \tag{5.2}$$

where Q is measured in cubic meters per second.

Moving water also has kinetic energy, $KE = \frac{1}{2}mv^2 = \frac{1}{2}\rho Axv^2$, and the kinetic power of water in motion is therefore $P = \frac{1}{2}\rho Av^3$. This is the source of energy for underwater turbines, which take advantage of tidal motion and river flow, and undershot waterwheels. Not all of this power can be extracted, however; the water must flow away from the bottom of the waterwheel or dam and therefore still has some residual kinetic energy. If the velocity changes from v_i to v_f, then the power generated is

$$P = \frac{1}{2}\rho A(v_i^3 - v_f^3). \tag{5.3}$$

The total power available from water in a dam of height h is therefore

$$P = \rho Qgh + \frac{1}{2}\rho A(v_i^3 - v_f^3), \tag{5.4}$$

where v_i is the velocity of the water as it reaches the dam and v_f is the velocity as it flows away from the bottom of the dam.

EXAMPLE 5.1

The Three Gorges Dam

Completed in 2006, the main section of the Three Gorges Dam in China is about 100 m tall, and the flow rate of water is about 1.1×10^4 m^3/s. How much power can be generated from the hydraulic head?

Using Equation (5.1), we have $P = \rho Avgh = \rho Qgh = 1.1 \times 10^{10}$ W, or about 11 GW. Running continuously for a year, the dam will generate about 0.3 quad per year. When the entire dam complex is complete, the dam will produce about twice this much electrical energy.

5.2 ■ Wind

Wind is the fastest-growing renewable source, rising by 70% in the United States since 2000. The cost of wind-generated energy in the United States is now about 6 cents/kWh (1 kWh = 3.6×10^6 J). With a 1.8-cent federal subsidy, this brings the cost down to less than 5 cents/kWh, which makes it economically competitive with other sources of electricity.[1] The energy payback time for a windmill, the operating time needed to generate enough energy to build a new one, is about 14 months for a windmill operating at 25% capacity. The cost of wind energy has dropped by a factor of 10 in the past 30 years and is projected to drop by another 50% from the current cost in the next 10 years. At the current rate of growth, 6% of the U.S. electricity demand will be met by wind generators in 2010. Europe is in the lead with wind-generated electricity, but China has recently taken the lead in building windmills (and photovoltaic solar cells). Seventy-two percent of the global wind energy installations for 2005 occurred in Europe, although since then the United States has taken the lead in new installations [5]. Denmark now generates nearly 20% of its electricity using wind, and Germany, which has almost three times the wind-generating capacity as the United States, already generates 6% of its electricity from wind. China is expected to surpass the rest of the world in wind-generated electricity in the very near future. Globally, electricity generated from wind energy has doubled every 3 years since 1990, making it by far the fastest-growing energy resource [1]. It is estimated that 10% of the world's electricity will be generated by wind within the next 10 years.

Estimates of the available global wind energy are quite large and depend heavily on assumptions about wind turbine design, location, and height. The available global annual wind energy is estimated to be about 36,000 quad, with about 11,000 quad located over land or in shallow water [1, 3, 6, 7]. If only winds below 100 m over land or close to shore with speeds in a range from 5 m/s to 25 m/s are included, the estimate is about 350 quad [3], which is almost seven times the total global electrical energy use (54 quad in 2006). The Energy Information Administration estimates the wind potential for the United States to be about 90 quad.

As with any energy resource, there are obstacles to the implementation of wind power. As can be seen in Table 5.1, wind resources are not necessarily located where they are most useful. The Patagonia region of South America

TABLE 5.1

Location of wind resource by region [1, 8, 9]. Winds in these regions average 5 m/s or greater at a height of 80 m above the surface of the earth.

Region	Location
Asia	East coast, inland mountains
Africa	North coast
Australasia	West and south coasts
Europe	North and west coasts, some Mediterranean regions
North and Central America	Western coastal regions, northeastern coast, central plains
South America	South (Patagonia)

has enormous wind energy resources but has the lowest population density in South America. Wind resources are also widely dispersed; the energy per square meter of land is low, 2 to 3 W/m^2. Even in regions where wind speeds are high, in order to keep wind turbines from blocking wind from each other they must be spaced far apart, unlike other electricity generation devices, which can be concentrated in a small area. A spacing of five turbine diameters reduces the power of the second turbine by about 30%, whereas a spacing of 10 diameters reduces the power by about 10% [10].

In order to take advantage of the fact that wind speeds increase with the distance from the surface of the earth, modern commercial wind turbines are very tall. A typical wind turbine built today might have its hub higher than 80 m off the ground, with a rotor diameter of 100 m or more, and generate 5 MW of electric power. Figure 5.2 shows the blade of a small wind turbine, only 20 m long. Because of the enormous size of these wind turbines, additional problems such as noise, TV and radio interference, threat to migrating birds, and aesthetic considerations have been cited against the use of wind power.

Probably the most significant difficulty in the use of wind power is the fact that wind is an intermittent source, varying both seasonally and daily. Some locations also experience changes in average wind speed by as much as 30% from year to year [3]. Modern turbines need a wind speed of 3 to 4 m/s to begin generating electricity. As a result, wind energy cannot be guaranteed to be available at the time of day or year when it is most needed. For instance, air conditioning use is heaviest in the summer, when wind speeds are typically lower in many locations. On the other hand, some applications such as charging electric car batteries, some industrial processes, and heating water are well suited to a resource that is intermittent. Variable winds can also result in the generation of

FIGURE 5.2

Single wind turbine blade, 20 m long, from a small wind turbine.

more electricity (e.g., during storms) than can be handled by the electrical grid. Large wind turbines currently being installed typically reach maximum generation capacity at a wind speed of 15 m/s and are designed to cease generation at wind speeds greater than 25 m/s.

The use of a widespread network of wind turbines helps average out local variability in wind speed, but total reliance on wind turbines for electricity is probably not possible. Some regions in Europe successfully use wind energy for 35% of the electricity generated without experiencing problems due to variable wind speeds. The power grid in the United Kingdom has coupled the implementation of wind energy with installation of large banks of emergency diesel engines, which can come online to make up for a shortfall in energy supply if wind speeds drop. The exact percentage of the global electricity use that can comfortably be supplied by wind without fears of interruptions due to wind variability is not known. However, it is clear that wind energy will have to be coupled with some combination of base load sources, backup energy supplies, and energy storage in order to be effective. Options for energy storage are considered in Chapter 6.

There are two basic designs for wind turbines: horizontal axis and vertical axis. Most of the following applies to both types of turbines. In both cases the wind passes over a rotor blade, creating a lift on the blade in a manner very similar to the lift on an airplane wing. The fact that the blade causes the wind to lose some of its kinetic energy results in a drag or friction force, which is also present on airplane wings. The blade's motion is the result of these two forces. A detailed explanation of how the lift and drag forces act on a wing or wind turbine blade can be found in reference [11].

As in the case of hydropower, the kinetic energy of the wind is $KE = \frac{1}{2}mv^2$, where m is the mass of a hypothetical "block" of wind traveling at speed v. The power contained in a "block" of wind with vertical surface area A and length x moving at velocity v is then

$$P = \frac{1}{2}\rho A v^3, \tag{5.5}$$

where now ρ is the density of air in kilograms per cubic meter. Not all this energy can be captured, because the air does not stop at the turbine blade; it will exit the region behind the turbine with a lower velocity. This is why placement of wind turbines is important; a turbine located downwind from another turbine will encounter lower wind speeds.

Because the wind turbine blades apply a resistance force to the wind and air compresses with increased pressure, it is more convenient to work with mass flow in kilograms per second rather than the flow rate, Q, in cubic meters per second, as was done for hydropower. Although the volume and pressure of a parcel of air may change as it passes a turbine, the total mass of air must remain fixed. The mass flow is the mass of air passing the turbine blades per second and can be defined as

$$\delta m = \rho A v_{ave}, \tag{5.6}$$

where $v_{ave} = \frac{1}{2}(v_i + v_f)$ and v_i and v_f are the wind speeds before and after passing the turbine. The change in kinetic energy per second of this parcel as it passes the turbine is $\frac{\Delta KE}{time} = \frac{1}{2}\delta m\left(v_i^2 - v_f^2\right)$, which is the energy captured by the turbine per second. Putting Equation (5.6) into this expression gives the power generated by the turbine as

$$(5.7)\quad P = \frac{\Delta KE}{time} = \frac{1}{2}\rho A v_{ave}(v_i^2 - v_f^2).$$

After some rearranging, substituting the expression for the average velocity into this expression gives

$$(5.8)\quad P = \frac{1}{4}\rho A v_i^3\left(1 + \frac{v_f}{v_i} - \left(\frac{v_f}{v_i}\right)^2 - \left(\frac{v_f}{v_i}\right)^3\right)$$

for the power generated.

To find the maximum power possible as a function of the change in wind speed, we take a derivative of this equation with respect to $\frac{v_f}{v_i}$ and set the result equal to zero. This gives $\frac{v_f}{v_i} = \frac{1}{3}$ for maximum power. Putting this into Equation (5.8) gives

$$(5.9)\quad P_{max} = \frac{16}{27}\left(\frac{1}{2}\rho A v_i^3\right).$$

Because the total power available in the wind is $\frac{1}{2}\rho A v_i^3$ from Equation (5.5), the ratio 16/27 = 59% gives the upper limit to the efficiency of a wind turbine, which occurs when the wind speed drops by two thirds after passing the turbine. Equation (5.9), then, gives the maximum power that can be extracted from the wind and is known as the Betz limit, after its discoverer, Albert Betz. It should also be noted that, for the technical reasons stated earlier, most wind turbines are designed for a maximum power output at 15 m/s and a cutoff at 25 m/s. This means the rate of energy captured does not actually scale as the velocity cubed, as might be expected given Equation (5.5) for the power contained in the wind. When these wind speed limits and other factors such as friction are included, real wind turbines turn out to have efficiencies of about 25% [10]. Of course, this low efficiency is not as relevant as in the case of the internal combustion engine because in the present case the "fuel" is free and renewable.

Two additional considerations of wind power are the height of the wind turbine and the average wind speed. Near the earth's surface the wind speed scales with height to the 1/7 power. This means that if the wind at 10 m has a speed of v_{10m}, then the wind at a height of 60 m will be $v_{60m} = (60\ m/10\ m)^{1/7} v_{10m} = 1.3\ v_{10m}$. Consequently, the hub of a wind turbine should be positioned as high as practically possible, and the blades must be able to withstand a significant difference in force from one end to the other, because the force experienced by the blade varies with height and is proportional to velocity cubed.

The fact that the power scales as the cube of velocity (within the designed operating speeds) also makes wind turbine power strongly sensitive to fluctuations in speed, as mentioned earlier. Suppose the average wind in a particular location is 7 m/s but the wind fluctuates so that it has velocities of 3 m/s, 7 m/s and 11 m/s for equal amounts of time. The average power will be proportional to $[(3\text{ m/s})^3 + (7\text{ m/s})^3 + (11\text{ m/s})^3]/3 = 1.6(7\text{ m/s})^3$, or $1.6v_{ave}^3$. The effect of having the higher speed for one third the time boosts the power output by 1.6 more than the power if the speed stayed constant at the average 7 m/s.

A final point about wind turbine power is the problem of blade soiling. The collection of insects and dirt on the leading edge of a turbine blade can reduce the power collection by as much as 50%. This requires periodic cleaning of the blades, no small task for a blade 50 m long attached to a hub 100 m off the ground.

EXAMPLE 5.2

Wind Turbine Energy

Suppose you are considering constructing a wind turbine in a region where the wind has an average speed of 3 m/s at a height of 5 m. You decide to build a turbine with a 50-m-diameter rotor with its hub 30 m high. What is the maximum possible power that can be generated? What would this number be after the blade becomes soiled? What would be a realistic estimate of the power generated, assuming 25% efficiency?

At a height of 30 m the average wind speed will be about $v_{30m} = (30\ m/5\ m)^{\frac{1}{7}} v_{5m} = 1.3v_{5m} = 3.9$ m/s. The area of the rotor will be $\pi r^2 = 1963\text{ m}^2$. Using Equation (5.9) we have $P_{max} = \frac{16}{27}\left(\frac{1}{2}\rho A v_i^3\right) = \frac{16}{27}\left(\frac{1}{2}(1.3\text{ kg/m}^3)1963\text{ m}^2(3.9\text{ m/s})^3\right) = 45$ kW. Replacing the Betz limit of $\frac{16}{27}$ with 25% efficiency gives 19 kW. Severe blade soiling cuts these numbers roughly in half.

For comparison, the annual residential electrical energy use in the United States is slightly more than 11,000 kWh per household, which would necessitate an average generating capacity of about 1.3 kW per residence. Total per capita electrical use in the United States is about 13,000 kWh (five times the world average), which would necessitate a generating capacity of about 1.5 kW per person.

5.3 ■ Biomass

Biomass is a broad generic term applied to any organic material that can be used as a fuel or energy source. Plants grown specifically for conversion to fuel and industrial byproducts such as used cooking oil and wood chips from lumber production are included. Biomass can be loosely classified as either wood or wood products, such as charcoal and black liquor (processed wood pulp); biogas; organic waste; or biofuels such as ethanol and biodiesel. Organic waste that can be used as an energy source consists of wood residue (sawdust), crop residue (straw, stalks), animal manure, sewage sludge, municipal waste, and landfill gas. Crops currently under intense development for biofuels include switchgrass, corn, and soybeans in the United States, *Miscanthus* in Europe, and sugar cane in Brazil.

In 2006 wood and wood products supplied a total of about 20 quad of energy globally, and biomass provided about 0.6 quad of electricity. Most of this wood use was for cooking and heating in Africa and rural Asia. Ethanol and biodiesel production was about 1.4 quad in 2006, about 2.5% of the global fuel used for transportation [1]. In the United States in 2006, biomass, including wood and biofuels, made up about half of the nonhydro renewable energy used and 3.4% of the total energy used [12]. Brazil currently fuels half its vehicles with pure ethanol from sugar cane and the rest with a mixture of 22% ethanol and 78% gasoline [10].

The chemical or biological process for converting a biomass feedstock into a biofuel depends on the biomass. For example, the sugars in sugar cane and sweet sorghum can be converted into ethanol and butanol by fermentation, an anaerobic biological process. However, maize (corn) and cassava are mostly starch, so to be used as a fuel the starch must first be converted to sugar by a saccharification process, also biological. Switchgrass, poplar, *Miscanthus*, corn stover, bagasse (sugar cane residue), forest trimmings, yard waste, and construction debris are composed mostly of cellulose, which can be either gasified at high temperature or treated by pyrolysis or the Fischer–Tropsch process (described in Chapter 6) to form complex hydrocarbons from which fuel can be extracted. Cellulosic fermentation for these stocks using special strains of bacteria is also under investigation, as are bacteria that can convert waste biomass directly into complex hydrocarbons. Palm oil, soy oil, and commercial waste oil can be treated by a chemical process called trans-esterification to form biodiesel fuel. Alternatively, all these primary feedstocks can be burned to produce steam for electricity generation. The most recent investigations of the use of biomass involve breeding or genetically modifying algae, which can generate hydrogen directly for use as a fuel.

Based on satellite data it is estimated that the total energy contained in plant material on the surface of the earth is about 2,850 quad [1]. On one hand, this represents mostly uncultivated plant material; the annual productivity, at least in some regions, could probably be augmented significantly by intensive cultivation. Using increased fertilization, hybrid plants, improved pesticides, and modern farming methods, the United States was able to increase crop yield of corn by almost a factor of five over the past 80 years. On the other hand, as we have recently seen, diversion of food crops for biofuels can significantly raise the price of food [13]. Preservation of natural habitats, such as the national parks of many countries and important bioreserves such as the Amazon basin, would also preclude the use of the entire biomass of the earth for fuel.

The main obstacle to an increased use of biofuels is the low overall efficiency of plant photosynthesis (see [14] for an explanation of the limits of plant growth). Although the photosynthesis process itself is very efficient, plants are at best only 8% efficient (and less than 1% in most cases) in converting solar energy into a form that can be harvested, as noted in Table 2.1. Intensely managed energy fuel crops such as tree plantations and sugar cane grown with modern techniques are estimated to generate between 0.4 and 1.2 W/m^2, whereas crop residues and wood from unmanaged forests provide half this much [3, 15]. For comparison, commercially available photovoltaic (PV) solar cells, discussed later in this chapter, are currently 10% to 25% efficient. The solar energy hitting the earth's surface varies significantly from location to location and time of day, but a reasonable average for mid-latitude regions is about 200 W/m^2, so the energy return from a solar cell is about $200\ W/m^2 \times 0.25 = 50\ W/m^2$, a factor of 50 higher than the best fuel crop.

As discussed more completely in Chapter 6, the energy needed for the various conversion processes mentioned here, such as pyrolysis and the Fischer–Tropsch process for converting biomass into a fuel, is significant. Currently the most common conversion process is fermentation, which converts plant sugars into ethanol. The efficiency of this conversion process is particularly dependent on the biomass feedstock. For example, it takes an input of about one energy unit in the form of fuel for farm machines, fertilizer, and conversion processes to produce 1.2 energy units of ethanol from corn, an energy gain of only 17%. This is because corn (and other biomass feedstocks such as wood byproducts and switchgrass) have significant amounts of cellulose, which does not ferment easily. Plants that are composed of a larger fraction of sugar produce a higher energy return on energy invested (EROEI). The EROEI for soybeans converted to biodiesel is approximately 3.5, and it is 8 for sugar cane converted to ethanol. The high energy balance for the production of ethanol from sugar cane makes the use of ethanol in Brazil successful; the low energy return of ethanol from corn makes it much less attractive as a biofuel. Cellulosic fermentation, which is currently under investigation, may improve the energy return for some plants with higher cellulose levels. Genetically modified plants may also improve these energy returns, but they will need to improve significantly to compete with existing solar power technology.

EXAMPLE 5.3

Energy from Fuel Crops

The total arable land in the United States is about 1.6×10^6 km^2. Using 1.2 W/m^2 for a plant crop, how much power can be generated in the United States by plants if all arable land is used? Assuming this figure can be maintained continuously all year, how much energy can be generated in a year? Use the energy return rates for corn, soybean, and sugar cane to find the energy content of the fuel that can be produced this way if the input energy comes from the fuel stock itself. Compare your final result to the energy used for transportation in the United States in 2004, which was 27.8 quad.

The power output is 1.6×10^6 km^2 $\times$ 1.2 W/m^2 = 1.9×10^{12} W. Multiplying by the number of seconds in a year gives the number of joules available per year as 6.1×10^{19} J = 57.7 quad if all the energy could be harvested with no energy invested.

If the crop is corn, with an energy return rate of 1.2 energy units per input unit, we have $0.2/1.2 \times 57.7$ quad = 9.6 quad of fuel, about a third of what was used in 2004. For soybeans we get $2.5/3.5 \times 57.7$ quad = 41.2 quad. If sugar cane could be grown in the United States (it cannot except in a few areas), the fuel obtained from using all the arable land to make fuel would yield 50.5 quad.

Example 5.3 shows that the total transportation needs of the United States could possibly be met with biodiesel made from soybeans but at a high environmental and social cost. The estimates shown assume that all arable land, including national forests and parks, is used for fuel and none for food production. The Energy Information Administration estimates that about 5% of the arable land in the United States could be diverted to fuel production without compromising food production or land set aside for other purposes such as national forests (by comparison, Brazil currently plants 4% of its arable land in sugar cane) [10]. Using the figures in the example, 5% of the U.S. arable land dedicated to fuel crops would produce only 0.5 quad if planted with corn and 2 quad if planted with soybeans. Although making ethanol from corn is energetically a factor of three less efficient than making biodiesel from soybeans, ethanol use in the United States has grown by a factor of 3.4 since 2000, and current use is 13 times greater than that of biodiesel [12].

Carbon dioxide emission will be discussed in Chapter 7; however, it is significant that one advantage biofuel claims to have over fossil fuels is that plants undergoing photosynthesis absorb carbon dioxide to make sugar. The process produces oxygen:

$$6CO_2 + 6H_2O \rightarrow C_6H_{12}O_6 + 6O_2. \quad (5.10)$$

Although both the conversion of sugar to ethanol and the burning of the ethanol in a car engine eventually return the CO_2 to the atmosphere, the net carbon dioxide flux is approximately zero. However, recent research has questioned this assumption by pointing out the additional CO_2 added to the atmosphere during the planting phase [16]. Whether biofuels are truly carbon neutral remains to be seen.

The low energy balance, the question of carbon dioxide emissions, and the growing question of competition between plants used for food versus fuel has made biofuels much less attractive, especially for corn. The only benefit of using plants to convert solar energy into fuel as compared to solar PV is that the fuel produced is portable, making it useful for transportation purposes, and it can be easily stored. In regions with long growing seasons and large areas of unused fertile land, such as Brazil, where more efficient plants such as sugar cane can be grown, biofuels can make a significant contribution to energy for transportation. In many other locations, however, the manufacture of biofuels will probably remain marginal because of fundamental limitations in plant efficiency. In contrast, the use of waste biomass, which currently constitutes half the world's biomass feedstock, does make sense because this energy would otherwise be wasted. Municipal garbage has an energy content of about 10 MJ/kg, which is about a fourth that of gasoline per kilogram. The United States produces about 228 Mt of municipal waste a year, which amounts to more than 2 quad of energy that could be obtained from garbage.

5.4 Solar

If you did Problem 2.9 in Chapter 2 you should have calculated 1,362 W/m^2 as the blackbody radiation reaching the top of the earth's atmosphere coming from the sun, which is very close to the measured value of approximately 1,366 W/m^2, known as the solar constant, *S*. The daily average power striking the upper atmosphere of the earth varies with latitude and time of year. Near the equator the average is between 500 W/m^2 and 550 W/m^2 for a 24-hour period [4].[2] These figures are also the approximate energy flux for much of the northern hemisphere during the summer. In the winter this flux drops to between 150 W/m^2 and 200 W/m^2 at a latitude of 40° because of the change in angle between the sun and the surface of the earth.

The formula for calculating the power arriving at the earth's surface as a function of location and time of year, called the insolation, *I*, is given by

$$I = S\cos(Z), \tag{5.11}$$

where

$$Z = \cos^{-1}(\sin\phi\sin\delta + \cos\phi\cos\delta\cos H). \tag{5.12}$$

Here ϕ is the latitude of the location; δ is the declination, which varies in the northern hemisphere from 23.5° on the summer solstice (June 21 or 22) to –23.5° on the winter solstice (December 21 or 22); and H is the hour angle. The hour angle is given by $H = 15° \times (time - 12)$, where time is the hour of the day, counting from midnight.

EXAMPLE 5.4

Insolation at a Given Latitude

Calculate the insolation for a latitude of 60° at noon on the equinox (September 21 or 22).

The hour angle is $H = 15° \times (time - 12) = 0$ for noon. On the equinox the declination is exactly halfway between 23.5° and –23.5° and so equals zero. Using Equation (5.12), we have $Z = \cos^{-1}(\sin 60 \sin 0 + \cos 60 \cos 0 \cos 0) = 60°$. The insolation for a latitude of 60° at noon on the equinox is $I = S\cos(Z) = 1366\ W/m^2 \cos(60) = 683\ W/m^2$.

The earth's surface is spherical, but the target presented by the earth to rays coming from the sun is circular. Therefore, we expect the total energy per second striking the earth to be $\pi R^2 S$. Because this is spread out over the area of a sphere during a 24-hour period, the daily average energy reaching the top of the earth's atmosphere per second averaged over all latitudes is

$$\frac{\pi R^2 S}{4\pi R^2} = \frac{1}{4}S = 342 \text{ W/m}^2. \quad (5.13)$$

As we saw in Chapter 2, the peak frequency of this radiation depends on the surface temperature of the sun and occurs in the visible part of the spectrum. Not all the solar energy arriving at the earth's upper atmosphere reaches the surface, as we will examine in detail in Chapter 7. Certain frequencies are blocked by water vapor and other gases in the atmosphere. An additional effect is that at low angles (high latitudes or sunset and sunrise) the sunlight passes through a larger amount of atmosphere, reducing the intensity by as much as 15%.

The total reflectivity of an object is given by a unitless number known as the albedo, a. The earth's albedo is about 0.30, indicating that 30% of the incident energy is reflected back into space [6]. The daily amount of solar radiation reaching the earth's surface on average for all latitudes is then

$$\frac{1}{4}(1 - a)S = 250 \text{ W/m}^2. \quad (5.14)$$

The flux reaching the surface is also very dependent on the local climate, varying by as much as 70% for areas that have significant average amounts of cloud cover, compared with arid high-altitude locations. The equator gets more power, as noted earlier, and the poles get less because of the angle of the sun's rays relative to the earth's surface. For middle and low latitudes, the areas where most of the world's population lives, the average daily energy flux reaching the earth's surface can be assumed to be about 200 W/m^2 if cloud cover is minimal.

EXAMPLE 5.5

Solar Rooftop Energy

Estimate the power that can be collected from a typical rooftop in the United States. How does this compare with the 1.5-kW average per capita electrical use?

The average square footage of a house in the United States is about 2,000 ft^2, which is very roughly 200 m^2. If every living space with this area were a single story with a roof of the same size, the energy captured would be 200 W/m^2 × 200 m^2 = 40 kW, assuming all the energy is captured. A 20% efficient PV solar cell array would collect 8 kW. If we suppose the living spaces are all in apartment buildings with four single-person dwellings per rooftop, this lowers the energy captured to 2 kW per person, still more than the current average total per capita electrical use (and significantly larger than the residential use).

In the preceding discussion we used the 24-hour average solar flux at the earth's surface, 200 W/m^2. However, it should be pointed out that the actual flux depends not only on weather conditions but also on the angle of the sun and the latitude where it is observed. The flux of the sun at the surface of the earth at noon on the equator is about 1,000 W/m^2. A fixed horizontal solar panel of area A will receive a power of $P = (1000\ \mathrm{W/m^2})A\cos\theta$, where θ is the angle between the vertical and the sun, which depends on latitude and time of year. Solar collector systems are often rated in terms of their available peak power, which usually assumes an incoming solar flux of 1,000 W/m^2 at noon with the panel directly facing the sun.

From these figures it should be clear that solar energy is plentiful. Based on an average flux of 200 W/m^2, the total solar energy reaching the world's land mass is 6.5×10^5 quad annually, with another 1.3×10^6 quad reaching the surface of the ocean. Using 20% efficient solar cells and transmission, storage, and conversion losses of 50%, an area of solar cells about 750,000 km^2 would be needed to supply the entire world's energy needs. Using the same assumptions, an area about the size of North Carolina covered with solar cells would supply the entire electrical supply of the United States. The percentage of a country's land area needed to supply all the country's energy is shown in Table 5.2 for a few countries.

Several European countries, the United States, and Japan are pursing the idea of locating PV solar panels on commercial and residential rooftops and paying to have the energy fed back into the grid. In 2000, for example, Germany introduced a guaranteed price payback, good for 20 years, for electricity fed back into the electrical grid by locally owned solar cells. This incentive resulted in a sixfold increase in solar energy use in the next 3 years.

Currently there are two main ways to collect solar energy: PV and solar thermal. Large- and small-scale applications for both are under intensive research. Each has advantages and disadvantages, which are discussed in the following sections.

TABLE 5.2

Annual energy use of select countries and the percentage of land mass of the country covered with solar cells that would supply all the annual energy use. The figures assume current solar cell technology, with 20% efficient cells and 50% storage and transmission loss, and are based on [15].

Country	Annual Energy Use 2005 (quad)	Percentage Land Surface Needed for All Solar
Argentina	2.7	0.2
China	46.6	4.7
Denmark	0.88	3.7
Egypt	2.8	0.4
France	11.2	3.7
Ghana	0.1	0.1
Japan	22.4	10.8
Russia	17.1	0.3
United Kingdom	9.8	7.8
United States	98.8	1.9

5.4.1 Photovoltaics

As mentioned in Chapter 2, a solar cell is essentially a photocell running in reverse. Figure 5.3 shows a schematic of the photocell process, which should be compared to Figure 2.5 for a light-emitting diode. If they have the right energy, incoming photons can create an electron–hole pair at the junction of p- and n-type semiconductors. Free electrons are formed on the n-side by boosting electrons over the band gap into the conduction zone, where they can move as a current. On the p-side, boosting electrons out of the valence band leaves holes, which act as positively charged particles that are also now free to move. As more electrons are created on the n-side of the junction, they repel each other and flow away from the junction. Positive holes also flow away from the junction on the p-side. The flow of electrons constitutes an electric current, and the energy of the moving electrons can be used in an external device, represented by the resistor, R. In the figure the conventional current, I (in amperes), flows to the right because conventional current is labeled to flow opposite the direction of the electron motion.

Recall from Chapter 2 that the energy of a photon is given by $E = hf = hc/\lambda$, where h is Planck's constant, c is the speed of light, and f is the frequency. The band gap limits the efficiency of a solar cell because only photons with

FIGURE 5.3

Schematic of a generic solar cell. Photons (electromagnetic radiation) cause electrons to jump across the band gap at the junction of a p–n diode. See Chapter 2 for an explanation of band gap.

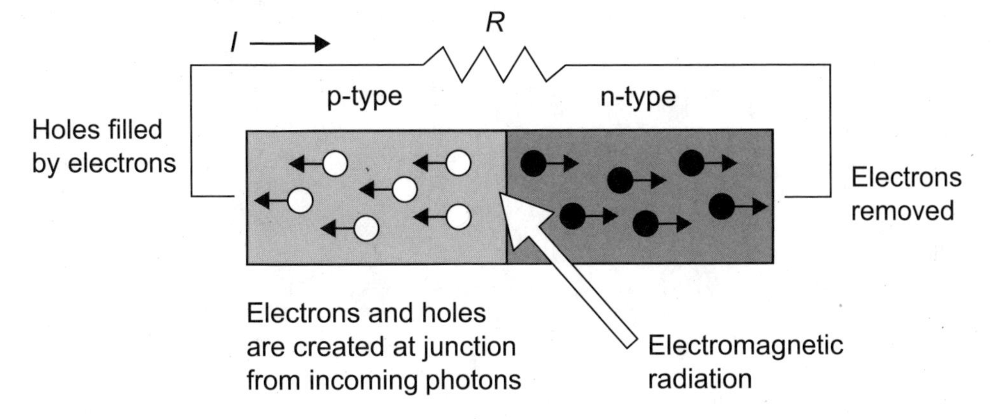

frequencies high enough to have energies large enough to boost an electron over the band gap are useful. In other words, only photons with $E > E_{band}$, where E_{band} is the energy of the band gap, can be captured. Photocells also cannot use the additional energy of photons with energies much larger than the band gap; once the electron has enough energy to cross the band gap, the rest of the photon's energy is absorbed as random kinetic energy (heat). Most PV cells in use today are designed to capture visible photons because the atmosphere is most transparent to these frequencies, even though a significant amount of the incoming electromagnetic energy is outside the visible spectrum.

Efficiency for PV cells is defined as the energy output in the form of electrical work divided by the radiant solar energy absorbed. Table 5.3 lists different types of PV cells under development and the efficiency of laboratory prototypes. Commercial silicon photocells, which constitute more than 90% of the market, have efficiencies of about 16%. In addition to losses due to the failure to capture photons with energies lower or much higher than the band gap, some energy is lost by electron–hole recombination before they have a chance to move away from the junction. This loss can be calculated from the open circuit voltage, the voltage across the PV cell when it is not connected to a circuit. The ratio of the open circuit voltage to the band gap voltage is the percentage of energy lost to recombination, which turns up as heat.

In multijunction cells a combination of materials is used so that more than one band gap is present. In this way photons of a larger range of energies can be captured by the different band gaps in separate parts of the cell. Single crystal

TABLE 5.3

Laboratory efficiencies of various prototype photovoltaic cells [17].

Type	Best Efficiency (%)
Multijunction	
Three junction	40.7
Two junction	30
Single junction	
Single crystal	25
Concentrator	28
Thin film	24.7
Silicon crystalline	
Single crystal	27.6
Multicrystalline	20.3
Thick film	16
Thin film	
Copper indium selenide	19.9
Cadmium telluride	16.5
Amorphous silicon–hydrogen	12.1
Nano-, micro-, poly-silicon	16
Multijunction polycrystal	15.8
New photovoltaic designs	
Dye sensitized (Graetzel) cells	11.1
Organic cells	5.4

cells are cells manufactured from a single monolithic crystal, which requires an elaborate chemical process. Making cells from a conglomerate of smaller crystals or from amorphous silicon decreases manufacturing costs. Amorphous cells are also flexible, unlike crystalline cells, which is an advantage in some applications. In dye and organic cells photons are first absorbed by a dye or organic compound to free electrons, which then pass into an n-type semiconductor. These types of solar cells currently have the lowest efficiencies but are also the cheapest to manufacture and can also be used in applications where mechanical flexibility is desired. Concentrating solar cells use lenses or mirrors to concentrate solar energy on the solar cell (it should be pointed out that this cannot increase the solar energy per square meter, but it can reduce the cost of a solar cell arrangement because a smaller area of solar cells captures the energy from the larger area of the lens or mirror). As seen in Table 5.3, combinations of all these varieties of solar cells are under investigation.

As is the case for any energy technology, a certain amount of energy has to be invested for manufacturing before the technology can be used to produce energy. As noted for bioethanol produced from corn, the energy output is 1.2

units of energy for one unit of invested energy. For solar cells a more useful figure is the energy payback time. It takes about 2.2×10^9 J to make a square meter of single-crystal silicon PV cells and 1.6×10^9 J for multicrystalline silicon [18]. These cells have efficiencies of about 12%, so, using 250 W/m^2 for the incoming radiation, it will take

$$(2.2 \times 10^9 \text{ J/m}^2)/(0.12 \times 250 \text{ W/m}^2) = 7.3 \times 10^7 \text{ s} = 2.3 \text{ yr} \quad (5.15)$$

of cell operation to make up the energy used to create the photocell. Although PV cells drop in efficiency over time, the projected lifetime for most commercial designs is between 25 and 35 years. If this PV cell lasts 30 years, it will produce another 2.6×10^{10} J of energy, 10 times the amount needed to manufacture it. Energy payback times for PV cells have dropped significantly over the past 20 years, and newer types of PV cells are expected to have energy payback times of less than a year.

The major disadvantages of PV solar are the cost and, like windmills, the fact that the driving energy source is intermittent and dispersed. Although the cost of PV solar electricity has been decreasing for the past 25 years, PV electricity still costs about five times more per kilowatt-hour than electricity generated from a modern gas turbine system and four times more than wind. It is projected that the cost of PV electricity will be competitive with electricity from coal plants within the next 5 to 10 years [15]. A major remaining problem, then, is storing energy for use when these resources are not available. Clearly, widespread use of solar cells or wind will also require the implementation of energy storage devices, a topic that is discussed in Chapter 6. One last obstacle to PV solar cell use is that they create direct current. This means additional electronics are needed to convert the direct current into alternating current for most electronic equipment in use today, increasing the cost of using this source.

5.4.2 Thermal Solar

Passive solar storage systems, such as using heat from the sun to warm rocks or water for use after the sun goes down, have been in use since the dawn of civilization. Modern applications of passive and active solar thermal storage for buildings are considered in Chapter 6. Here we give a brief overview of concentrating thermal solar energy to drive various steam engines, which are in turn used to generate electricity.

Solar thermal electric systems use arrays of mirrors, either flat or parabolic, to concentrate the sun's energy, raising the temperature of a working fluid to between 300°C and 1,000°C. Collections of mirrors in a parabolic trough can be used to heat the fluid as it flows through a pipe at the focus of the parabola. The sun's energy can also be focused on a single target in a tower, using arrays of flat mirrors or a parabolic dish mirror that track the movement of the sun. Water, various gases, and molten salts have been investigated as working fluids for these applications. The heated fluid can be used to run a traditional steam engine to generate electricity, or the heat can be used directly to run a Stirling engine. Because of the high temperatures possible, the Carnot efficiency for these processes is high.

To date several solar thermal electric systems have been built, and more are planned. Currently the costs of building such a system make them economically unattractive, but cost is projected to go down as technical details and manufacturing processes improve. The Luz International company went bankrupt in 1991 after federal tax breaks were removed but not before it constructed nine solar plants with a total generating capacity of 354 MW. See Problems 5.35 and 5.36 at the end of this chapter for further information.

5.5 ■ Tidal, Geothermal, and Other Renewable Energy

In this section we consider various other possibilities for renewable energy. Although in some cases the available global energy flux is substantial, for a variety of reasons these resources have yet to be developed into viable energy resources. The current consensus seems to be that it will be some time before technical advances permit these resources to play a significant role in the global energy supply. Further details about the current status of development of these resources can be found in references [19] and [20].

5.5.1 Tides

One means of harnessing tidal energy is to collect water at high tide to form a hydraulic head, as is done for dams, and then capturing the energy using hydropower technology as the water flows back out at low tide. The height difference between low and high tide varies greatly around the world. The Bay of Fundy in Canada has the largest tides in the world, varying by 17 m every 6 hours (Figure 5.4). It is estimated that, by damming the mouth of the bay, approximately

FIGURE 5.4

Low tide in the Bay of Fundy. At high tide the water reaches the top of the sculpted part of the rocks.

4.2×10^{16} J = 0.04 quad could be generated annually [1]. Other places that have tides of this magnitude are the Severn Estuary and Solway Firth in the United Kingdom and the Gulf of Khambhat in India. The largest tidal plant built to date is on the La Rance Tidal Basin at Saint Malo, France, which generates 7.7×10^{15} J = 0.01 quad a year from tides that vary by about 7 m.

Damming tidal basins has a significant environmental impact on local estuaries, and the alternative, anchoring turbines underwater to capture the flow of tidal current, has gathered more interest. This mechanism captures the movement of tidal currents using turbines similar to wind turbines to collect the kinetic energy of the water as it moves with the tides or ocean currents. A number of experimental tests have been made, and there are plans to put prototypes in place in the near future [1]. The motion of the tides very gradually slows down the earth's rotation, and the total tidal energy that exists globally can be calculated from the rate at which the earth's rotation is slowing [10]. Based on this calculation there is about 90 quad of tidal energy available globally (see Problem 5.41).

Calculating how much of the global energy can actually be captured from local tidal currents is more complicated than the calculation of the Betz limit for wind turbines. In that calculation it was assumed that there was little resistance to the air changing volume due to pressure differences as it passed the turbine. However, water does not easily change volume, and if the turbine is located in a narrow channel the water flow will interact strongly with the sides and bottom of the channel. This makes any calculation of extractable energy highly dependent on the geological terrain at the site of the turbine. Estimates of the energy that could be extracted from the best 28 sites in the world add up to about 10 quad [3].

5.5.2 Waves

The total energy flux of all the waves on the globe is quite large, but the energy near the coast is thought to be on the same order of magnitude as the global tidal energy [3, 10]. As in the case of tidal energy, wave energy can be predicted more accurately than wind or changes in solar energy flux due to weather variations. On the other hand, wave energy has a very low density because it is spread over the two thirds of the globe that is covered with water.

EXAMPLE 5.6

Wave Energy

How much wave energy is available? The average energy per unit surface area carried by a water wave of sinusoidal shape is given by

$$\frac{E_{ave}}{Area} = \rho g y^2 / 2, \tag{5.16}$$

where ρ is the density of the liquid, g is the acceleration of gravity, and y is the amplitude of the wave. Multiplying both sides by the wavelength, λ, and the frequency, f (or the speed, because $f\lambda = v$), gives the power per meter width of the wave:

$$\frac{P_{ave}}{l} = \rho g v y^2 / 2. \tag{5.17}$$

Waves in the ocean near the coast have an average height of about a meter, a frequency of about 0.1 Hz, wavelengths of about 10 m and speeds of about 1 m/s. This gives a power of $P_{ave}/l = 5$ kW/m, or 5 kW per meter of coastline.

The total coastline of the east coast of the United States is about 3,300 km, so the power available at the east coast is about 5 kW/m × 3,300 km = 1.65×10^{10} W. If all this could be captured (100% efficiency) over the period of a year, this would be about 0.5 quad per year.

An amazingly large number of different devices and mechanisms have been proposed and tested for extracting wave energy [20]. They can generally be classified into three basic types. In overtopping or surge mechanisms, water is collected in a basin from waves that break over a barrier or surge through a one-way gate. Once trapped, the water has a hydraulic head, and the energy can be extracted as in a dam. Oscillating water column devices use the oscillating level of waves to compress a working fluid by using a one-way valve. Heaving and pitching mechanisms use the kinetic energy of the wave directly. In these devices, floating objects heave up and down due to wave action, and the force of the up-and-down motion can be captured by a stationary set of levers attached to the floats. Waves also cause pitching from side to side of a floating object, and this mechanical motion can be converted into the rotation of a crankshaft. At present it is not clear which of these devices, if any, may eventually provide significant amounts of energy.

5.5.3 Geothermal

The center of the earth has a temperature of about 4,500°C due to gravitational pressure and radioactive materials in the earth's core. The temperature decreases the farther from the center of the earth so that at the top of the mantle on which the continental crust floats, the temperature ranges from 500°C to 1,000°C, compared to average surface temperatures of less than 40°C. This large temperature difference would make for a very efficient heat engine with basically unlimited energy; however, it is technically difficult if not impossible to access. The earth's crust is 5 to 10 km thick under the ocean and about 30 km thick under the continents. Although there have been a few attempts to drill down to the base of the crust, thus far none have succeeded. Wells of 10 km depth are technically feasible but very costly.

In non–geologically active locations in the continental crust, there is a temperature increase of about 25°C to 30°C per kilometer depth, with the result that a 3-km-deep well has a temperature of about 100°C at the bottom. As mentioned in Chapter 2, the constant temperature of the earth at shallow depths, maintained by this temperature gradient, has been used in applications for heat pumps, which are the fastest-growing use of geothermal energy.

The flux of energy coming from the core of the earth amounts to about 0.06 W/m^2 leaving the earth's surface. Multiplied by the surface area of the earth, this amounts to more than 900 quad per year, which is close to one estimate of how much geothermal energy is ultimately available [6]. In theory it would be possible to extract this energy by circulating water or other fluids in extremely deep wells. Depending on assumptions about the available technology, in particular how deep a well can be drilled, estimates of geothermal energy resources range as high as 960 quad and as low as 15 quad for hydrothermal sources [10]. If wells are ever drilled all the way to the mantle beneath the crust, this estimate would rise by roughly a factor of 50.

The earth's crust is segmented into large pieces called tectonic plates, which slowly move around the surface of the earth, grinding against each other. In some regions they slide under one another, and in other locations they pull apart from each other. The boundaries of these plates constitute regions of high geological activity, and in these locations hot magma from the mantle can come closer to the surface, causing volcanoes and geological hot spots. In these regions, hot molten rock at temperatures of 1,000°C may be within 2 or 3 km of the surface. If water from the surface reaches these heat sources, it becomes steam under high pressure. This steam sometimes reaches the earth's surface in the form of geysers. The earliest use of geothermal energy dates to the early 1900s, when steam from geological activity was used to run generators to make electricity.

Geologically active locations are generally classified as either high- or low-temperature fields and can be further classified based on the types of rock formation present and whether water is present. For efficient electricity generation the rock must be above 150°C. Most facilities generating electricity from high-temperature fields have wells 1 to 2 km deep and well temperatures of 250°C to 350°C. Steam from existing ground water or water pumped into these wells, which becomes steam, is used in a Brayton cycle steam turbine to generate electricity (see Chapter 3). Currently about 0.5 quad of geothermal energy is used globally per year: about 0.2 quad for generating electricity and 0.3 quad for direct heating use [3].

Technical and economic difficulties currently limit the use of geothermal energy; however, a significant number of new projects designed to capture this energy source are under way. As mentioned earlier, wells as deep as 10 km are technically possible, but the costs are prohibitive. This limits large-scale geothermal energy to more shallow wells located near tectonic plate boundaries and local hotspots. Because of the geological features underlying the country,

Iceland is able to supply about 25% of its electricity and more than 85% of its heating needs from geothermal sources, making it a notable example of the use that can be made of geothermal energy.

5.5.4 Thermoclines and Salt Gradients

The ocean, which covers two thirds of the surface of the earth, captures a large amount of solar energy, with the result that for most of the ocean in the tropics there is a temperature difference of more than 20°C between the surface and depths of 1 km. The amount of accumulated energy is about 60,000 quad. Because of the vast amount of energy stored in the tropical ocean, ocean thermal energy conversion received a significant amount of interest in the 1970s, after the steep rise in oil prices at that time. However, as we saw in Chapter 3, thermal efficiency is very low for heat engines in which the temperature difference is small. For a temperature difference of 20°C the Carnot efficiency is $\eta = \left(1 - \frac{T_C}{T_H}\right) = 7\%$. This would still leave an extractable amount of energy of 4,200 quad; however, this energy is spread over very large areas of the ocean, which makes it technically difficult to extract. Additional losses in efficiency for real extraction processes make it doubtful that this source will contribute significantly to the global energy supply in the foreseeable future.

The solar effect resulting in a natural temperature difference of the tropical ocean can be enhanced artificially on a smaller scale by using a salt gradient. In a reservoir or pond with a high salt concentration near the bottom, visible solar radiation can penetrate to the salty layer, but infrared radiation from the hot salt solution is trapped by the upper layers of less salty water. Using this technique, it is possible to generate a temperature difference as large as 80°C, which increases the Carnot efficiency to the point of being useful [10]. Experimental systems are in use in Israel near the Dead Sea and in the United States at El Paso, Texas. Salt and temperature gradient engines have also been investigated for use with solar collector arrays [20].

5.5.5 Energy from Waste and Recycling

As we saw in Chapter 1, the increase in global population has been responsible for a significant increase in the amount of municipal and other waste disposed of each year. This provides one compelling reason to recycle discarded material: The sheer volume of waste makes it difficult to find sufficient landfill space. A second reason has to do with the energy content of these byproducts of civilization. Biomass from municipal waste with an average energy content of 10 MJ/kg has been mentioned previously as an energy source with the potential to provide 2 quad of energy per year in the United States. However, some of the components of municipal waste have much higher energy content. Plastic has an energy content of 42 MJ/kg, paper 18 MJ/kg, and wood scrap 15 MJ/kg, and these items constitute more than 50% of municipal waste by weight [21, 22]. These figures, when compared with the energy content of crude oil at 42 MJ/kg, make it clear that "mining" municipal waste for energy is a serious possibility. Currently in the United States about 15% of municipal waste is burned using energy recovery systems for either heat or electricity, according to Department of Energy figures.

A second energy consideration related to recycling has to do with the initial energy cost of manufacturing a product and the energy needed to produce the same product using recycled stock. The energy savings is very dependent on the material being recycled. It takes some 300 MJ to produce a kilogram of aluminum from ore but less than 20 MJ to produce a kilogram of aluminum from recycled scrap, a 93% savings in energy. Using scrap copper saves about 75% of the energy compared with mining and refining copper ore. Manufacturing a kilogram of iron from ore takes about 28 MJ, but only 1.3 MJ is needed to produce iron from scrap iron. Using recycled paper as stock for a paper mill results in about a 40% savings in energy. Recycling glass containers, on the other hand, results in less than a 30% energy savings over manufacturing glass directly from unprocessed sources [22].

The case of plastic recycling is complicated by the fact that at least seven different types of plastics have official classification numbers. For many purposes these different types cannot be mixed when used as recycled stock. For example, a single polyvinyl chloride (PVC) bottle mixed into a batch of polyethylene terephthalate (PET) bottles being recycled will ruin the outcome because PVC will ignite below the melting temperature of PET [23]. If the plastics are separated properly there can be as much as a 70% energy savings over making the plastic directly from petroleum products, but separation is labor intensive, which makes it costly. The lack of a sufficient market for recycled plastic has resulted in a net decline over the past 20 years of plastic recycling in the United States. Further details about performing a lifetime cost analysis (LCA) that accounts for the total energy involved in making, transporting, using, and disposing of a consumer product can be found in reference [24].

5.5.6 Other

Human beings are very creative, so it is not surprising that there have been a large number of other proposals for capturing various energy fluxes present in the environment. Capturing thermal energy from underground nuclear explosions, flying kites in the upper atmosphere to capture high-speed winds, launching solar collecting satellites that would beam energy to the earth's surface using microwaves, capturing ice from the poles to use as a cold reservoir for a heat engine, capturing the energy released when freshwater and seawater mix, and collecting the energy in lightning have all been proposed. Except for the investigations of water mixing and lightning (see Problems 5.47 and 5.48), none of these proposals have advanced past the planning stage. Because of the technical difficulties involved, it seems doubtful that any of them will play a significant role as an energy source in the foreseeable future.

5.6 Summary

Table 5.4 summarizes pertinent facts about the various renewable energy sources discussed in this chapter. As mentioned in the sections for each source, estimates of the available resource depend strongly on assumptions about what to include. For example, there are enormous amounts of energy in wind, but probably only winds over land out to within a few hundred kilometers of shore and within 1 km of the earth's surface can reasonably be tapped.

TABLE 5.4

Estimates of annual global renewable energy resources. The figures are based on references [1], [3], [4], and [10].

Source	Estimated Total Energy (quad)	Estimated Recoverable (quad)	Current Use (quad)
Solar	810,000 over land	30,000	2.2 thermal, 0.2 photovoltaic
Wind	11,000 over land	200–1,800	0.4
Biomass	2,850	12–1,200	21
Hydroelectric	300–900	45–60	2.7
Tidal	90	10	0.01
Wave	60	15–30	0.01
Geothermal	13,000	30	0.5

For other sources, such as geothermal, economic factors are a significant deterrent. Drilling down to the base of the earth's crust would make huge amounts of energy available, but drilling to those depths is economically unattractive at present. Still, several conclusions can be drawn from the data.

Clearly the estimated recoverable solar energy is nearly two orders of magnitude greater than any other source using current technology. The total energy reaching the earth from the sun is about 7,500 times the total global energy use today. The estimated recoverable solar energy striking land surfaces is 65 times the present global consumption. Cost and energy storage are the main factors limiting further implementation of this vast resource. Wind and biomass offer substantial amounts of energy for the near future. As mentioned earlier, hydroelectric is nearing saturation in the developed world but has excellent potential for expansion in South America, Asia, and Africa. Tidal and wave energy will play only a minor role in global energy in the foreseeable future, probably only in specific coastal areas where tides or waves are larger than the global average. Geothermal remains an unproven source awaiting further technological development.

When considering the future energy supply, the economic costs and technical challenges should be considered in addition to the available resource. Technical and economic factors can be summarized by the estimated cost to build, operate, and fuel a plant per watt of electrical power generated. These expenditures are often given as the levelized cost, which is defined as the yearly revenue needed to recover all expenses over the lifetime of the plant. Capital cost, the cost to build the plant, may be given as the overnight capital cost (the cost in present-day dollars to build the plant per kilowatt-hour) or the levelized cost (the cost averaged out over the lifetime of the plant including finance charges). Table 5.5 shows the National Academy of Science's projection of levelized electricity cost by source assuming completion in the year 2012 [25]. The capacity factor, the ratio of actual output over its normal operating lifetime to the maximum possible output, is included. Also shown are the typical plant sizes and the years from the start of construction until completion [26]. A few nonrenewable sources are included for comparison. The cost figures are in 2007 dollars and are averages over regional variations in construction and fuel costs.

As seen in the table, onshore wind and hydropower plants are already competitive with coal-fired plants and are cheaper than advanced technology gas turbine plants. Electricity from nuclear reactors, biomass, and geothermal plants are close to the cost of a traditional coal plant. It should also be pointed out that the costs shown are average costs; local fuel and transmission costs may change the figures significantly in a given region. As mentioned previously, building a

TABLE 5.5

Typical electrical power plant size, time needed to construct the plant (lead time), capacity factor, and projected 2012 levelized costs in 2007 dollars per megawatt-hour [25–27]. O&M cost includes operating and maintenance not including fuel.

Source	Size (MW)	Lead Time (years)	Capacity Factor (%)	Capital Cost ($)	Fixed O&M ($)	Fuel Cost ($)	Transmission Cost ($)	Total Cost ($)
Coal	600	4	85	57	3.7	23	3.5	87
Gas combined cycle			87	20	1.6	55	3.8	80
Gas combustion turbine	230	2	30	36	4.6	80	11	132
Nuclear	1,350	6	90	82	10	9	4	105
Biomass and landfill gas	80	4	83	64	9	25	4	102
Geothermal	50	4	90	77	22	0	5	104
Hydropower	500	4	40–60	57	7	0	5	69
Wind, onshore	50	3	36	73	10	0	8	91
Wind, offshore	50	3	33	171	29	0	9	210
Solar thermal	100	3	31	220	21	0	10.6	252
Photovoltaic	5	2	22	342	6.2	0	13	362
Energy-efficient building								0–60 (saved)

PV plant is a factor of four more expensive than obtaining wind energy. These numbers make it clear why wind is the fastest-growing renewable resource and why gas and coal are still the most popular sources of electricity. The capacity factors in Table 5.5 indicate how much electricity can actually be generated from a plant with a given rating after taking into account time offline for maintenance and intermittent energy sources such as wind or sunlight. Nuclear power has the highest capacity factor because nuclear power plants can usually run for many years with only very short maintenance periods when the plant is offline. The intermittent nature of wind and solar energy lowers their capacity factors.

As a final comparison, Table 5.5 shows the construction cost of an energy-efficient building for an equivalent energy savings (energy that doesn't have to be generated). As can be seen, this figure is close to the cost per kilowatt hour for producing electricity from a gas-fired plant [27]. Conservation can be as cost effective as building new power plants. Many industries are beginning to take advantage of this point.

As we have repeatedly seen, projections of the future use of a given energy source are fraught with uncertainties. The levelized cost of various plants shown in Table 5.5 does not take into account possible changes in fuel costs, plant efficiency, and pollution, three other factors related to the total cost of using a particular energy source. In 2007 the cost for obtaining natural gas was $4.84 per million Btu ($1.055 \times 10^9$ J) and $9.58 for crude oil, whereas coal was only $1.04 per million Btu. The low cost of coal and the low construction cost of gas-fired power plants have made these two choices the most cost effective in the past, as can be seen from the total costs shown in Table 5.5. However, a change in the cost of natural gas or a carbon tax on fossil fuels (a possibility we will consider in Chapter 8) could quickly and significantly affect the desirability of building more of either of these types of plants, regardless of construction cost. On the other hand, most renewable sources are less sensitive to fuel costs and depend more on initial or capital costs. As pointed out in Chapter 4, the time needed to bring new technology online is about a decade and power plants have a lifetime of about 40 years. If renewable sources are to replace traditional fossil fuel energy sources, these new technologies need to be developed now.

Questions, Problems, and Projects

For all questions,

- Cite all sources you use (e.g., book, article, Wikipedia, Web site). Before using Wikipedia as a source, please read the policies of research and citing in Wikipedia: http://en.wikipedia.org/wiki/Wikipedia:Researching_with_Wikipedia and http://en.wikipedia.org/wiki/Wikipedia:Citing_Wikipedia. For Wikipedia you should verify the information in the article by reading the sources listed.
- For all sources, briefly indicate why you think the source you have used is a reliable source, especially Web resources.
- Write using your own words (except for direct quotes, which should be enclosed in quotation marks and properly cited).

1. As mentioned in the text, estimates of available renewable resources can vary by as much as a factor of 10. Find several estimates of the potential hydropower available and compare them. Which do you think are more accurate and why?

2. Given the figure of 1.3 kW per home as the electrical power needed in the United States, how large a dam and water flow would be needed to supply a typical family with all their electrical needs, assuming 50% efficiency of collection, transmission, and storage? Repeat this calculation for your own annual electrical use.
3. Find the flow rates of any local streams or rivers in your area. Using Equation (5.3) for the power available from the kinetic energy, find the available hydroelectric power from these resources.
4. Using a reliable source, write an explanation of the pros and cons of large dam projects such as the Three Gorges project in China. Include ecological and health issues, carbon emissions, and potential political problems.
5. Using reliable sources, investigate the question of seismographic problems caused by the weight of water behind a large dam. Give some examples where it is believed that geological problems were caused by building and filling a large dam. Include a discussion of silting, for example in the Aswan Dam in Egypt.
6. Go to the EIA Web site (http://tonto.eia.doe.gov/country/index.cfm), pick two countries, and compare their energy profiles (e.g., imports, exports, consumption, historical changes in consumption, types of fuel used). How do these countries compare to the same figures for the world? Write a summary of your findings.
7. Find several estimates of the potential wind energy available, either locally or globally, and compare them. Which do you think are more accurate and why?
8. Based on reliable sources, write a summary of the problems with implementing wind energy. For each problem, find as many proposed solutions as you can.
9. One highly publicized problem with wind energy is the threat to migrating birds. Proponents of wind energy point out that about 1,000 times more birds are killed by domestic house cats and 5,000 more birds die from flying into windows than are killed by wind turbines. Using reliable sources, look at both sides of the argument and build a case either supporting or opposing the argument against wind turbines because of the potential damage to migrating birds. Take into account a projected increase in the number of wind turbines over the next few years.

10. From a reliable source, find a discussion of the plan to put wind turbines of the shore of Cape Cod in the United States. What are or were some of the concerns of the residents of in this area? Compare these concerns with the claims of the proponents of the idea. Who do you think has the better case and why?
11. Find a description of each of the following types of wind turbines and list the pros and cons of each design type and how they work, their efficiencies, and so on: horizontal axis, Darrieus (eggbeater), Savonius.
12. China is the fastest-growing market for wind energy. Using reliable sources, summarize the current status of wind energy in China and compare it with similar figures from Europe and the United States.
13. An amazing variety of wind turbine blade shapes have been proposed over time. Find as many different designs as you can and make a copy or a sketch of each. Which can be classified as horizontal and which are vertical axis turbines?
14. The National Renewable Energy Laboratory publishes wind maps for various regions at the following Web site: http://www.nrel.gov/wind/international_wind_resources.html. Download one of these maps for your favorite country or state and examine the map. Summarize the data, noting where wind velocities are sufficient for wind turbines and how far these areas are from large population areas. Note any other interesting data.
15. Derive Equations (5.8) and (5.9).
16. Make a table of wind speeds as a function of altitude up to 100 m, given that the wind speed at a particular location is 3 m/s at an altitude of 5 m.
17. Repeat Example 5.2 for the case where the wind speed is only 2 m/s at 5 m and the wind turbine diameter is 25 m.
18. Give the figure of 1.3 kW per household as the electrical energy needed, design a wind turbine system for a family living in your area. Consult the wind speed charts online at http://www.nrel.gov/rredc/ for your area. Adjust the height and rotor size to achieve the 1.3 kW figure per residence, assuming 25% efficiency.
19. Wind energy has a density of 2–3 W/m^2. How much area would be needed to supply all the U.S. electricity from wind? How large a wind farm would be needed to build the equivalent of a typical 500-MW coal power plant?

20. Using reliable sources, investigate any problems a country or region has faced as a result of relying heavily on wind energy. Include comments about recent problems with wind energy in Denmark and Texas.
21. Repeat the calculation in Example 5.3 for world figures. Go to the Central Intelligence Agency Factbook (https://www.cia.gov/library/publications/the-world-factbook/) to find arable land and the Energy Information Administration (http://www.eia.doe.gov) to find global transportation use.
22. Following Example 5.3, how much energy could be produced by 20% efficient solar cells if placed over the same area? How does this compare with fuel crops? (Note that solar cells do not need to be placed on fertile crop land as biofuel farms must be; the same area of desert could be used.)
23. Using reliable sources, write a summary of (a) the process by which plant sugars can be turned into ethanol and (b) the process that turns starch into a usable fuel. Comment on the efficiency, feedstock, and estimates of future use for both processes.
24. Using reliable sources, write a summary of the current state of development of cellulosic fermentation, hydrogen-producing algae, and bacteria that convert biomass directly into petroleum-like products.
25. Using reliable sources, write a summary of the current uses of biogas. Include the use of local sources such as methane from composting toilets and animal manure and of larger sources such as methane from landfills.
26. Ethanol use in the United States is currently 13 times the use of biodiesel. Using reliable sources, come up with a list of reasons why this is so. Include political considerations, such as the effectiveness of the corn lobby.
27. The actual impact of growing biofuels on the price of food is a very complex issue that has been hotly debated. The United Nations blames 65% of the spike in food prices in 2008 on the cost of energy, whereas the U.S. Secretary of Agriculture has said the effect is much less. Using reliable sources, make two arguments: one that the impact of biofuels on global food prices is minimal, the second that biofuels play a key role in food prices. Evaluate the arguments and decide which one is more plausible.

28. Using reliable sources, investigate the suggestion that some species of *Jatropha* would make a good fuel crop.
29. Repeat Example 5.4 for noon at your latitude and the current date (assume the declination changes uniformly during the year from solstice to solstice).
30. Fill in the steps to show that an area the size of North Carolina covered with solar cells would be needed to supply all the U.S. energy needs. How does this area change if the photocells are 15% efficient? What about 35% efficient? How does this compare with your wind energy calculation in Problem 19?
31. Using a reliable source, write a summary of the 100,000-roof project started in 2000 in Germany. What are some successes and failures of the project? Report on similar projects in other countries, such as the million-roof program in the United States.
32. Calculate the energy per year that could be captured by 25% efficient solar collectors, assuming a 50% transmission and storage loss if 10% of the area of the state you are living in were covered in solar collectors. Assume the daily average insolation is 200 W/m^2. How does this compare with the 100 quad total annual energy use in the United States?
33. Find the solar flux for your area and, using the area of your house, calculate how much solar energy you could collect with 20% efficient solar cells mounted on your roof, assuming a 50% storage loss. Is this enough to supply all your electricity needs?
34. Using reliable source, write a short explanation of how a solar or thermal chimney works.
35. A number of large-scale solar collecting initiatives have been tried but are no longer operating. Using reliable sources, write a summary of two of the following projects, including the reasons for shutting down the project: Eurelios, Solar One, Solar Two, CESA-1, Themis, MSEE, Phoebus, Rancho Mirage, and King Abdul Aziz City.
36. Summarize the status of any large-scale solar collecting project that is still in operation. Use reliable sources.
37. Using reliable sources, find out and report on the status of the plan of the Japan Aerospace Exploration Agency or the Solaren project to build a satellite to collect solar energy in space and beam it down to Earth.

38. About what wavelength of photon is most efficiently absorbed by silicon with a band gap of 1 eV? Amorphous silicon has a band gap of about 1.7 eV. What wavelength photon is absorbed by amorphous silicon? Where are these wavelengths in relation to the peak of the visible spectrum?

39. Using a reliable source, write a summary of how each of the types of PV cells listed in Table 5.3 works and advantages and disadvantages of each.

40. What is the energy payback time for amorphous PV cells, which require 4.3×10^8 J per square meter to manufacture, if they are 6% efficient?

41. From a reliable source, find the rate at which the earth is slowing due to tidal motion. Using the change in the earth's rotational kinetic energy per year, calculate the approximate amount of tidal energy available. *Hint*: Rotational kinetic energy is $\frac{1}{2}I\omega^2$, where $I = 2mr^2/5$ is the moment of inertia of a solid sphere, and ω is the angular velocity.

42. For the heat flux coming from the earth's core of 0.06 W/m^2, how much area would be needed to supply all the U.S. energy needs? What percentage of the surface area of the United States is this? How does this compare with wind and solar?

43. Using reliable sources, summarize the current state of the implementation of tidal, wave, geothermal, thermocline, and salt gradient technology. Describe ongoing projects and estimations of the likelihood of success and project the future of the resource.

44. Using reliable sources, write a summary of the following mechanisms for capturing wave energy: oscillating water column, the Pelamis, the Wave Dragon, the Archimedes wave swing, Biowave, Superbuoy, Seawave slot cone, the Edinburgh duck, and Salter's duck. There are many others.

45. Repeat Example 5.6 for the world's coastline. Based on this assessment, do you think wave energy will play a significant role in world energy use in the future? Explain your conclusion.

46. Using reliable sources, write a brief summary of the history of ocean thermal energy conversion (OTEC) and its current status as an energy source.

47. The change in entropy that occurs when freshwater is mixed with saltwater releases a significant amount of energy. Two projects, one in the Netherlands and the other in Norway, have examined the possibility of capturing the energy released when fresh river water mixes with saltwater in the ocean. Using reliable sources, report on the status of these projects. What is the potential energy supply?
48. Globally there are about 100 lightning strikes per second. Each strike carries about 500 MJ of energy. How much energy is available annually (in quad) from lightning? Given the available resource, would it make sense to investigate the technical problems that must be overcome to harness lightning as an energy source?
49. Using reliable sources, investigate two or three of the more imaginative proposals for renewable energy mentioned in Section 5.5.6. State the technical barriers that would have to be overcome to use each proposed source. Evaluate the likelihood that any of these ideas will come to fruition.
50. Using reference [24] (or equivalent), pick two consumer products and report on the lifetime cost analysis (LCA) of each. How does the energy used during the normal lifetime of each item compare with the energy needed to make, transport, and dispose of it at the end of its usable life? How are the LCAs of each item different from each other?
51. Using reliable sources, find the chemical names of the various types of plastics (PETE or PET, HDPE, V, LPDE, PP, PS, and number 7), what they are used for, sample physical properties, energy content, the percentage of each in municipal solid waste, how much of each is recycled, and any other interesting information you come across.

REFERENCES AND SUGGESTED READING

1. World Energy Council, *2007 Survey of Energy Resources* (http://www.worldenergy.org/documents/ser2007_final_online_version_1.pdf).

2. International Hydropower Association (http://www.hydropower.org/).

3. V. Smil, *Energy at the Crossroads* (2005, MIT Press, Cambridge, MA).

4. B. Sorensen, *Renewable Energy: Its Physics, Engineering, Environmental Impacts, Economics and Planning* (2000, Academic Press, New York).

5. American Wind Energy Association, *American Wind Energy Association Annual Wind Industry Report, Year Ending 2008*, 2009 (http://www.awea.org/publications/reports/).

6. J. W. Tester, E. M. Drake, M. J. Driscoll, M. W. Golay, and W. A. Peters, *Sustainable Energy: Choosing among Options* (2005, MIT Press, Cambridge, MA).

7. A. V. DaRosa, *Fundamentals of Renewable Energy Processes* (2005, Elsevier, Burlington, MA).

8. National Renewable Energy Laboratory, *Wind Energy Resource Atlas of the United States* (http://www.nrel.gov/rredc/).

9. C. L. Archer and M. Z. Jacobson, "Evaluation of Global Wind Power," *Journal of Geophysical Research* 110 (June 2005):D12110.

10. D. Hafemeister, *Physics of Societal Issues* (2007, Springer Science, New York).

11. J. S. Denker, *See How It Flies: A New Spin on the Perceptions, Procedures, and Principles of Flight* (http://www.av8n.com/how/#mytoc).

12. Energy Information Administration, *Annual Energy Review 2007* (http://www.eia.doe.gov/emeu/aer/pdf/aer.pdf).

13. United Nations, *Economic and Social Survey of Asia and the Pacific 2008: Sustaining Growth and Sharing Prosperity* (http://www.unescap.org/survey2008/download/01_Survey_2008.pdf).

14. T. R. Sinclair, "Taking Measure of Biofuel Limits," *American Scientist* 97(5) (Sept.–Oct. 2009):400.

15. P. B. Weisz, "Basic Choices and Constraints on Long-Term Energy Supplies," *Physics Today* 57(7) (July 2004):47.

16. T. Searchinger et al., "Use of US Croplands for Biofuels Increases Greenhouse Gases through Emissions from Land-Use Change," *Science* 319 (Feb. 2008):1238.

17. L. Kazmerski and K. Zweibel, *Best Research-Cell Efficiencies*, Power Point Presentation, National Renewable Laboratory (http://www.nrel.gov/pv/thin_film/docs/bestresearch-cellefficienciesall.ppt).

18. National Renewable Energy Laboratory, *PV FAQs: What Is the Energy Payback for PV?* (http://www.nrel.gov/docs/fy05osti/37322.pdf).

19. J. M. Deutch and R. K. Lester, *Making Technology Work: Applications in Energy and the Environment* (2004, Cambridge University Press, Cambridge).

20. G. Boyle, *Renewable Energy: Power for a Sustainable Future* (2004, Oxford University Press, Oxford).

21. National Renewable Energy Laboratory (http://www.nrel.gov/).

22. "The Truth about Recycling," *The Economist* (US Edition) Technology Quarterly, June 9, 2007.

23. R. E. Hummel, *Understanding Materials Science* (1998, Springer-Verlag, New York).

24. M. F. Ashby, *Materials and the Environment: Eco-Informed Material Choice* (2009, Butterworth-Heinemann, Burlington, MA).

25. L. R. Glicksman, "Energy Efficiency in the Built Environment," *Physics Today* 61(7) (Dec. 2008):35.

26. U.S. Energy Information Administration (Department of Energy), Report DOE/EIA 0554 (2008) (http://www.eia.doe.gov/emeu/international/contents.html).

27. National Academy of Science, *Electricity from Renewable Resources: Status, Prospects, and Impediments*, 2009 (http://www.nap.edu/catalog.php?record_id=12619).

Notes

[1] At least three different costs can be quoted for electrical power: the cost per kilowatt-hour to construct and maintain the power plant, the operating cost to generate the electricity per kilowatt-hour, and the price the consumer pays for the electricity. Average consumer cost for electricity in the United States is about 10 cents/kWh. The cost to industry is slightly lower at 7 cents/kWh. The generating cost for coal is about 1 cent/kWh and about 3 cents/kWh for natural gas and petroleum, not including taxes or proposed carbon emission trading costs.

[2] This figure is less than half the solar constant because of the glancing angle of solar radiation arriving at a location that is experiencing sunset or sunrise.

CHAPTER 6

Energy Storage

The role each energy source plays in powering society depends on variables such as availability, development costs, and politics. But regardless of these factors, energy is generally not useful to modern society unless it can be stored. Energy storage is particularly important for a world using renewable energy because most of these sources are inherently intermittent and often of low energy density. In this chapter we review various means of energy storage, both for transportation needs and for daily and longer-term load leveling.

For transportation, fuels must be lightweight while containing significant amounts of energy so as not to put energy burdens on moving the fuel in addition to the vehicle and contents. When transportation is not an issue, other forms of energy storage can be considered. As we know from Chapter 2, transforming energy from one form to another can never result in more energy than was there to begin with and usually results in a decrease in the usefulness of that energy for doing work. The efficiency limits imposed by the second law of thermodynamics on energy conversions were considered in Chapter 3. The limitations of these laws on conversion processes for specific fuels are considered in this chapter.

The information in previous chapters now puts us in a position to consider the combined efficiency for producing a particular fuel and using it for various transportation choices. This is often called the total well-to-wheels efficiency or, if the energy to manufacture and dispose of the vehicle is included, the life cycle efficiency.

Hydrogen as a fuel and its production are discussed in this chapter because, as we shall see, it is not a primary energy source but rather is manufactured from other energy supplies and used as an energy carrier. This chapter also considers energy storage on a larger scale, for example for load leveling in electricity distribution, a factor that will be important for many of the renewable energy sources considered in the previous chapter.

6.1 ■ Energy Content of Fuels

Table 6.1 shows the energy content of several common substances, some of which are in use today as fuels. Gasoline carries a large amount of energy per kilogram, so it is not surprising that it is a very popular fuel source for vehicles where weight is a factor. In fact, it is only considerations of carbon emissions, treated in the next chapter, and the quantity of remaining world oil resource that are motivating the current discussion of replacing this fuel in vehicles.

At first glance hydrogen also looks appealing as a portable fuel source because it is a very lightweight substance with high energy content. As noted in the table, however, it also occupies a large volume at standard temperature and pressure (STP), greatly lowering the energy per volume of this fuel. The fact that a sturdy storage tank is needed to transport liquid or compressed hydrogen also makes it less desirable, although alternative storage mechanisms currently under investigation such as nanotubes or gas–solid adhesion may solve this problem. Other difficulties of using hydrogen as a fuel are mentioned later in this chapter. Similar storage problems are found with other fuel gases such as propane, natural gas, and compressed air, but these problems are not insurmountable; vehicles using these fuels are already on the market.

The choice of lead batteries as a portable fuel for vehicles can be seen to be problematic from the data in Table 6.1 because of their very low energy density. All-electric cars exist, but an electric car using lead batteries is possible only because of the very high efficiencies of electric motors as compared to gasoline engines, and even then the typical range of a battery power car is only about 160 km. More energy-dense lithium batteries or other alternatives may make a shift to all-electric vehicles more feasible in the near future. A more detailed discussion of batteries as energy storage devices is found later in this chapter.

The figures in Table 6.1 also explain the current interest in biodiesel and ethanol as a fuel for vehicles, although as we shall see later, there are problems with these fuels as well. The reason we do not use wood, straw, or vegetables as fuel for automobiles should also be clear from the chart; their low energy per weight

TABLE 6.1

Energy content per weight and per volume for various substances [1].

Fuel	Energy Content (MJ/kg)	Energy Density (MJ/l)
Uranium-235 in fission reactor	88,250,00	1,500,000,000
Hydrogen (STP)	114.0	0.01
Hydrogen (liquid)	10 (with tank)	9.4
Hydrogen (compressed gas)	5 (with tank)	1.5 (150 atm)
Liquefied petroleum gas (propane)	49.6	25.3
Gasoline	46.9	34.6
Diesel	45.8	38.7
Plastic (polyethylene)	42.2	42.6
Crude oil	42.0–43.0	37
Plant oil (biodiesel)	38–42	~33
Butter	30	27
Ethanol	29.6	22
Coal	22–32	~72
Natural gas (STP)	20.0	0.03
Cereal grains, bread	~15	~26
Air-dried wood	12–15	2–3
Straw, dung	12–15	2–3
Meat	5–10	5–10
Potatoes	4	6.4
Beer	1.8	1.8
Carbon fiber flywheel	0.8	0.1
Fruit and vegetables	0.6–1.8	0.6–1.8
Lithium batteries	0.5–2.5	1.0
Compressed air	0.1 (with tank)	0.1
Lead batteries	0.1	0.15

STP = standard temperature and pressure.

makes them an unlikely fuel source for transportation, particularly when coupled with the low second law efficiencies of heat engines. With half as much energy per kilogram as gasoline, coal was once used as a fuel for steam locomotives, but with the advent of diesel fuel, a lighter-weight alternative, coal-fired trains have mostly been abandoned except in countries where there is no alternative.

Finally, in the table we see why diet plans such as Weight Watchers generally recommend eating fruits and vegetables as opposed to meat and potatoes; one can more easily fill up on vegetables while consuming fewer calories. However,

TABLE 6.2

Fuel energy versus approximate power density [2].

Source	Energy Content (MJ/kg)	Power Density (W/kg)
Supercapacitors	0.05	5,000
Flywheels (conventional)	0.05	400
Flywheels (advanced)	0.1	3,000
Lithium ion battery	0.3	150
Nickel–zinc battery	0.2	150
Lead battery	0.1	70
Ethanol	29.6	100
Gasoline	47.0	100
H_2 fuel cell	10.0	50

athletes are encouraged to load up on pasta and other carbohydrates made from cereal grains for the obvious reason that these fuels have high energy content.

In addition to the energy content per weight or volume, it is also important to realize that not all sources can provide energy at the same rate, thus presenting further difficulties in using a certain fuel for a given purpose. Although gasoline has a high energy content per kilogram, the rate at which that energy can be used is lower than that of some other energy storage mechanisms, as can be seen in Table 6.2. The power needed to overcome friction and drag forces for an automobile traveling at low constant speeds is about 8,000 W (approximately 10 hp) for an average car at normal speeds, but the power needed for reasonable acceleration is about 20 times higher. Electric motors can produce more power at low speed than gasoline engines. This coupled with the fact that newer batteries (such as lithium ion and nickel–zinc batteries) can deliver more power (the same energy in a shorter time span) explains the current move to hybrid cars, which use an electric motor for acceleration and a smaller, more efficient gasoline motor for constant-speed highway travel.

6.2 ■ Energy Needed for Transportation

About 28% of the total energy used in the United States is used for transportation. Globally this figure is about 20%, and nearly all of this energy comes from petroleum products. In this section we consider the power needed to accelerate a typical car, the power needed to keep it moving at constant speed against drag forces, and the power needed to travel uphill. Any combination of motors and fuel sources must be able to provide the necessary power to perform these functions, and this limits what alternative fuels and motors can be used for transportation.

Aerodynamic drag forces on a vehicle moving with velocity v are proportional to velocity squared:

$$(6.1)\quad F = \rho A c_d v^2/2,$$

where A is the frontal area of the vehicle and ρ is the density of air.[1] The drag coefficient, c_d, is a unitless number that is a function of the exact shape and roughness of the surface of the vehicle. Values for the drag coefficient are determined empirically and range from about 0.35 for most cars currently on the road to 0.15 for research vehicles. Power is defined to be the energy expended per time, but an alternative expression for power is $P = Fv$.[2] From this we see that the power needed to overcome drag forces is

$$(6.2)\quad P_{drag} = \rho A c_d v^3/2.$$

This is the energy per second needed to keep the car in constant motion against drag forces and accounts for the largest fuel use in an average vehicle.

Rolling resistance is proportional to the mass and speed of the car:

$$(6.3)\quad P_{\mathrm{rolling}} = c_r m g v,$$

where m is the mass of the car, g is the acceleration of gravity, and c_r is the coefficient of rolling friction, which is typically about 0.01. Because rolling resistance is proportional to speed but drag is proportional to speed cubed, rolling resistance plays a bigger role at low speed but drag forces are more significant at higher speeds. Rolling resistance forces on a moving vehicle amount to about a 15% loss of energy on average for the typical car in ordinary circumstances.

Power for hill climbing depends on the steepness of the hill, which can be given by the angle of incline, θ. In this case the engine must overcome gravitational potential energy, $U = mgh$, where h is the height of the hill. The power needed to do this at constant speed is

$$(6.4)\quad P_{hill} = mgv\sin\theta,$$

where we have used the fact that $h/t = v\sin\theta$ for a car climbing a hill of angle θ with speed v.

Power needed for acceleration as a function of instantaneous speed is

$$(6.5) \quad P_{acc} = Fv = mav,$$

where P_{acc} is the acceleration and v is the instantaneous speed. From this we see that more power is needed to accelerate from a high speed than from a lower speed. In other words, a 5-mph increase in speed, accomplished over the same time period, requires more power if you are already traveling at 70 mph compared to an initial speed of 50 mph. This is in addition to the extra power needed to overcome rolling resistance and drag, which depend on speed and speed cubed, respectively. We also see from these equations that power needed for acceleration, hill climbing, and rolling depend on the mass of the vehicle, which explains why fuel energy density is a relevant factor.

EXAMPLE 6.1

Speed, Acceleration, and Gas Mileage

Estimate the horsepower needed for constant speed, acceleration, and hill climbing for a typical car.

For a typical car of frontal area 2 m^2 and mass 1500 kg traveling at 20 m/s (about 45 mph), the power loss to rolling resistance from Equation (6.3) is $P_{rolling} = c_r mgv = 2744$ W = 3.7 hp for $c_r = 0.01$, and the power loss to drag forces, according to Equation (6.2), is $P_{drag} = \rho A c_d v^3/2 = 720 \text{ W} \approx$ 1 hp. A car may have as much as 200 hp but needs only about 5 hp to travel at a constant speed on level ground at 45 mph.

It should be clear from Equation (6.2) that doubling the speed increases the power needed by a factor of eight. For a car traveling at 30 m/s (67 mph), the power needed to overcome drag is 3.2 hp, and at 40 m/s (90 mph) the power needed is 7.7 hp; increasing your speed by 10 m/s more than doubles the power needed. Gas mileage scales with energy used per mile, and this is the reasoning behind the lowering of speed limits after the oil crisis of the 1970s; decreasing the speed limit from 70 mph to 55 mph uses nearly 50% less fuel per mile traveled. Due to factors involving engine design, most cars have their best fuel economy at about 45 mph, so the best way to improve your gas mileage is to slow down to 45 mph.

Suppose we want to accelerate the same car described above at a rate of 5.4 m/s^2 (equivalent to zero to 60 mph in 5 s). According to Equation (6.5) the power needed for acceleration depends on the speed as well as the acceleration. So to have this acceleration at an instantaneous speed of 20 m/s requires 1.5×10^5 W = 203 hp, not including drag and rolling friction. To achieve this acceleration at 30 m/s requires 2.3×10^5 W = 304 hp, not including drag and friction. The maximum force that a car engine can supply is limited by the size of the engine, but the drag forces increase with speed. This is analogous to the terminal velocity of a skydiver; a constant downward force due to gravity accelerates the diver until air resistance, which increases with speed, equals the downward force, and the acceleration drops to zero. This means that for a car there will be a maximum upper speed; acceleration cannot continue forever. Notice that lowering the acceleration also decreases the amount of power needed, which increases gas mileage. Jackrabbit starts should be avoided to improve gas mileage.

Finally, suppose we want to drive this car up a hill with a 5% incline (4.5°) at 20 m/s. Equation (6.4) gives $P_{hill} = mgv\sin\theta = 2.1$ kW = 29 hp. From these calculations it should be clear that acceleration is a significant power demand on a car engine. As we saw in Figure 2.6, cars have increased in horsepower by 40% in the past 30 years, in large part because of popular demand for better acceleration.

Which of the factors calculated in Example 6.1 have the biggest effect on gas mileage depends on how the car is driven. As mentioned previously, drag forces are more important at high speed. Acceleration uses a lot of power, but generally cars do not accelerate during much of the time they are on the road. For normal driving, the energy needed to overcome drag forces at high speed plays the biggest role in determining gas mileage.

EXAMPLE 6.2

Hybrid Cars

For ordinary cars the kinetic energy of forward motion is dissipated as heat in the brake pads when the car stops. Hybrid vehicles capture this energy using a generator that slows the car down while recharging a large battery. These systems capture about 50% of the kinetic energy of the car when braking. What is the gasoline savings if the car makes 100 stops from a speed of 20 m/s?

For a 1,400-kg car traveling at 20 m/s, the savings in energy is $E = 0.5 \times \frac{1}{2}mv^2 = 140$ kJ. As we saw from Table 6.1, gasoline has an energy density of 34.6 MJ/l. An engine with 15% efficiency delivers 0.15×34.6 MJ/l = 5.2 MJ/l to move the car. Dividing this into 140 kJ gives a savings of 0.027 l, or 0.01 gal per stop. Stopping 100 times saves about a gallon of gas.

As shown in the previous example, hybrid cars are expected to perform better in stop-and-go traffic because of the regenerative braking system. This coupled with the use of electric motors for acceleration, a smaller gasoline engine, and the fact that the gasoline engine is turned off when the car is stopped makes a hybrid car more fuel efficient than ordinary cars. Some hybrids also use a smaller gasoline engine operating on the Atkinson cycle, a more efficient cycle than the usual four-stroke engine. Most newer hybrid cars are also carefully designed to have lower drag coefficients, a feature that is explored in Problem 6.7 at the end of this chapter.

6.3 ■ The Energy Cost of Energy

In addition to considerations such as energy density and power density, the energy cost to extract and convert the fuel to a usable form must be factored into a choice of fuel. As we noted in Chapter 2, any time energy is converted from one form to another, there is generally a loss of energy, which can be expressed as conversion efficiency. As mentioned in Chapter 5, the energy return for corn ethanol is only 1.2; in other words it takes 1 gal of gasoline in the form of fertilizer, tractor fuel, and processing energy to make 1.2 gal of ethanol fuel. If this additional energy cost is not taken into account, the figures in Table 6.1 can easily be misleading. The first stage in this calculation is the conversion of the raw fuel into a usable form; Table 6.3 lists several such conversion processes. From the table we see that approximately 15% of the energy of crude oil is lost in converting it to gasoline for use in an automobile. Fifteen percent of the energy in natural gas is lost in compressing it into tanks for commercial use. Some of these conversion processes are discussed later in this chapter.

It should be emphasized that, unlike the figures in Table 6.1, these numbers for efficiencies can change with new technology (possibly with the exception of plant photosynthesis, which may be limited by nature). For example, solar cell efficiencies have increased over time, and this trend is expected to continue. There are physical limits to efficiencies imposed by the second law of thermodynamics, but we have not yet reached those limits for most conversion processes, including modern heat engines. The figures in Table 6.1 also do not include the energy lost in extracting the raw source before it is converted to a fuel. As we saw in Chapter 5, for primary extraction techniques the additional energy needed for withdrawing crude oil from the ground is about 5% of the total energy captured. This energy extraction cost increases for oil wells over

TABLE 6.3

Conversion efficiencies for various fuels [1]. Note that these processes represent different types of conversions and do not include the energy needed to extract or create the source. Crude oil to gasoline is a chemical conversion, whereas photovoltaic is a conversion of electromagnetic waves to electricity.

Process	Conversion Efficiency (%)
Natural gas to compressed	85
Crude oil to gasoline	85
Natural gas to electricity (gas turbine)	65
Natural gas to hydrogen	60
Coal to gasoline (Fischer–Tropsch)	50
Natural gas to electricity (conventional)	35
Grid electric to hydrogen	24
Solar to electric (photovoltaic)	15–25
Soybeans to biodiesel	40
Corn to ethanol	20
Plant photosynthesis to biomass	<8

time; as more oil is extracted, more energy-intensive methods, such as steam injection, are needed to remove the remaining oil. If the energy cost for extracting and producing the fuel becomes higher than the energy obtained from the fuel, it makes little sense to produce the fuel.

Recall from Chapter 2 that for multistep energy conversion processes, efficiencies are multiplied. So even though the efficiency of converting soybeans to biodiesel is approximately 40%, the first step in making biodiesel fuel, terrestrial plant photosynthesis, is very inefficient, bringing the overall efficiency of the fuel much lower. Soybeans are less than 1.5% efficient in converting sunlight to plant material, so the overall efficiency of converting sunlight to soybeans to biodiesel is $0.40 \times 0.015 = 0.006 = 0.6\%$. The energy investment in the form of fertilizer and farming equipment needed to grow the soybeans is not included in this figure. Biodiesel efficiency should be compared with solar cells, which are currently about 15% to 25% efficient in converting solar energy into a usable form. Even including the energy needed to make the photocell, the overall efficiency is higher because of the low conversion efficiencies of plants. From a calculation based only on energy efficiencies, electric cars powered with batteries charged with solar energy come out far ahead of cars with internal combustion engines running on biofuels, particularly when the energy overhead for growing soybeans and manufacturing solar cells is included.

Given this, why should biodiesel be considered as a possible transportation fuel? In some instances it may be advantageous to create (or convert to) a fuel that is easily transportable, even if there is a net energy loss. Because of the low energy density of batteries, all-electric vehicles have limited driving ranges, whereas cars running on biodiesel are not very different from cars running on gasoline. As another example, although more energy can be gained by directly burning coal, straw, or other biomass, converting these energy sources into liquid fuel for transportation is useful in terms of convenience because coal- or straw-burning vehicles are impractical. In addition, many types of biomass such as used cooking oil, corn stalks, sawdust, and other agricultural byproducts would otherwise be left unused but can be converted into fuel (biodiesel).

Currently the most common process for converting coal and biomass to a liquid fuel is the Fischer–Tropsch process, invented in Germany in 1920 and used during World War II to supply the German army with fuel. This process begins with coal or other carbon sources such as biomass, which are first heated and partially combusted to give carbon monoxide:

(6.6) $C + H_2O \rightarrow H_2 + CO$ (from coal),

or

(6.7) $2CH_4 + O_2 \rightarrow 4H_2 + 2CO$ (from methane).

The carbon monoxide is then combined with hydrogen to make liquid hydrocarbons:

(6.8) $(2n + 1)H_2 + nCO \rightarrow C_nH_{(2n+2)} + nH_2O$,

where n is a whole number, and the formula indicates some combination of complex hydrocarbon molecules as the final result.

The final product is an assortment of liquid hydrocarbons, some of which can be extracted as a fuel in a process similar to oil refining. The overall efficiency of the process is about 50% because of the input energy needed to make process occur. Once again, if coal or biomass could be burned directly in an automobile, it would not be a good idea to waste 50% of the energy converting it into a liquid fuel. It is the convenience of having an easily handled liquid fuel with high energy content that makes the process attractive. It is also the case that some countries, such as the United States, have large resources of coal but smaller oil resources. The conversion of coal to gasoline, though wasteful in energy terms, would make the United States less dependent on oil imports. Additional factors, such as availability and carbon dioxide emission, will also influence the decision and desirability of using coal to make gasoline, as we will discuss in Chapter 7.

6.4 ■ The Hydrogen Economy

Because of its high energy content per weight and the fact that it can be used in highly efficient fuel cells, hydrogen has recently attracted a great deal of attention as fuel for transportation. This has led some proponents to talk about a future "hydrogen economy" in which hydrogen replaces gasoline in many applications. In this section we examine the challenges inherent in this idea, starting with a look at potential sources of hydrogen.

The first challenge of using hydrogen as a fuel is that although it is the most abundant element in the universe, on Earth it is found almost exclusively in compounds such as hydrocarbons and water. Because there are no naturally occurring sources of pure hydrogen, it must be manufactured, which requires a net input of energy (for this reason we are treating hydrogen in this chapter rather than in Chapter 4 or 5, on energy sources). Currently most commercially available hydrogen is derived from natural gas using a process called steam reforming.[3] In this process steam at high temperatures (about 800°C to 1,000°C) is reacted with methane (natural gas) to yield hydrogen and carbon monoxide:

$$\text{(6.9)} \quad \text{energy} + CH_4 + H_2O \rightarrow CO + 3\,H_2 .$$

Steam reforming has an overall efficiency of about 60% in converting natural gas to hydrogen when the energy needed to make the process occur is included. It should be noted that here, unlike the case for converting coal to fuel, we have started with a portable fuel (methane) and used 40% of the energy in that fuel to create a different portable fuel (hydrogen). Energy-wise this process makes sense only if the process of using the hydrogen to do work (e.g., to propel a vehicle) is significantly more efficient than using natural gas directly without converting it to hydrogen. Converting methane into hydrogen and using the hydrogen in a fuel cell (discussed in Chapter 3) is more energetically advantageous than burning the methane directly in a car engine, as can be seen with a rough calculation.

EXAMPLE 6.3

Hydrogen Fuel Cell Cars

How does the efficiency of hydrogen from steam reforming of methane, used in a fuel cell, compare with an internal combustion engine using the methane directly?

Using an electric motor at 80% efficiency coupled with a 60% efficient fuel cell using hydrogen produced by steam reformation would have a total efficiency of 80% × 60% × 60% = 29%. This is significantly more efficient than burning the methane in an ordinary internal combustion engine, which is about 15% efficient. On the other hand, fuel cells can also run directly on methane, with efficiencies approaching those of hydrogen fuel cells, in which case conversion of methane to hydrogen for use in a fuel cell is much less desirable.

Electrolysis is another way to generate hydrogen. In electrolysis direct current is run through water, which causes water to separate into hydrogen and oxygen in gaseous form. Hydrogen is formed at the anode (positive electrode), and oxygen forms at the cathode (negative electrode):

$$\text{(6.10)} \quad \text{electrical current} + 2H_2O \rightarrow 2H_2 + O_2.$$

Generally the electrodes are made of platinum, which acts as a catalyst, and if the electrodes are sufficiently separated the two gases can be trapped before they recombine to form water. Some of the electrical energy used in this process acts to heat the water rather than produce hydrogen. If the hydrogen is collected and the output energy from burning the hydrogen is measured, the efficiency of the entire process (output thermal energy obtained by burning the hydrogen divided by input electrical energy) is approximately 60%, not including the energy needed to generate the electricity.[4]

EXAMPLE 6.4

Hydrogen from Steam or Electrolysis?

Which is more energy efficient: making hydrogen from steam reforming or electrolysis using traditionally generated grid electricity?

As we saw in Chapter 2 and 3, the best modern gas turbine plants have efficiencies of 65% for converting natural gas to electricity, whereas conventional plants using coal to make electricity have typical efficiencies lower than 40%. Using a gas turbine to generate electricity, which is used to make hydrogen, has a total efficiency of 65% × 60% = 39%, which is less than the 60% to make hydrogen directly from natural gas by steam reforming. For conventional electricity from coal the efficiency is only 24%, so making hydrogen from grid electricity does not make good energy sense unless coal (which cannot be conveniently used for transportation) is the only available fuel stock.

As Example 6.4 shows, steam reforming is currently a more efficient way to make hydrogen than electrolysis using traditionally generated electricity, assuming there is a supply of methane. Hydrogen supplied from natural gas will be affected by available natural gas stores, which, as we saw in Chapter 4, may be limited. Although hydrogen fuel cells emit no pollutants, both steam reforming and electrolysis using grid electricity are carbon-emitting processes, which also must be considered. On the other hand, hydrogen made from electricity coming from a photovoltaic cell could be a much more sensible idea because solar energy is essentially an unlimited source, and the energy needed for solar cell manufacturing is small. Research is under way to combine the light-gathering process of photovoltaic devices with the electrolytic process into a single device with higher efficiency that generates hydrogen in a single step [3]. Hydrolysis using electricity from hydroelectric or nuclear plants is another possibility. From these rough calculations of efficiency for the production and use of hydrogen we see that the choice to use hydrogen as a fuel will probably also mean a choice to use fuel cells and electric motors for transportation. Further comparisons of various combinations of fuel source, conversion processes, and engine efficiencies are given in the next section.

The amount of electricity needed for electrolysis decreases as temperature increases because some of the energy needed to break the hydrogen–oxygen bond comes from thermal energy rather than electrical energy. In fact, above 2,500°C water breaks down into hydrogen and oxygen without the addition of any electric current. Because thermal energy can be created more efficiently than electrical energy, this process, called high-temperature electrolysis, can be more efficient than electrolysis at room temperature. Using natural gas or coal combustion as a heat source does not provide a gain in efficiency over the direct use of these traditional fossil fuels. However, if the heat is provided as waste heat from these or other renewable sources, the combined efficiency can be higher, similar to the case of cogeneration we considered in previous chapters. Heat from nuclear reactors, solar thermal energy, and geothermal energy are all potential heat sources that could be used to achieve higher electrolytic efficiencies in the manufacturing of hydrogen.

Hydrogen directly from plant photosynthesis is another potential hydrogen source. As we have seen, the photosynthetic process in most plants is about 1% efficiency for turning solar radiation into stored plant energy, and 8% is considered an upper limit. This stored energy is normally in the form of starches and sugars, which can be further processed into usable fuel with corresponding energy conversion losses. However, there is another, more primitive form of life called algae in which the photosynthetic process can have a slightly higher efficiency. Because of the enormous number of these organisms and the fact that they have been around much longer than terrestrial plants, algae are responsible for most of the oxygen in the atmosphere today. Most algae live in the sea, where they come in many different forms (everything from seaweed to single-cell organisms), and in their natural state most do not have efficiencies higher than 1%. Through gene manipulation, however, it has been possible to create modified versions of algae that have efficiencies of up to 15% and produce hydrogen as a byproduct. These new life forms are still in the laboratory, and it remains to be seen whether they can be used to manufacture the quantities of hydrogen needed to replace conventional fuels [3].

Storage and distribution of hydrogen is another consideration in a hypothetical hydrogen economy. As mentioned earlier, hydrogen at STP has a large volume per unit of energy, necessitating a more sophisticated containment device than a gasoline tank. To compress or liquefy hydrogen and put it in a tank requires energy, and the container itself reduces the energy content per kilogram for stored hydrogen. As noted in Table 6.1, liquefying hydrogen reduces the energy per kilogram of hydrogen by a factor of 5 relative to gasoline, and compressing hydrogen gas lowers it by a factor of 10. Because hydrogen is a very small molecule, containers of hydrogen and hydrogen supply pipelines must have much tighter seals than for methane or gasoline, and the high pressures needed to contain hydrogen during transportation contribute to this problem. In the past the ability of hydrogen to seep through very small openings has made it useful as a trace gas to find leaks in vacuum systems. Exposure of high-strength steel to hydrogen also eventually weakens the steel, which can lead to failure of pipes under high pressure.

The risks of an accidental explosion or fire from hydrogen as compared to other fuels such as methane or gasoline depend on the circumstances in which the fuel is being handled [4]. Transportation through pipelines, transportation by tanker truck, manufacturing, fueling a vehicle, and hydrogen carried as a vehicle fuel all are associated with different risk factors. On one hand, hydrogen diffuses much faster than most other flammable gases or liquids, so spills are less dangerous because the gas dissipates rapidly. It is also true that the autoignition temperature (the temperature at which it will ignite if no spark is present) of hydrogen is higher than that of gasoline and methane. On the other hand, hydrogen is flammable at a wider range of concentrations than either gasoline vapor or methane and has a much lower minimum ignition energy (the energy a spark must have to ignite it). The potential for explosive combustion is also higher for hydrogen than most other fuels. Hydrogen flames are invisible because they produce no soot or smoke, making them more dangerous than gasoline or methane. Careful risk assessments should be done before hydrogen is introduced to the public as a vehicle fuel (see Chapter 8 for further comments on risk assessment).

Because hydrogen is not an energy resource but must be produced from other energy sources, many authors have questioned the desirability of a hydrogen economy [5, 6, 7]. Whether creating hydrogen (e.g., using renewable resources) and then using it as a transportation fuel will end up being more efficient than, for example, the all-electric vehicles using high-capacity batteries now under development remains to be seen.

6.5 ■ Total Well-to-Wheels Efficiency

As we have frequently mentioned, the total efficiency of a process is found by multiplying the efficiency of each step in the process. This is often called a total life cycle energy analysis or, for transportation applications, a well-to-wheels analysis. In this section we examine the implications of this principle in designing a transportation system with maximum energy efficiency, taking into account not only the engine selected but also the efficiency of the steps needed to make the fuel, as discussed in the previous sections. Figure 6.1 shows the result of a series of calculations done to compare the efficiencies of various combinations of engines and energy sources [8–10]. It should be emphasized that these figures are for existing technology and do not indicate the theoretical maximum upper efficiencies possible for the processes involved. The values also assume traditional sources for the fuels used. For example, electricity is assumed to be generated conventionally from a mix of 50% coal-fired plants, 18% natural gas, 20% nuclear, and the rest from other sources such as wind and hydroelectric. Hydrogen is assumed to come from steam processing of natural gas. The energy

FIGURE 6.1

Total well-to-wheels greenhouse gas emissions (GHG) in grams per kilometer driven and energy use for several choices of alternative vehicles (data from the GREET simulation available from Argonne National Laboratory [9]). *Feedstock* refers to the energy needed to extract and transport the raw fuel, *fuel* refers to the energy needed to convert the feedstock to a usable fuel, and *vehicle operation* refers to the energy actually used to operate the vehicle. The following vehicles are included: internal combustion engine (ICE), advanced technology engine (SIDI), hybrid engine (HE), flexible fuel vehicle using 85% ethanol (FF), liquid natural gas engine (LPG), fuel cell (FC), and all-electric. Fuels considered are hydrogen (H2), diesel, gasoline, and liquid natural gas (LPG).

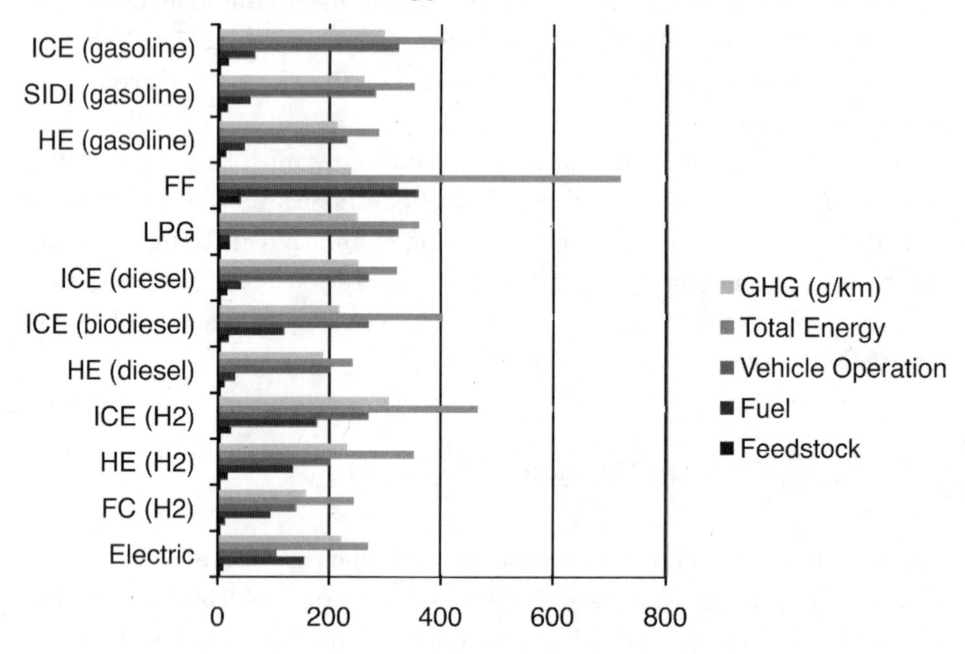

cost of manufacturing the vehicle and disposing of the vehicle at the end of its useful life and the vehicle lifetime are assumed to be the same for each case and so are not included in Figure 6.1. Although many of these parameters will change, the results shown are useful for making choices about transportation alternatives in the near future.

In Figure 6.1, energy used in vehicle operation, measured in joules per kilometer, is a proxy for gas mileage and makes it easier to compare various different combinations of transportation from a consumer viewpoint. From a societal perspective, total energy per kilometer and carbon emissions are more significant. From the figure we see that hybrid technology improves gas mileage (energy used in vehicle operation), total well-to-wheels energy use, and the emission of greenhouse gases over conventional internal combustion engines for gasoline, diesel, and hydrogen fuels. As also noted in the figure, hybrid diesel, hydrogen fuel cells, and all-electric vehicles have similar total efficiencies and emissions. These vehicles seem to be reasonable choices for further development, at least from an energy and emission perspective. Vehicles built to run on various mixtures of alternative fuels such as biodiesel and ethanol are the net losers in this comparison in terms of energy used in vehicle operation, total energy used, and emissions. Because electricity generated for the grid can be done on a larger scale and in a more controlled way, it is expected that plug-in hybrids (not shown), which use a battery that is recharged at night by plugging into the grid, will bring further improvements in total fuel economy and emission control. Again, should fuel supply sources change (e.g., by a switch to electricity generated by photocells), these results could change significantly. Problem 6.32 includes a link to the spreadsheet model that produced the data in Figure 6.1. The reader is encouraged to download the software and experiment with different fuel and engine factors to get a better feel of the sensitivity of these data to various parameter choices.

EXAMPLE 6.5

New Electricity Needed for All-Electric Cars

How many more power plants would have to be built in order to supply the necessary electrical energy if all vehicles of all types in the United States switch to all-electric vehicles driven on the road?

From Chapter 2 we know that the United States uses about 27 quad of energy for transportation per year. If we assume vehicles are about 20% efficient, then the actual energy needed to move the vehicles is 5.4 quad. As a rough first approximation, assume the electric engines and batteries used in the electric cars have a net efficiency of 50%. So the electrical energy that must be generated is 5.4 quad/0.5 = 10.8 quad.

Suppose a typical power plant is 1,000 MW and can run 80% of the time. During a year this power plant can supply 1,000 MW $\times$ 1 year $\times$ 0.80 = 2.5×10^{16} J = 0.024 quad. To supply all the transportation energy in this way would take approximately 10.8 quad/0.024 quad = 450 new power plants. The current generating capacity of the United States is about 950,000 MW, or 950 power plants with 1,000 MW capacity, so this would mean approximately half again as many power plants would need to be built.

If we assume the 10.8 quad electrical energy is generated with 50% efficient power plants, the total primary energy (in the form of coal and natural gas) used for transportation would drop from 27 quad to 21.8 quad.

We will discuss greenhouse gas emissions in Chapter 7; however, it should be clear that choices in fuel efficiency for transportation are linked to carbon emissions. The all-electric vehicle referred to in Figure 6.1 is assumed to use traditional grid electricity, which is generated predominately from fossil fuels. If the additional power plants needed in Example 6.5 are fossil fuel plants, carbon emissions for switching to all-electric vehicles are worse than for hydrogen fuel cells using steam-reformed hydrogen. However, if this electricity can be generated from renewable sources such as wind and solar, then the carbon emission for both cars drops to near zero.

One reason plug-in hybrids were not included in Figure 6.1 is that the energy used is highly dependent on how the car is driven. If the car operates only over short distances so that it functions mostly on its battery-operated electric engine, then the results will be more like those of an all-electric car. If instead the plug-in hybrid is driven mostly over long distances, then it operates more like an internal combustion engine car. It is also very difficult to arrive at a miles per gallon figure for electric and plug-in hybrids because the electricity used can be generated by many different types of power plants. In the following example the equivalent miles per gallon and carbon dioxide savings for a plug-in hybrid vehicle are estimated assuming typical driving times and current technologies.

EXAMPLE 6.6

MPG for a Plug-In Hybrid

Empirical data taken from hybrid cars converted to plug-in hybrids and efficiency calculations indicate that the energy needed per mile to run only on the electric engine is about 120 Wh/mi to 150 Wh/mi [11, 12]. The average daily round-trip commute for drivers in the United States is about 30 miles, for a total annual mileage of 11,000 miles. Assuming half the trip is made using only electricity from the grid, what is the equivalent fuel economy, annual carbon dioxide emissions, and cost compared with a hybrid car that gets 45 mpg and a normal internal combustion engine car that gets 30 mpg?

Electrical energy is used at a rate of 130 Wh/mi for 15 mi, so 130 Wh/mi $\times$ 15 mi $\times$ 3,600 s/h = 7.0×10^6 J of energy is needed for this part of the trip. If we assume this electricity is generated at 50% efficiency, then the actual energy used is 1.4×10^7 J. Gasoline contains 34.6 MJ/l of energy, so dividing this into the energy used gives a gasoline equivalent of 0.4 l.

For the other half of the trip we can assume the car operates like a hybrid car getting 45 mpg, so for 15 mi we need 0.33 gal or 1.33 l. The total gasoline equivalent used in the 30-mile trip is 1.73 l, or 0.46 gal, which gives a gas mileage of about 65 mpg.

Burning gasoline releases 10.7 kg of carbon dioxide per gallon. Carbon emissions for the 11,000-mile year is given by (11,000/30 mpg) × 10.7 kg/gal = 3,923 kg of carbon dioxide for the conventional car, 2,615 kg for the 45-mpg hybrid, and 1,757 kg of carbon dioxide for the plug-in hybrid.

At 10 cents/kWh, the electric part of the trip in the plug-in costs (130 Wh/mi ÷ 1 KWhr/$0.10 = $0.01, or $0.20 for the 15 miles. At $2/gal the other 15 miles costs $0.67, for a total of 87 cents. For the hybrid the cost is $1.30, and for the conventional car the cost is $2.00.

It should be clear from Example 6.6 that the claims in the media of more than 100 mpg for plug-in hybrids are suspect. The actual figure will be highly dependent on assumptions about driving conditions. Most plug-in hybrids under development today are projected to have a 20-mile all-electric range, which would yield a gas mileage similar to that of the example because two thirds of the commute could be done in all-electric drive. The calculation is also sensitive to the figure used for the electrical equivalent of a gallon of gasoline. If the gasoline were actually used to generate electricity with an internal combustion engine at 20% efficiency, the plug-in hybrid would have the same fuel efficiency as the hybrid car. Costs for each scenario are also highly dependent on energy pricing. Increased electrical demand as the result of more plug-in hybrid cars being built will probably raise the cost of electricity. One further point is that new power plants are generally more efficient than older plants, so equivalent gas mileage will go up over time for a plug-in hybrid.

Notice that the energy needed to build the vehicle and the energy needed to dispose of the vehicle after its useful life have not been included. A report from Argonne National Laboratory estimates that it takes about 3.8 million Btu to manufacture a midsized car, about 1.2 million Btu to paint it, and about 1.5 million Btu to dismantle it for disposal or recycling [12]. This totals nearly 6.5 million Btu, or 6.8×10^9 J. Using 34.6 MJ/l for gasoline from Table 6.1, this would be about 200 l, or 52 gal of gasoline. Clearly the well-to wheels-energy used in a normal vehicle lifetime is significantly more than the energy needed to manufacture it, so our decision to ignore construction energy costs is justified.

6.6 ■ Large-Scale Energy Storage Systems

Much of our discussion of fuels has focused on the storage of energy for use in vehicles. It is also desirable to store energy on a larger scale to level out daily and seasonal variations in energy used for homes and industry. This will become extremely important if solar and wind energy are ever to figure significantly in energy consumption for the obvious reason that these resources do not produce a constant energy supply. On a shorter time scale, load-leveling energy storage systems for the electrical grid are needed so that the grid can respond quickly to changes in demand. Various technical solutions for dealing with this problem such as pumped hydroelectric, flywheels, and supercapacitors are considered here.

Figure 6.2 shows the sources that the Tennessee Valley Authority (TVA) Power Company uses to supply electricity to one of its service regions (Raccoon Mountain) over a typical 24-hour period. The selection of the various sources at different times as shown in the figure is a result of the performance of the source. Electricity from nuclear reactors is cheap to produce, once the plant has been built, but cannot be turned on or off very quickly, so it makes sense to use it as a constant base supply. Coal plants also generate electricity cheaply and so are typically used as base supplies but can also respond to changes in demand. Hydroelectric

FIGURE 6.2

Typical daily electrical consumption by source for a 24-hour period, based on figures from the Tennessee Valley Authority [13]. These figures are for a large region and represent multiple plants for each source.

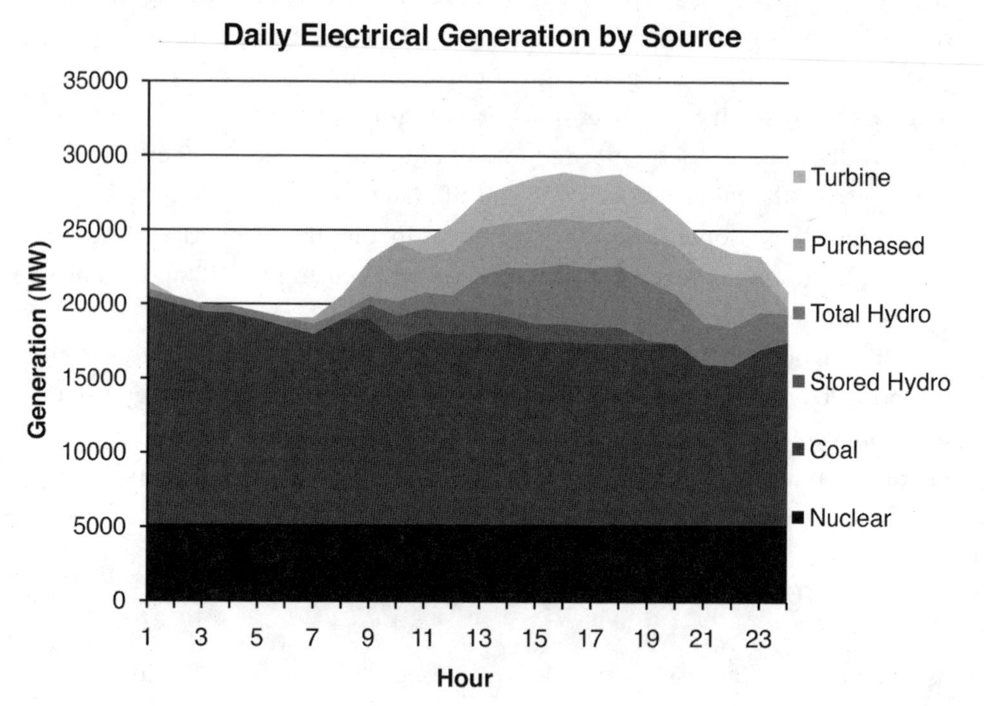

is actually the cheapest method for making electricity, but because it can respond very quickly to changes in demand it is often used at peak demand times. The stored hydroelectric energy shown is generated from water pumped back into the lake behind a dam. In this case the energy to move the water back uphill is produced by coal in off-peak hours (between hours 22 and 7). Pumped hydro actually is a net energy loss (because of efficiencies in the pumps and generators less than 100%) but provides additional energy when it is needed during the day. Turbines run on either natural gas or fuel oil but are an expensive source of electricity and so are used only at peak need. Notice also that a significant amount of energy is imported from surrounding power systems via the electrical grid for this region. This is one of the important features of electricity as an energy carrier; the electric grid allows the rapid transfer of energy to other locations as demand changes.

Short-term load balancing in the electric grid can also be facilitated by creating a so-called smart grid for the distribution of electricity. In this scenario, industrial processes using large amounts of electricity (e.g., aluminum processing) would be in constant electronic communication with power-generating stations. The system would allow suppliers and industrial customers to cooperate automatically to determine the best schedule for heavy electrical use, reducing use during peak demand and increasing use during off-peak hours. This would be especially helpful for using intermittent sources efficiently; industries could agree to use wind and solar electricity when it is most available in return for a discount. Consumer appliances such as thermostats, clothes washers and dryers, water heaters, and refrigerators might also participate in this electric load leveling plan. A "smart" refrigerator, built with a computer chip and sensors that monitor its internal status, would communicate with the grid and, for an agreed-upon price break to the consumer, cycle on and off less often during peak demand while still maintaining a sufficiently low temperature to preserve its contents. Similarly, washers and dryers, water heaters, and plug-in hybrid cars could communicate with the grid so that they use electricity only when a surplus generating capacity is available. Additional benefits of a smart grid include greater stability and the possibility of a less centralized system that would be less prone to catastrophic failure or terrorist attacks.

There are also seasonal changes in energy use during the year. Air conditioners are the largest users of electrical energy during the summer, and as can be seen in Figure 6.3, they contribute to a significant increase in electrical power consumption in the United States. Most heating in the United States is provided by oil or natural gas, but some electricity is used for heating, and this is reflected in the figure as well.

FIGURE 6.3

Annual variation in electrical energy use for the United States [14].

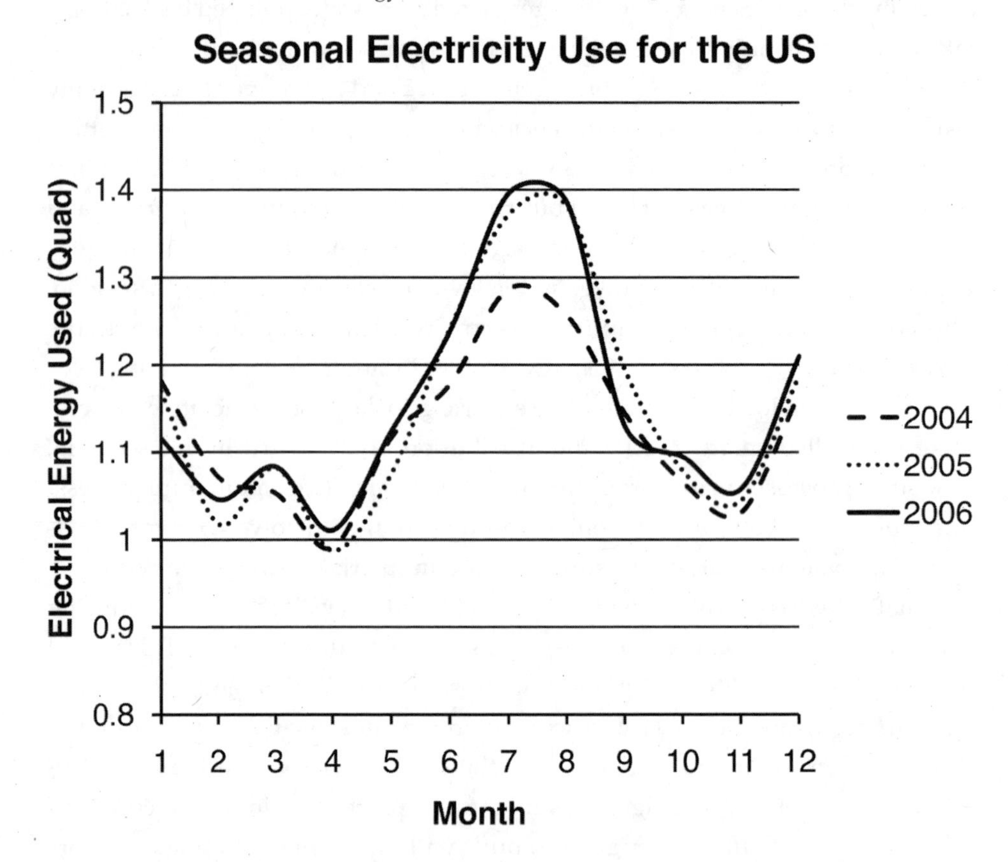

At present, load leveling on a seasonal scale is not being practiced. However, as can been seen from the TVA example in Figure 6.2, daily load leveling is already in use and is expected to increase. Pumped hydroelectric is the most common form of load leveling in use today, but alternatives are being investigated. A comparison of various methods for storing energy is shown in Table 6.4, with the energy content of gasoline as a comparison. These figures reflect current technology and could change with new technological breakthroughs. Several of these storage methods, including pumped hydro, are discussed in more detail in the following sections.

As noted in the table, none of these storage mechanisms can match the energy stored per kilogram in gasoline. However, for load leveling in electrical generation the available power and cycle efficiency are much more significant than energy density. Conventional coal- or gas-fired electric plants generate between 1 MW and 1500 MW of electrical power (MWe), depending on their

TABLE 6.4

Characteristics of various energy storage devices in existence today (with gasoline as a comparison) [2, 15, 16]. Cycle life is the number of times the device or system can be discharged and recharged with energy before significant deterioration in the efficiency of the process occurs.

Storage Type	Energy (kJ/kg)	Potential Size (MWe)	Cycle Efficiency (%)	Power (W/kg)	Cycle Life
Gasoline	47,000				
Pumped hydroelectric	1	100–1,000	65–80		50,000
Compressed air	100	50–1,000	40–50		10,000
Flywheels		1–10	95		50,000
Steel	30–120				
Carbon fiber	~200				
Thermal					Infinite
Water (40°C–100°C)	259				
Rocks (40°C–100°C)	40–50				
Rocks (200°C–400°C)	160				
Iron (200°C–400°C)	30				
Salt compounds	300				
Batteries					
Lithium ion	~580	0.5–50	99	1,800	1,200
Nickel cadmium	~350		70–90	150	1,500
Nickel hydride	~290	0.5–50	66	250–1,000	1,000
Lead storage	~140	0.5–100	70–80	180	500–800
Superconducting magnet	100–10,000	10–10,000	85		Infinite
Supercapacitors	18–36	1–10	95		100,000

MWe = equivalent megawatts electrical energy produced.

size, and typical nuclear plans generate about 500 MWe. These figures are for individual generating units; typically several units operate at the same site to provide much larger energy resources. A typical home in the United States uses energy at a rate of 1.3 kW. In order to be useful as an energy storage mechanism for load leveling, the process should be scalable to 1–10 MW in size, whereas for home use, storage devices can be as low as 1 kW in capacity and still be useful. As we can see from the table, lead–acid batteries can be used to store energy on the megawatt scale, more than sufficient for storing solar energy for use in a home.[5] The other batteries listed were developed for portable applications such as computer batteries and are only now being considered for larger-scale applications.

A second consideration for storage devices is the cycle efficiency: the quantity of energy lost in the process of storing and then reclaiming the energy. For example, in the case of pumped hydroelectric storage, 20% to 35% of the energy is lost each time energy is stored for use later on, depending on the efficiency of the individual components of the system used. Obviously, the cycle efficiency should be as high as possible because each time the device is used some energy is lost. Cycle life is the number of times the device or system can be discharged and recharged with energy before significant deterioration in the efficiency of the process occurs. Details of several technologies listed in the table are given in the following sections.

6.6.1 Solar Thermal Energy Storage

Thermal storage of energy from the sun has a long history, as can be seen in Figure 6.4. Applications include solar heating of rocks and slabs of concrete for

FIGURE 6.4

White House Ruins dating from about 1200, Canyon de Chelly National Monument. Like many early Native American settlements in the Southwest, the structures were built into a south-facing canyon wall to collect solar energy for the cool desert nights.

warmth at night, thick adobe walls to help equalize daytime and nighttime temperatures, and water tanks painted black to absorb energy from the sun during the day. The urban heat island effect, which raises the average nighttime temperature of some cities by as much as 3°C, is in part attributable to the heat reservoirs created by large amounts of concrete found in cities.

The Trombe wall is a common architectural design that was created in 1956 and is designed to equalize the day and night temperatures inside a building. The basic concept is shown in Figure 6.5. During the winter, when the sun stays closer to the horizon, solar radiation shines directly through a glass window and heats a large thermal mass, generally concrete or some other material that has a high heat capacity painted black. At night air circulates past the wall (by natural or forced convection) to provide heat to the building. During the summer the sun is higher in the sky, and a roof with a large overhang prevents the sun from shining directly onto the wall. Reflective curtains can also be used to block the heat gain by the Trombe wall during the summer. Several modifications such as earthen walls and underground containers of stones have also been used.

FIGURE 6.5

Diagram of a Trombe wall.

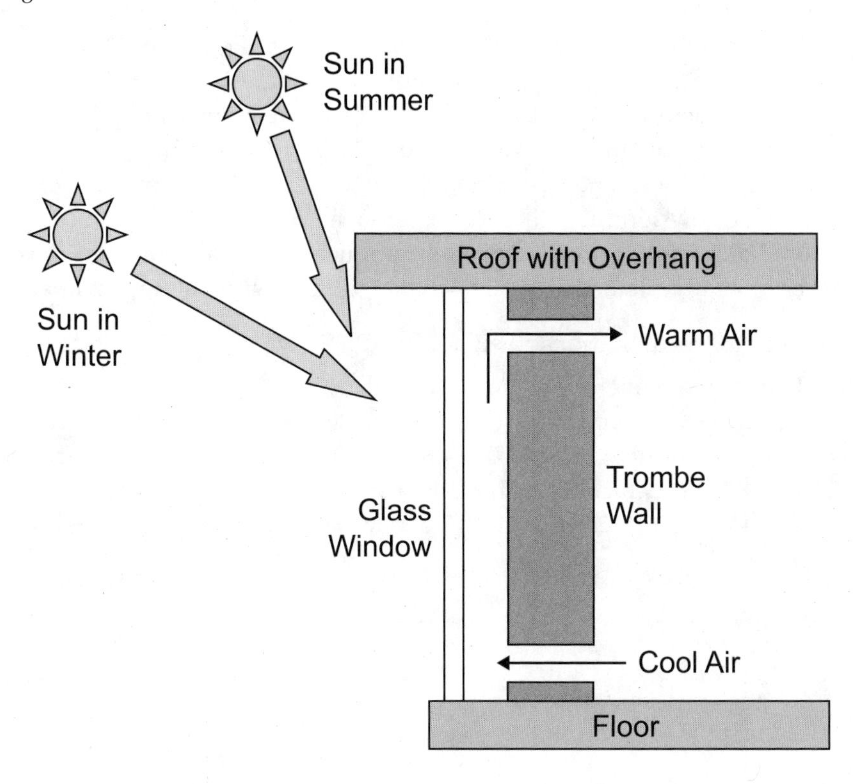

EXAMPLE 6.7

Thermal Energy Storage

In this example we will use the equations for heat flow from Chapter 2 to calculate the heat gain from a hypothetical Trombe wall made of concrete. We assume the wall is 2 m × 5 m and 20 cm thick and is 50% efficient in collecting energy (the net albedo is 0.5) for an 8 hour period. We also want to calculate how long it will take the wall to radiate this heat back into the room; if the heat is lost too quickly or too slowly the wall will not be effective in transferring the heat captured during the day to warm the building at night.

Using the daily average solar flux from Chapter 5 of 200 W/m^2 for an 8-hour period at 50% efficiency, we can store 0.5 × 200 W/m^2 × 2 m × 5 m × 8 h × 3,600 s = 2.88 × 10^7 J. This is about 27,000 Btu. For comparison, it is typically recommended that a 180-ft^2 room have a furnace with a rating of about 10,000 Btu to heat it.

The density of concrete is about 2,400 kg/m^3, and it has a specific heat capacity of 1,000 J/kg K. Using Equation (2.5) the temperature increase of the concrete during the day is $\Delta T = Q/mC$ = 2.88 × 10^7 J/(2,400 kg/m^3 × 2 m × 5 m × 0.20 m × 1,000 J/kg K) = 6.0°C, assuming it does not lose any heat. So a wall originally at the temperature of 21°C (70°F) would heat up to 27°C (81°F) during the day if heat loss into the room is ignored.

We can now use Newton's law of cooling, Equation (2.27), to find the rate of cooling at night. Equation (2.27) is $T(t) = T_S + (T_o - T_S)e^{-(\lambda t)}$, where the initial temperature of the cooling wall is T^o. We want the inside ambient temperature to remain constant at temperature T_S. This is not guaranteed, but we can still get an approximation of how long it will take for the wall to cool off by assuming T_S does not change much. In many applications the parameter λ is experimentally determined; however, we can get an approximate value from the definitions for Equation (2.27), where $\lambda = \frac{UA}{mC}$ and U is the U-factor for concrete.

We can get an approximate value for U from the R-factor of concrete per inch, which is 0.1 ft^2°Fhr/Btu = 0.0176 Km^2/W per inch (from Table 2.3). The U-factor for the 20-cm wall (converting inches to centimeters) is then 0.0254 m/(0.0176 Km^2/W × 0.20 m) = 7.20 W/Km^2. This gives $\lambda = \frac{UA}{mC} = 1.5 \times 10^{-5}/s$. For an 8-hour night we have

$e^{-(\lambda t)} = e^{-(1.5\times10^{-5}/s\times3600s\times8\ hr)} = 0.65$. This means the wall will cool overnight to 65% of the value it had at the end of the day. Here we have assumed that during the 4 hours between day and night (the two 4-hour periods of dusk and dawn) the wall neither gains nor loses energy. Obviously, optimizing some of these parameters (e.g., the material of the wall) would allow a greater heat transfer from daytime to nighttime.

Thermal energy storage can also be used in the summer to reduce air conditioning and refrigeration needs, and in fact it is one of the oldest forms of thermal energy storage. Long before modern refrigeration, snow and ice were often stored in insulated buildings during the winter for use in the summer. In a modern application, ice or cold water is made during off-peak hours (normally at night) and used for cooling during summer days. Stanford University and the Alabama Power Company headquarters in Birmingham, Alabama both use electrical energy during off-peak hours to generate ice: 4,000 metric tons per night in Stanford's case and 440 metric tons in Alabama [17, 18]. This ice is used to cool buildings during the day in the summer. Using this system, Stanford reduced its peak energy demand by one third, saving approximately $500,000 per year.

6.6.2 Solar-Assisted Chemistry

The idea of using solar energy to assist the electrolysis process was mentioned earlier. Recently other solar-assisted chemical reactions have been proposed that could provide a transportation fuel or energy storage mechanism [19]. The following reaction of methane and carbon dioxide requires 248 kJ/mole and occurs at 700°C, heat that could be provided by solar energy:

$$(6.11)\quad 248\ \text{kJ} + CH_4 + CO_2 \rightarrow 2H_2 + 2CO.$$

This reaction produces hydrogen fuel and consumes carbon dioxide but has the disadvantage of also producing carbon monoxide. Three options are available at this point. Using catalysts, the products in Equation (6.11) could be reacted to form methanol, CH_3OH, which is a portable fuel. The addition of steam under the right conditions can form additional hydrogen and CO_2. But perhaps the best option would be to generate hydrogen and carbon monoxide during the day, cool these products, and save them for use at night. The reaction in Equation (6.11) can be run in reverse, giving off 248 kJ/mole, which can be used to generate steam for a steam engine and generator combination. In this way, solar energy is stored in chemical form for use at times when solar energy is not available.

6.6.3 Batteries

Battery advancement has been rapid in the past few years, and intense research is still under way. Each battery cell contains two chemical reactions, one in which electrons are given off (occurring at the anode, which becomes negatively charged) and one in which electrons are absorbed (occurring at the cathode, which is positively charged). Electrons at the anode are at a higher potential energy than electrons at the cathode and so will flow from anode to cathode through an external circuit. This flow of electrons constitutes an electrical current, I, measured in amperes, and can be used to do electrical work. The anode and cathode are in chemical contact with each other by means of an electrolyte, which may be a liquid or a solid permeated with a liquid.

The chemical reactions occurring in a lead–acid battery are as follows. The electrolyte is sulfuric acid (H_2SO_4) and water. At the cathode, which is solid lead oxide, a reduction reaction occurs with an activation energy of 1.685 V:

$$(6.12)\quad PbO_2 + SO_4^{2-} + 4H^+ + 2e^- \leftrightarrow PbSO_4 + 2H_2O.$$

At the anode, which is solid lead, an oxidation reaction occurs, giving off the two electrons needed at the cathode:

$$(6.13)\quad Pb + SO_4^{2-} \leftrightarrow PbSO_4 + 2e^-,$$

with an activation energy of 0.356 V. As we saw in Chapter 3, activation energies can be calculated from the Gibbs free energy for the reaction. The positive activation energies, expressed in volts, indicate that the two reactions will generally proceed to the right as long as there is sufficient material left at the electrodes and enough electrolyte. The combined energy of this reaction is a little over 2 V per pair of electrodes. Individual cells, each containing this reaction, can be combined to form a battery; six such cells would make a 12-V battery. Commercial car and boat batteries are available in 6-V, 12-V, and 24-V sizes, all multiples of the 2-V primary reaction.

The reactions in Equations (6.12) and (6.13) are reversible, so if current from some external source is introduced to the battery and supplies electrons to the cathode and removes them from the anode, the reaction will run in reverse, thus recharging the battery. One difficulty in recharging batteries is that the forward reaction occurs with maximum efficiency for the case of large surface areas of the cathode and anode being exposed to the electrolyte. As the reactions proceed, the lead and lead oxide dissolve into the electrolyte. When the battery is recharged, the lead precipitates out of solution, but it may not precipitate out back onto the anode and cathode, and if it does, it may not redistribute itself in the same configuration. For this reason, after discharging and charging many times, most batteries cannot be fully recharged to their original state. As mentioned previously, charging a battery entails a small heat flow into the battery; discharging gives off heat. Likewise, some hydrogen gas is given off in the recharging part of the cycle, which, aside from being potentially dangerous, reduces efficiency. These energy losses indicate that the efficiency cannot be 100% for a discharge–recharge cycle.

The reactions shown for the lead–acid battery are typical but are not the only possible chemical reactions that produce electrons [20]. A simple grade school science fair project is to make a battery using a zinc strip as the anode and a copper strip as the cathode and a potato or a lemon as the electrolyte. Hundreds of other chemical combinations have been found to produce useful current flows, including some plastics. Fuel cells and flow cells, which were discussed in Chapter 3, are actually types of batteries but with the advantage that the materials at the cathode and anode are continually being replaced. Most of these chemical reactions are not easily reversed, in which case the battery cannot be recharged. The most common rechargeable batteries in use today are nickel cadmium, nickel metal hydride, and rechargeable alkaline.

Commercial batteries are often rated by ampere-hours. An ampere, which measures current, is the number of coulombs of charge per second traveling past a point in a circuit. So an ampere-hour (current times time) is really the amount of charge stored in the battery. Another useful rating for a battery is the power rating, in watts, which indicates how fast the energy can be extracted from the battery. As can be seen from the properties listed in Table 6.4, different reactions have different power ratings, cycle efficiencies, and lifetimes. These properties (and cost) determine which battery is appropriate for which application. A calculation of the number of batteries needed to store enough energy for the United States over a 1-week period is given as an exercise (Problem 6.45).

EXAMPLE 6.8

Charge Stored in a Battery

How much available energy does a 100-Ah, 12-V battery contain? If it is rated at 1.0 kW, how long will it last? What is this maximum current?

An ampere is a coulomb per second, so 100 Ah × 3,600 s/h = 3.6 × 10^5 C. A volt is a joule per coulomb, so at 12 V this represents 4.3 × 10^7 J. If this is being used at the rate of 1.0 kW, we have 4.3 × 10^7 J/1.0 kW = 43,000 s = 11.9 h. The maximum current flow can be determined from $P = IV$ and is 83 A.

As a final note, it should be mentioned that most electrical applications use alternating current (AC), whereas batteries produce direct current (DC). Converting DC to AC requires additional equipment, raising the cost of a battery storage system. The conversion process itself is not 100% efficient, so battery storage systems have to be designed for capacities larger than the amount needed by the application for which it is used.

6.6.4 Flywheels

Kinetic energy in joules is calculated by $KE = \frac{1}{2}mv^2$, where v is the velocity of the object in meters per second and m is the mass in kilograms. However for rotating objects a different equation is often useful. The moment of inertia is an indication of how much mass an object has and how far the mass is located from the axis of rotation and is measured in kg m^2. A formal definition, useful for calculating the moment of inertia for symmetrically shaped objects, is $I = \int_0^R r^2\, dm$, where R is the radius of the object and dm is an incremental piece of the object's mass. The moments of inertia for various objects, calculated using this formula and assuming uniform density, are listed in Table 6.5. It should be noted that the value of the moment of inertia depends on where the axis of rotation is; a rod rotating around its length has a different moment of inertia than a rod rotating around one end.

The amount of kinetic energy stored by a rotating object is conveniently expressed using its moment of inertia, I, and its angular velocity, ω, using an equation similar to the one for linear kinetic energy (in fact, this equation can be derived from the kinetic energy formula):

$$KE = \frac{1}{2}I\omega^2. \quad (6.14)$$

TABLE 6.5

Moments of inertia for several regular solids.

Description and Axis	Figure	Moment of Inertia
Rod of length L (around end)	L	$I = mL^2/3$
Rod of length L (around middle)	L	$I = mL^2/12$
Solid sphere	r	$I = 2mr^2/5$
Solid disk (around axis)	r	$I = mr^2/2$
Cylindrical shell (around axis)	r_2 r_1	$I = m(r_1^2 + r_2^2)/2$

Here, angular velocity is measured in radians per second, which can be converted from revolutions per minute (rpm) using the fact that one revolution is equivalent to 2π radians. The units of rotational kinetic energy will remain joules.

In most applications for energy storage, a large disk or wheel-shaped flywheel is spun at high speed to store energy for later use. Covering for short-term fluctuations in electricity supply for power plants, as demonstrated in Example 6.9, is a typical application. The axis of the flywheel is connected to a device that can act as both an electric generator and an electric motor. Energy is added to the flywheel when the device acts as a motor; energy is retrieved by using the device as a generator. Generators and electric motors were discussed in Chapter 3.

EXAMPLE 6.9

Energy Stored in a Flywheel

How much energy can be stored in a solid disk of radius 1 m with a mass of 2,000 kg spinning at 1,050.0 rad/s (10,000 rpm)? If the device is to be used for load leveling for 15 minutes, what power is available?

Using the moment of inertia from Table 6.5, we have $\frac{1}{2}I\omega^2 = \frac{1}{2}\left(\frac{1}{2}2000\text{ kg}(1\text{m})^2\right)\left(1050\frac{\text{rad}}{\text{s}}\right)^2 = 551\text{MJ}$ of energy. Notice that, because radius and angular velocity are squared in these equations, increasing these parameters has the largest effect on energy storage. Doubling the rotation speed increases the stored energy by a factor of four, as does doubling the radius.

If we divide the energy stored by 15 minutes, we get 0.6 MW for the available power. Using 10 units like this would make 6 MW of power available for the 15-minute time span. This is sufficient to smooth out short-term fluctuations in power supply due to changes in demand or while other sources can be brought online.

A limitation on the amount of energy that can be stored in a flywheel is the strength of the material of which the flywheel is made, as the sample calculation in Example 6.10 shows. First-generation flywheels were made of steel, but current versions are made of carbon fiber, which has a higher tensile strength and can rotate at speeds up to 50,000 rpm. To achieve these speeds the flywheel is rotated in a vacuum chamber to eliminate air friction, and physical contact with the surroundings is further reduced by using magnetic bearings. A second, related limitation to the use of flywheels is that, to ensure safety at these high rotation rates in the event of a failure, high-performance flywheels must be enclosed in protective shielding, which generally adds significant weight to the device. For nontransportation applications this is not a serious limitation, and the device can be located underground for additional safety.

EXAMPLE 6.10

Limitations of Flywheel Speed

As a sample calculation of the speeds and forces involved in a flywheel, consider a rotation rate of 10,000 rpm, which gives an angular velocity of 1,047.2 rad/s $\left(\omega = 10000\ \frac{\text{rev}}{\text{min}} \times \frac{2\pi\ \text{rad}}{60s} = 1047.2\ \text{rad/sec}\right)$. The speed of the outer rim is given by $v_T = r\omega$, so the edge of a 2-m-diameter flywheel traveling at 10,000 rpm is 1,047.2 m/s or more than 2,342 mph, three times the speed of sound.

The centripetal force needed to keep a small piece of the flywheel of mass m moving in a circle of radius r is given by $F = mv_T^2/r$. So a 100-g piece of the rim of this flywheel needs a force of $F = \frac{0.1\ \text{kg} \times (1047.2\ \text{rad/sec})^2}{1\text{m}} = 109{,}663N$ in order to keep moving in a circle. The yield strength of steel, the force per area under which it will permanently deform, is 250×10^6 Pa, or $250 \times 10^6\ \text{N/m}^2$. If we imagine that a rod or spoke of iron with 1 cm^2 cross-sectional area extends from the center to hold the 100-g piece in place at the rim, the available force without deformation would be $250 \times 10^6\ \text{N/m}^2 \times 0.0001\ \text{m}^2 = 25{,}000$ N. This tells us that a steel flywheel with a 1-m diameter will not be able to withstand the rotational forces on it at 10,000 rpm.

6.6.5 Capacitors

A capacitor is an electronic device that stores electrical charge. The basic idea is that if there are two surfaces close together but electrically insulated, positive charge on one surface will attract negative charge on the other surface so that a stable configuration of charge can be stored. Connecting the two surfaces via an external circuit allows the stored charge to do useful work as a current flow. Capacitance, in farads, is defined as the ratio of the amount of charge stored to the voltage applied, or

$$C = Q/V. \tag{6.15}$$

A 1-farad capacitor holds 1 C of charge when 1 V is applied across its terminals. It should be emphasized that capacitance is a number that is fixed by the structure of the capacitor; applying more voltage will increase the amount of charge stored, but the capacitance, because it is the ratio of the two, remains unchanged.

The capacitance for storing charge in a system of two flat surfaces of plates of area A separated by a distance d is

$$c = \frac{\varepsilon A}{d}, \tag{6.16}$$

where ε is a constant called the permittivity that depends on the material between the two surfaces. For parallel plate capacitors the separation between the plates is constant and the electric field, defined as the force on a small test charge, $\vec{E} = \lim_{q_t \to 0} \frac{\vec{F}}{q_t}$, is uniform between the two surfaces. Capacitors can obviously be made to hold more charge by increasing the surface area, decreasing the distance between the surfaces, or changing the material between the surfaces. In most applications a large surface area is obtained in a smaller space by wrapping the two surfaces, separated by insulating material, into a cylinder.

The permittivity of a vacuum is $\varepsilon_o = 8.85 \times 10^{-12}$ F/m and is the lowest permittivity. The atoms or molecules of some materials, called dielectrics, will react to the electric field between the two capacitor plates by separating their charge. The molecule remains neutral overall but has a bit more positive charge oriented toward the negative capacitor plate and an equal amount of negative charge oriented toward the positive plate, as shown in Figure 6.6a. This reorientation of the molecules requires energy, and therefore a capacitor with a dielectric inserted between its plates stores more energy. As the charge is removed from the capacitor, the molecules reorient, giving up stored potential energy, which is available for higher-voltage current flow.

Capacitors have many applications besides energy storage; for example, they can be used as low-frequency filters for analog signals and as elements to isolate direct current flow in a circuit. The stored energy is in the form of a uniform electric field that points from the positive side to the negative side of the capacitor when it is charged. Electric fields, E, measured in N/C, cause forces on charges: $q\vec{E} = \vec{F}$. Electrical work then becomes $W = -\int_{x_i}^{x_f} \vec{F} \cdot d\vec{s} = -q\int_{x_i}^{x_f} \vec{E} \cdot d\vec{s} = -qEd$ in the

FIGURE 6.6

Cross-section of a standard parallel plate dielectric capacitor and an ultracapacitor.

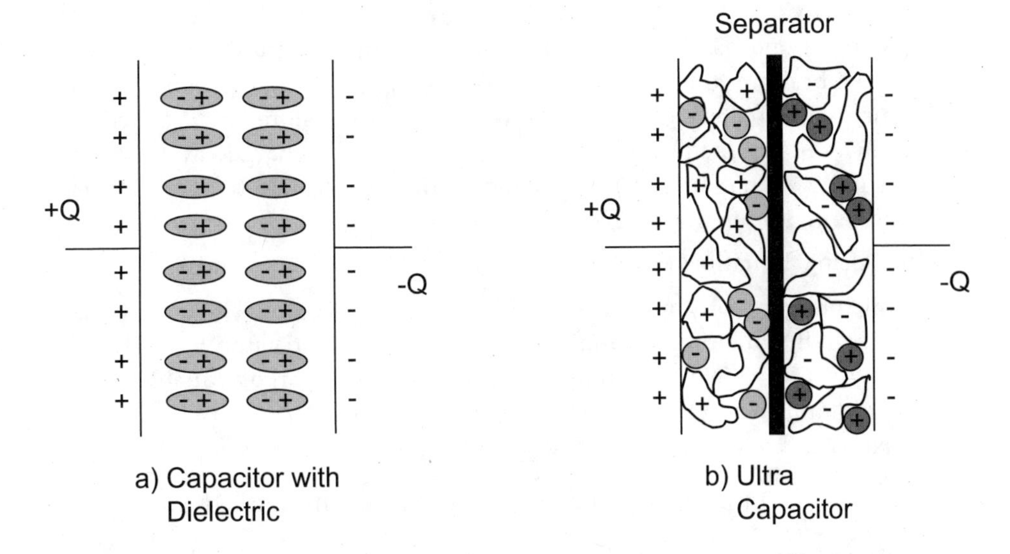

case of a uniform field. Electric field is difficult to measure directly, so a more useful expression for energy stored in a capacitor involves electrical potential energy, $U = qV$. To find the work needed to charge the capacitor to a total charge Q, which is equal to the potential energy stored in the capacitor, we may write

$$(6.17)\quad W = U = \int_0^Q V\,dq = \int_0^Q \frac{Q}{C}dq = \frac{1}{2}\frac{Q^2}{C} = \frac{1}{2}CV^2 = \frac{1}{2}VQ,$$

where we have used the definition of capacitance in Equation (6.15).

As mentioned earlier, the capacitance increases with the surface area available to hold charge. In an electrochemical double-layer capacitor, sometimes called an ultracapacitor or supercapacitor, clever use is made of a solid material that is porous and therefore has a large internal surface area. A schematic diagram is shown in Figure 6.6b. The effective surface area is made much larger than in a normal parallel plate capacitor, enabling the device to hold up to a thousand times more charge. Because they can react more quickly than batteries to changes in current flow, supercapacitors are being considered for energy storage during regenerative braking in automobiles.

EXAMPLE 6.11

Capacitors

As mentioned in Section 6.1, a hybrid car might capture about 140 kJ while braking to a stop from 20 m/s. Suppose instead of storing this energy in a battery (as is done now), the energy is stored in a capacitor. How much capacitance is needed if the capacitor is charged at a voltage of 100 V? What area of capacitor is needed if the dielectric constant is $10\varepsilon_o$ and the separation is 0.001 m?

Using Equation (6.15) we have $W = 140 \text{ kJ} = \frac{1}{2}CV^2$ and $C = 28$ F, much larger than capacitors used in electronic circuits, which are typically in the microfarad range. Fully charged, this capacitor would have $2\ W/V = Q = 2{,}800$ C of charge, equivalent to the charge available from a 0.78-amp-h battery (1 amp-h × 3,600 s = 3,600 C).

Using Equation (6.14), the area of the capacitor would be $A = \frac{dC}{\varepsilon} = \frac{dC}{10\varepsilon_o} 3.16 \times 10^8 \text{m}^2$. Obviously, enclosing this much area in a convenient package would be a significant engineering problem.

6.6.6 Compressed Air

Several power plants around the world now use compressed air energy storage (CAES) systems to store energy during off-peak hours. The air can be pumped into tanks or underground geological formations and used later in engines that use compressed air to do work against a piston, much as steam engines use expanding water vapor. Efficiencies are currently about 50% for these conversion processes. The energy per kilogram compares favorably with that of a storage battery, and several car manufacturers in Europe and India have working prototypes of automobiles running only on compressed air as a fuel.

To calculate the energy that can be stored in a tank of air, we start with Equation (2.2) for the work done in compressing a piston: $W = \int_{V_i}^{V_f} PdV$. Using the ideal gas law, $PV = nRT$, where n is the number of moles of gas, R is the ideal gas constant, and T is the temperature, we can change this equation to

$$(6.18) \quad W = nRT \int_{V_i}^{V_f} \frac{dV}{V} = nRT\ ln\left(\frac{V_f}{V_i}\right),$$

where we have assumed the process was isothermal (the temperature did not change). Because we really want the work done by changing the pressure in a tank of constant volume rather than changing the volume at constant pressure, we can use the ideal gas law again at constant temperature to write $P_iV_i = nRT = P_fV_f$, which gives $\frac{P_i}{P_f} = \frac{V_f}{V_i}$. Using this relation, Equation (6.18) becomes

$$(6.19)\quad W = nRT\ ln\left(\frac{V_f}{V_i}\right) = P_fV\ ln\left(\frac{P_i}{P_f}\right).$$

Equation (6.19) gives the work done in filling a tank of constant volume V from some initial pressure, P_i (for example atmospheric pressure) to a new, higher pressure, P_f. The work done in filling the tank will be a negative number, but the same number with a positive sign is the energy stored in the tank. The efficiency of this process depends on the efficiency of the pumps used, but because the process is isothermal and therefore reversible, it could in theory be close to 100%. To achieve a high efficiency a large heat bath would be needed to store the heat expelled during compression to be used when the air is extracted (which is a cooling process). In practical applications, such as transportation, it may be more realistic to not carry the process out isothermally and accept a lower efficiency.

EXAMPLE 6.12

Energy Stored in a Tank of Compressed Air

Suppose the strength of a tank limits a compression ratio of 1,000:1 for air inside relative to outside pressure. Further assume that the volume of the compressed air tank is 75 l (0.075 m^3, a bit more than 30 gal). How would this compare to the equivalent energy stored in a 10-gal tank of gasoline?

The work done in compressing the gas is $W = P_fV\ ln\left(\frac{P_i}{P_f}\right) = 1000atm \times 1.0 \times 10^5\ Pa/atm \times 0.075m^3 \times ln\left(\frac{1}{1000}\right) = -5.2 \times 10^8$ J. If we assume 50% efficiency in compressing the gas and then extract this energy, we get 2.6×10^8 J = 260 MJ.

A 10-gal tank of gasoline stores (using Table 6.1) 10 gal × 3.785 l/gal × 34.6 MJ/l = 1,310 MJ. So this compressed tank of air contains about the energy in 2 gal of gasoline.

6.6.7 Pumped Hydroelectric

As we saw in Chapter 5, water stored behind a dam has the capacity to generate electricity. In Figure 6.3 we saw that daily electricity demand varies, and one solution to this problem is to use pumped storage. Water is pumped into a dam at night or other times when electricity demand is low. It can then be used to run generators when demand is high. There is some net loss of energy because the pumps and generators are less than 100% efficient, but overall losses are usually less than 30%. Although some energy is lost, the electric utility saves money by not having to build and maintain extra capacity for times of peak demand. This technique can be used even where the terrain is flat by using subterranean reservoirs. A cavern with a volume of 4.0×10^6 m^3 at a depth of 2 km can store $U = mgh = \rho Vgh = 7.8 \times 10^{13}$ J of energy. Over a 10-h period this can supply $U/t =$ 2.1 GW, equivalent to a medium-size generating plant. The energy is extracted in a manner similar to that of hydroelectric plants; as the water descends from a surface holding tank, the kinetic and potential energy is captured by turbines.

6.6.8 Superconducting Magnets

Superconducting magnetic storage (SMES) is a new concept that is being investigated for large-scale energy storage. As we have seen previously, normal current-carrying wires lose energy in the form of heat at the rate $P = I^2R$. Some materials, called superconductors, experience a loss of all electrical resistance below a certain temperature, called the critical temperature, which varies depending on the material. Because wires made of superconductors below their critical temperature effectively have no resistance, the power loss is zero. The idea behind using this phenomenon as an energy storage device is to generate a magnetic field in an electromagnet that has superconducting wires. The current will flow indefinitely with no resistance loss, storing energy in the magnetic fields of the magnet. Efficiencies for storing and retrieving the current are about 95%.

Energy per cubic meter stored in a magnetic field is

$$u = B^2/2\mu_o, \tag{6.20}$$

where B is the magnetic field in tesla and $\mu_o = 4\pi \times 10^{-7} Tm/A$ is the permeability constant. A 10-T field stores $u = B^2/2\mu_o = 4.0 \times 10^7$ J/m^3. Superconducting systems are currently available to smooth fluctuations in grid voltage. They can respond much faster to voltage changes than other storage systems, which makes them perfect for maintaining steady line voltages.

Current limitations to SMES are the fact that superconductors will not permit magnetic fields beyond a certain magnitude, and superconductivity stops if the temperature rises above the critical temperature. Type I superconductors lose their capacity for resistanceless current flow at fields of about 0.05 T. Fields of 40 T are possible with type II superconductors. There is also a significant energy investment in keeping these magnets below their critical temperatures. To date superconductors with temperatures as high as 130 K have been made in the laboratory, but commercial versions of these materials are not yet available. SMES has not yet been applied to the longer energy storage times needed to load balance daily or seasonal fluctuations in energy demand.

6.7 ■ Summary

In Chapter 2 we defined energy efficiency as the ratio of the energy benefit to the energy cost. As we have seen in this chapter, the overall energy efficiency of a transportation option must take into account not only the energy-to-weight ratio of the fuel and the efficiency of the motor but also the energy needed to produce the fuel and the emissions involved in each step. Biodiesel and ethanol from corn used in internal combustion engines are not a particularly energy efficient combination and may be able to contribute only marginally to a reduction in petroleum use. Although fuel cells have high efficiencies, the energy cost of manufacturing hydrogen is substantial and depends significantly on the method used. Safety issues also are a concern with any plan to switch to hydrogen as a fuel for transportation. It is entirely possible that small improvements in battery efficiencies and costs will, in the end, relegate fuel cells to nontransportation applications in which hydrogen from cogeneration can be used and safety is less of an issue.

Most of the applications for large-scale energy storage discussed in this chapter, with the exception of pumped hydroelectric and passive solar storage, are new developments. Even batteries have changed significantly in the last 20 years. These innovations are quickly finding applications. Given the fact that a majority of the renewable energy sources mentioned in Chapter 5 will necessarily be coupled with some kind of storage device, energy storage is an important issue. It is estimated that a volume of about 1 km^3 of common lead–acid batteries would be needed to store a 3-day supply of energy for all the current electrical needs of the United States [21]. Which of the energy storage mechanisms outlined in this chapter will end up being the most significant in the future energy supply stream remains to be seen. Probably some combination of all of them will be used, depending on the particular application.

The average electric power plant in the United States is about 50 years old [22]. Although new plants have been added to the system, the basic design of these plants and the grid that connects them have not been significantly modified. The implementation of a "smart grid" that monitors and adjusts electricity production and time of usage would greatly improve efficiency, reduce carbon emissions, and aid the implementation of renewable but intermittent resources.

Considerations of energy efficiency are the first step in selecting a potential energy solution for transportation, storage, or generation. Emissions, economic cost, and risk factors are other issues that influence these decisions and are considered in Chapters 7 and 8.

Questions, Problems, and Projects

For all questions,

- Cite all sources you use (e.g., textbook, Wikipedia, Web site). Before using Wikipedia as a source, please read the policies of research and citing in Wikipedia: http://en.wikipedia.org/wiki/Wikipedia:Researching_with_Wikipedia and http://en.wikipedia.org/wiki/Wikipedia:Citing_Wikipedia. For Wikipedia you should verify the information in the article by reading the sources listed.
- For all sources, briefly state why you think the source you have used is a reliable source, especially Web resources.
- Write using your own words (except for direct quotes, which should be enclosed in quotation marks and properly cited).

1. What volume of hydrogen (at STP), coal, wood, and potatoes would have to be carried in a car to have the same energy as a 44-l (10-gal) tank of gasoline? Comment on the feasibility of each fuel source for transportation.
2. Using reliable sources, write a brief description of the current research on nanotubes and gas–solid adhesion as methods of storing hydrogen. At what stage of development are these techniques?
3. Using reliable sources, write a brief summary of the "Points Plan" used in the popular diet plan created by Weight Watchers. Include comments

about the kinds of foods that have more points and which have fewer and compare them with the figures in Table 6.1.

4. Suppose you can harvest 5 metric tons of wood per acre from trees that have the energy content listed in Table 6.1. How many acres of trees would you need if you used wood as a fuel to drive 15,000 miles (the U.S. average) and got 30 mpg? Would this be feasible for the entire U.S. population?
5. Using reliable sources, define a Ragone plot and find some examples. What is this type of plot useful for?
6. For the next 2 weeks, try the following experiment. Fill your car up with gas and drive (very carefully) at 10 mph under the speed limit. At the end of the first week, fill up again and calculate your gas mileage. The second week, drive exactly at the speed limit and measure your gas mileage. What is the difference in gas mileage between the two weeks?
7. Use Equation (6.1) to calculate the drag force on your car and a friend's car. You may be able to find c_d, the drag coefficient for your car, on the Internet. If you can't, make an approximation based on the following values: square block (worst case), $c_d = 1.05$; pickup truck, $c_d = 0.70$; station wagon, $c_d = 0.60$; Ford Escort, $c_d = 0.40$; Toyota Prius, $c_d = 0.26$; teardrop shape (best case), $c_d = 0.04$.
8. Derive Equation (6.4) from $F = ma$ and the definition of work.
9. Verify with a sample calculation, the claim in the text that "decreasing the speed limit from 70 mph to 55 mph uses nearly 50% less fuel per mile traveled." Also look up the number of miles driven in 1973 in the United States and estimate the fuel saved by making this change in speed limit.
10. Repeat the calculation in Example 6.1 using the same assumptions for an empty school bus weighing 10,000 kg. How much power is needed for constant speed and how much for an acceleration half that given in the problem at a speed of 20 m/s? Repeat the calculation for a school bus loaded with 50 students, each weighing 50 kg. Repeat the calculation for the car loaded with four students. Can you draw any conclusions about the efficiency of bus transportation over transportation using cars?

11. How much savings in horsepower would be gained by cutting the mass of the car in Example 6.1 by 30%? (*Note*: Consult Problem 2.33 if you are worried about safety issues associated with lighter vehicles.)
12. For a round-trip drive from your home to campus and back, estimate the time you spend accelerating and the time spent driving at different speeds (e.g., highway, urban roads). Use these times and speeds and the power calculations in Example 6.1 to determine what part of your drive to campus uses the most energy. Assume a reasonable acceleration. Because this is a round trip, assume the energy used going uphill is compensated for by not using the same amount of energy going downhill on the way back. How does the total energy used change if you reduce your speed by 5 mph?
13. For the car in Example 6.2 equipped with regenerative braking, what is the gas savings for stopping from a higher speed, such as 30 m/s?
14. Using reliable sources, discuss the "hyper-milers," who get much higher than average gas mileage by changing their driving patterns. What techniques do they use to achieve these efficiencies? What is the maximum improvement in miles per gallon reported by these drivers?
15. As we saw in Chapter 3, fuel cells have high efficiency. Calculate the combined efficiency of using electricity from the grid (Table 6.3) and a fuel cell with 80% efficiency coupled to an electric motor with 90% efficiency. Comment on the prospects of using this combination of resources and conversions to solve the transportation energy problem.
16. Repeat Example 6.5 twice, first assuming 10% of the transportation needs are met with all-electric vehicles and then half. What percentage increase in generating capacity is needed for these cases?
17. Suppose a typical vehicle travels 10,000 miles per year. How many gallons of gasoline are saved per year if the gas mileage improves from 15 mpg to 20 mpg? How many gallons are saved if the mileage improves from 40 mpg to 45 mpg? From these calculations, what can you conclude about the importance of improving the mileage for low-efficiency vehicles compared with making superefficient vehicles?
18. Suppose a hybrid vehicle costs $4,000 more than a comparable car and gets 60 mpg compared to 30 mpg for the normal car. How many miles do you have to drive this vehicle to make up for the price difference, assuming gasoline is $4/gal? Suppose you drive 15,000 miles a year.

How many years does it take to make up the price difference? How sensitive are these calculations to the price of gasoline (redo the calculation for $2/gal and $8/gal)?

19. Repeat Example 6.6 assuming the plug-in hybrid travels 20 of the 30 miles as an all-electric vehicle. What is the equivalent mileage per gallon and annual carbon savings? How many barrels of oil per year are saved by switching from a conventional car to a plug-in hybrid using these assumptions?
20. Find and compare several discussions of plug-in hybrid vehicles (the Google RechargeIT.org project is one interesting place to start). What are some of the mileage claims being made? Do any of these sources state exactly how they calculate the gasoline equivalent of electricity? Which figures do you think are more reliable for the mileage of a plug-in hybrid and why?
21. Using reliable sources, discuss the status of any electric vehicle either available now or being planned for production in the near future. Compare the features of each model. What new technologies are being used? What claims are being made about gas mileage? Use a sample calculation similar to that of Example 6.6 to determine whether these claims are reasonable.
22. Find a discussion of the electric power outage that blacked out the mid-eastern part of the United States and parts of Canada on August 14, 2003. Summarize the event, its causes, and steps taken to reduce the possibility of another such crisis. What can be learned from this event about the distribution of electricity?
23. There have been five major electricity blackouts in the past 40 years in the United States. Using reliable sources, discuss these events and what has been learned from them. What caused them? How large an area was affected? How long were those areas affected?
24. Using reliable sources, find a discussion of direct solar to hydrogen solar cells and write a summary of your findings.
25. Using reliable sources, first provide a short discussion of what algae is, the different types of algae, and its history as earth's source of oxygen. Next, write a short summary of the current research in using algae to produce hydrogen, including pros and cons of implementing algae as a hydrogen production source.

26. Using reliable sources, find a discussion of the idea of cogeneration of hydrogen and electric power from a nuclear power plant. List the pros and cons of this process.
27. Find one reputable Web site or article in support of hydrogen as a future transportation fuel and one against it. Summarize both, evaluate their arguments, and state why you think they are reputable sources. Which source makes the better argument?
28. Find and read references [5], [6], and [7] on the proposed hydrogen economy. Summarize both points of view and analyze where they differ in outlook. State which of these points of view you think has the more accurate perspective and why.
29. A widespread urban myth of the early 2000s claimed that the Hummer was actually more fuel efficient than the Toyota Prius, a hybrid car. In a well to wheels comparison track down this rumor, find reputable sources supporting the two views, and draw a conclusion as to which side had the stronger argument.
30. Using reliable sources, write a brief summary of the all-electric vehicle, the EV1, produced by General Motors, or the Roadster, produced by Tesla Motors. Include comments about the controversy surrounding the EV1 and the latest news about the Roadster.
31. Summarize the data shown in Figure 6.1 in your own words. Which combinations of fuel and vehicle are most energy efficient? Which produce the least greenhouse gas? Compare the diesel internal combustion engine and the gasoline hybrid. What facts do you recall from earlier chapters that would explain these figures?
32. Go to the Argonne National Laboratory site and download the GREET spreadsheet program (http://www.transportation.anl.gov/modeling_simulation/GREET/index.html). Follow the included instructions and investigate different scenarios. For example, how do the numbers in Figure 6.1 change if electricity is generated entirely from renewable sources? How do the figures change if the hydrogen comes from photovoltaic sources instead of steam reforming of natural gas? What savings are gained by using grid-connected plug-in hybrid cars over hybrid cars? Report any interesting findings.
33. Using reliable sources, find information about the net energy needed for construction and disposal of a vehicle. How does the energy for

construction and disposal compare with the energy used in the normal operation of the vehicle?

34. Some sources claim the energy cost for building a car is about 10% of the energy used by the vehicle during a normal lifetime. Assume you drive 10,000 miles a year, the car gets 30 mpg, and you keep your car for 8 years. Using the energy density per volume for gasoline in Table 6.1, how much energy do you use? Does 10% of this figure come close to the value stated in the text for the energy used in manufacturing the car?

35. In the U.S. government's 2009 "Cash for Clunkers" program, an older car could be traded in for a model that had better gas mileage for a discount on the cost of the new car. Suppose a car normally lasts 8 years and gets 27.5 mpg (the CAFE standard). (a) Find the annual miles driven by the average driver in the United States and the gasoline consumed. (b) How much gas is saved if the mileage is increased by 5 mpg? (c) How does the savings compare with the energy cost of building the car (assumed to be about 10% of the fuel used in the 8 years)? (d) Did the "Cash for Clunkers" program make sense from an energy perspective? Explain your reasoning.

36. According to Figure 6.3, the United States used significantly more energy in June, July, and August in 2005 and 2006 than in 2004. Using reliable sources, investigate weather reports from those months and years and explain this difference.

37. A quad is roughly the energy equivalent to 3.6×10^7 metric tons of coal. According to Figure 6.3, approximately how many more metric tons of coal did the United States use for electrical generation in August 2006 than in April 2006? August 2005 compared to August 2006?

38. Using reliable sources, discuss the solar energy consequences of the choices made by early Native Americans in the southwestern United States to build homes into cliffs.

39. Using reliable sources, find a discussion of Swedish Thermodeck buildings. Summarize the various methods used to make these buildings more energy efficient. Be sure to mention the four means of heat transfer discussed in previous chapters.

40. Repeat Example 6.7, adjusting various parameters until you find a combination such that the Trombe wall cools off to 30% of its value in an 8-hour night.

41. For your house, find out how many Btus are needed to warm it in the winter. How much of this could be supplied by a Trombe wall such as the one in Example 6.7? Calculate the size of a Trombe wall that would supply all the heating needs for your house, using the same assumptions as in Example 6.7.

42. Suppose a company decides to cool its buildings by making ice at night and using the ice to absorb heat during the day. (a) How much heat could be absorbed by 6,000 mt of ice (*hint*: $Q = mL$)? (b) The power needed to make the ice depends on the coefficient of performance (C.O.P.), discussed in Chapter 3: $P = Q/(\text{C.O.P.} \times t)$, where Q is the heat removed from the water to make ice and t is the time needed. For a C.O.P. of 2.5 and an off-peak generation time of 15 h, how much power is needed to make the ice? (c) Compare your value with the energy needed to cool the building during the day, which, without using the ice, would take 3.0×10^6 W.

43. Using reliable sources, find a discussion of one of the rechargeable batteries now in use (nickel cadmium, nickel metal hydride, and rechargeable alkaline). Write the chemical equations, pros and cons, where and how they are used, and any other interesting information.

44. Given the battery properties listed in Table 6.4, which battery would you choose for a situation in which you needed a lot of current all at once? Which battery would be best for repeated use? Which battery would be best if you wanted to minimize energy loss for a charge–recharge cycle? Explain your answers.

45. The United States on average produces about 10^{12} W of electricity. (a) How much energy is this per day? (b) How much energy would we need to store to supply the U.S. electrical needs for a week? (c) How many standard 12-V lead–acid batteries (100-Ah rating) would be needed to store a week's worth of energy for the United States? (d) Assuming a volume of 0.01 m^3 per battery, what total volume of batteries would be needed? (e) Find the volume of the Empire State Building (or some other famous structure) and calculate how many such buildings would be needed to house this quantity of batteries.

46. How much energy can be stored in a 500-kg disk of radius 1.5 m rotating at 2,000 rpm? Repeat the calculation but double the mass. Repeat the calculation for 500 kg but double the radius. Repeat the calculation

for 500 kg and 1.5 m radius but double the rotation rate. Which parameter has the biggest effect on the stored energy?

47. Repeat the calculation of Example 6.9 for a similar flywheel rotating at 50,000 rpm. Find the yield strength of carbon fiber from a reliable source and calculate the maximum speed of a 2-m flywheel using assumptions similar to those of Example 6.10 (carbon fiber is one of the common choices for flywheel applications).
48. Repeat Example 6.11 for a supercapacitor with capacitance of 2,000 F.
49. What voltage would be needed for a supercapacitor of 2,000 F if it is to store the energy equivalent of 5 gal of gasoline? Does this seem reasonable (compare this to high-voltage power lines)?
50. Using reliable sources, summarize the current state of development of supercapacitors or ultracapacitors. What applications are being considered?
51. Using reliable sources, find a description of a car (or prototype) that runs on compressed air. Summarize the pros and cons of the car, any safety issues, when it will be for sale, and other interesting information.
52. Using reliable sources, write a description with comments of the air compression facilities at Huntorf, Germany; McIntosh, Alabama; or another commercial-scale venture using this method of large-scale energy storage. How much energy is being stored? For what is the energy used? What is the efficiency of the process?
53. Repeat Example 6.12 for the case of a compression ratio of 10,000 to 1. Also, examine the effect of tank size on the stored energy by doing the calculation for a tank the same size as the gasoline tank. Calculate the mass of air stored for Example 6.12 and its weight. How does this compare with the weight of the 10 gal of gasoline (assume the tanks weigh the same, even though the air tank will have to be heavier in order to withstand the pressure)?
54. Using Equation (6.19), calculate the energy stored in an underground cavity of volume 3×10^5 m^3 if the pressure is raised from 1 atmosphere (1.01×10^5 Pa) to 40 atmospheres. Assuming 50% efficiency, what fraction of the total energy used by the TVA system for one hour at peak output shown in Figure 6.2 could be supplied by this energy?

55. How much energy can be stored in a 40-T magnetic field? For this quantity of energy, how big a magnet would be needed to store 10 quad of energy (one third the current electrical use in the United States)?
56. Using reliable sources, write a short report on the current status of the "smart grid." How soon will consumer appliances become "smart"?
57. Using information from Chapter 5 and the storage mechanisms in this chapter, design a system that captures and stores enough energy to maintain a household that uses 1.3 kW of power. You can combine any renewable source (e.g., wind turbine, photovoltaic solar) with any storage mechanism (e.g., compressed air, SMES). List all the assumptions you make (e.g., efficiencies, energy flow).
58. Report on the various energy storage mechanisms used in the Van Geet off-grid house. (*Hint*: See http://www.nrel.gov/docs/fy04osti/32765.pdf.)

REFERENCES AND SUGGESTED READING

1. V. Smil, *Energies: An Illustrated Guide to the Biosphere and Civilization* (1999, MIT Press, Cambridge, MA).

2. V. Smil, *Energy at the Crossroads* (2005, MIT Press, Cambridge, MA).

3. G. W. Crabtree and N. S. Lewis, "Solar Energy Conversion," *Physics Today* 60(3) (2007):37.

4. T. Huld, "Compilation of Existing Safety Data on Hydrogen and Comparative Fuels," EIHP (European Integrated Hydrogen Project) report (http://www.eihp.org/public/documents/CompilationExistingSafetyData_on_H2_and_ComparativeFuels_S.pdf).

5. G. W. Crabtree, M. S. Dresselhaus, and M. V. Buchanan, "The Hydrogen Economy," *Physics Today* 57(12) (Dec. 2004):39.

6. "Special Issue: Toward a Hydrogen Economy," *Science* 305 (August 13, 2004).

7. R. Shinnar, "The Hydrogen Economy, Fuel Cells, and Electric Cars," *Technology in Science* 25 (2003):455.

8. M. Levine, J. Koomey, L. Price, and N. Martin, "Scenarios of U.S. Carbon Reductions: Potential Impacts of Energy-Efficient and Low-Carbon Technologies by 2010 and Beyond," 1997 (http://enduse.lbl.gov/projects/5lab.html).

9. A. Rousseau and P. Sharer, "Comparing Apples to Apples: Well-to-Wheel Analysis of Current ICE and FC Vehicle Technologies," Argonne National Laboratory Report, March 10, 2004 (see http://www.transportation.anl.gov/modeling_simulation/GREET/index.html for information about the modeling software).

10. A. Burnham, M. Yang, and Y. Wu, "Development and Applications of GREET 2.7: The Transportation Vehicle Cycle Model," Argonne National Laboratory Report, November 2006 (http://www.transportation.anl.gov/modeling_simulation/GREET/publications.html).

11. National Power Research Institute, "Environment Assessment of Plug-In Hybrid Electric Vehicles," Vols. 1 and 2 (http://my.epri.com/portal/server.pt?open=512&objID=243&PageID=223132&cached=true&mode=2).

12. The Google.org RechargeIT.org project (http://www.google.org/recharge/index.html).

13. Tennessee Valley Authority Reservoir Operation Study, Chapter 4.23 (http://www.tva.gov/environment/reports/ros_eis/index.htm).

14. Energy Information Administration (http://www.eia.doe.gov/cneaf/electricity/epm/epm_ex_bkis.html).

15. Bent Sorensen, *Renewable Energy: Its Physics, Engineering, Environmental Impacts, Economics and Planning* (2000, Academic Press, New York).

16. J. W. Tester, E. M. Drake, M. J. Driscoll, M. W. Golay, and W. A. Peters, *Sustainable Energy: Choosing Among Options* (2005, MIT Press, Cambridge, MA).

17. The Electricity Storage Association (http://www.electricitystorage.org/).

18. D. Hafemeister, *Physics of Societal Issues* (2007, Springer Science, New York).

19. R. M. Dell and D. A. J. Rand, *Clean Energy* (2004, Royal Society of Chemistry, London).

20. Woodbank Communications Ltd. in the UK has a very complete Web site called Electropaedia, on battery chemistry (http://www.mpoweruk.com/index.htm).

21. T. W. Murphy, "Home Photovoltaic Systems for Physicists," *Physics Today* 61(7) (Dec. 2008):42.

22. Department of Energy, *The Smart Grid: An Introduction*, 2008 (http://www.oe.energy.gov/SmartGridIntroduction.htm).

Notes

[1] This equation holds true for vehicles of most common shapes traveling at reasonable speeds in air. Other drag equations apply to objects traveling at very different speeds in different media.

[2] Recall that $W = \int_{x_i}^{x_f} \vec{F} \cdot d\vec{s}$, so $P = dW/dt = \int_{x_i}^{x_f} \vec{F} \cdot d\vec{s}/dt$. For a constant force and using $d\vec{s}/dt = \vec{v}$, we have $P = \vec{F} \cdot \vec{v}$.

[3] Steam reforming currently supplies 99% of the world's commercially available hydrogen.

[4] Higher efficiencies for electrolysis (up to 90%) are sometimes found in reports touting the use of hydrogen as a fuel; these figures often reflect the total chemical energy in both gases (oxygen and hydrogen) rather than the thermal energy that can be obtained by burning the hydrogen alone. At present, however, this additional energy is not considered to be in a usable form.

[5] Note that there are several safety issues such as gas emissions and heating during battery charging, which must be carefully monitored.

CHAPTER 7

Climate and Climate Change

Climate is a complex subject involving atmospheric chemistry, ocean circulation, and changes in the earth–sun system dynamics, and we cannot possibly cover all its subtleties here. (The reader is referred to more comprehensive sources for an in-depth treatment: See [1] and [2] for basic climate physics and [3], [4], and [5] for information about the more recent human-made changes to climate.) But to understand the longer-term environmental consequences of our energy choices, it is critical to study the fundamental concepts behind climate.

In this chapter we review the basic physical processes that affect the earth's climate. Weather is the day-to-day variation in temperature, humidity, and wind patterns in a given location. Climate consists of longer-term characteristics of the seasonal averages of these variables. As will be discussed in this chapter, changes in the earth's climate over the past few million years are well documented, and the natural mechanisms that cause these changes have been identified. From ice cores and ocean sediment data, we have especially detailed documentation of climate changes occurring over the past 650,000 years. In the past 250 years, however, through changes in land usage and emission of various industrial pollutants humans have begun to change some of the natural parameters that influence climate. In this chapter we will examine natural climate variations and then survey our current understanding of human-induced changes.

7.1 ■ The Natural Origins of Climate Variation

You may recall from Chapter 5 that the solar constant, S, is the annual average rate of energy reaching a square meter area at the top of the earth's atmosphere and is currently about 1,366 W/m^2. If this were the only influence on climate, the earth's surface temperature could easily be calculated as heat transfer between two blackbodies, as shown in Example 7.1. Although the earth and sun are not perfect blackbodies, their emission and absorption spectra are closely approximated by the Stephan–Boltzmann equation (Equation (2.18) from Chapter 2).

EXAMPLE 7.1

First Calculation of the Earth's Surface Temperature

Calculate the surface temperature of the earth assuming that there is no atmosphere and that the earth and sun are blackbodies.

As shown in Example 2.7, the energy reaching the earth per second from the sun can be calculated from the blackbody radiation given off by the sun and is about $S = 1{,}366\ W/m^2$. The surface of the earth is $4\pi r^2$, but only a circular cross-section of πr^2 faces the sun. If all the energy were absorbed, the incoming power would be $P_{in} = 1{,}366\ W/m^2 \times \pi r^2$ where r is the radius of the earth. In fact, the earth's albedo is about $a = 0.3$, which means the earth reflects about 30% of the incoming light. The actual incoming solar energy absorbed by the earth per second is then

$$(7.1) \quad P_{in} = (1 - a)S\pi r^2, \text{ where } a = 0.3.$$

The earth is also cooling by blackbody radiation given off into space, and the amount calculated from the Stephan–Boltzmann equation is temperature dependent. The earth will continue heating from incoming solar radiation until it reaches a temperature at which the outgoing blackbody radiation equals the incoming solar radiation. Only the side facing the sun absorbs energy, but the entire surface area of $4\pi r^2$ emits blackbody radiation (we are assuming that the rapid rotation of the earth distributes the average warming of the sun roughly equally over the entire surface). The energy leaving the earth is

$$(7.2) \quad P_{out}(T_E) = \sigma \varepsilon A T_E^4 = 5.67 \times 10^{-8} \frac{W}{m^2K^4} \times 1 \times 4\pi r^2 \times T_E^4,$$

where T_E is the earth's surface temperature. Here we have assumed, as a first approximation, that the earth without an atmosphere is a perfect blackbody with emissivity, ε, equal to 1.

At equilibrium we can set the energy gain from the sun equal to the energy lost due to radiation, which gives an approximate temperature of the earth's surface for the case of no atmosphere. We have

$$(7.3)\quad P_{in} = (1 - a)S\pi r^2 = P_{out}(T_E) = \sigma\varepsilon 4\pi r^2 T_E^4.$$

So for the energy balance between incoming energy and outgoing energy we have

$$(7.4)\quad (1 - a)S/4 = \sigma\varepsilon T_E^4,$$

or $0.7 \times 1366\ \mathrm{W/m^2} = 5.67 \times 10^{-8}\ \frac{W}{m^2K^4} \times 1 \times 4 \times T_E^4$, which gives $T_E = 254.8\ K = -18.3°C$.

The average surface temperature of the earth is actually about 15°C, which is 33°C higher than the value calculated in Example 7.1. Clearly other factors play a role in determining the surface temperature of the earth, primarily the atmosphere, as we will discuss. In the following section we will focus primarily on surface temperature, although it is only one element of climate. Other components of climate such as average precipitation, atmospheric and oceanic circulation patterns, cloud cover, and temperature in the upper atmosphere can also be calculated using standard physics equations, but these are not included here for the sake of brevity (see references [1] and [2]).

In addition to the net solar radiant energy flux, many other phenomena also affect the earth's surface temperature, and it is useful to divide these factors into two categories: climate forcing and feedback mechanisms. Modifications to the energy flux calculated in Example 7.1 as the result of various natural and human-made effects are called climate forcing mechanisms and are measured in watts per square meter. Depending on the mechanism, these forcings may act on the entire planet or only on select regions. A change in the amount and type of solar radiation reaching the earth's surface due to changes in the ozone content of the atmosphere is an example of a climate forcing mechanism that acts regionally near the South Pole. For comparison with other forcing mechanisms discussed in this chapter, seasonal variation of the solar constant from summer to winter for the equator is about 60 W/m^2 because of the change of angle at which the sun strikes the surface of the earth. For middle northern latitudes (e.g., 60° latitude), the difference is about 400 W/m^2, and for high latitudes the difference is 500 W/m^2 between summer and winter.

Changes which occur as the result of climate forcing but which also have an effect on climate are called climate feedback mechanisms, and they can be as large as or larger than forcing mechanisms. An example of a feedback mechanism is the amount of water vapor in the atmosphere. As we will see in the next section, water vapor plays a significant role in determining the earth's surface temperature. However, the amount of water in the atmosphere is determined by surface temperature, which can change as the result of forcing mechanisms. Ice coverage is another example of a feedback mechanism; a warming climate reduces the annual average ice cover, which reduces the earth albedo (reflectivity), causing further warming. In mathematical computer models of the climate, forcing mechanisms are input parameters, whereas feedback mechanisms are parameters that automatically adjust to the present conditions as the model goes forward in time.

In the subsections that follow we discuss various natural climate forcing mechanisms and the amount of forcing each mechanism is thought to cause. Various feedback mechanisms are also presented. If the rate of solar energy reaching the earth's surface changes, we expect the average surface temperature to change. It can also be the case that changes in the amount reaching a given region on the earth's surface will result in climate change because of the different absorption properties of land compared with ocean or ice, even when the total amount of energy reaching the surface does not change. The makeup of the earth's atmosphere also changes the rate at which energy reaches and leaves the earth's surface, thus influencing temperature. In this section we start with forcing and feedback mechanisms that operate over long time spans and then consider shorter-term influences. The much larger effect of the atmosphere and changes in atmospheric chemistry are left to the next section. The data behind the climate change observations discussed in this and the following section are treated in Section 7.3 (see also [5] and [6]).

One point that is easily misunderstood in reading the literature on atmospheric forcing is the distinction between natural forcing and changes in forcing. Because generally we are interested in variations from the established climate, changes in climate forcings are discussed more often than the naturally occurring forcings that lead to the standard climate. For example, water vapor and carbon dioxide play major roles in determining the surface temperature of the earth. However, most of the literature on global warming focuses on changes, both natural and human-caused, to the baseline energy forcing of carbon dioxide and water vapor feedback.

7.1.1 Changes in the Solar Constant

In Section 4.6.1 the mechanism for the energy generated by the sun was explained as a nuclear fusion process. In this process, because of the high pressures and temperatures in the interior of the sun, hydrogen ends up fusing to form helium (Equation (4.12)) while giving off energy. Over a time scale of billions of years the sun's core is slowly changing composition from hydrogen to helium. This change in chemical makeup caused by the nuclear fusion process has gradually changed the volume and irradiance of the sun over geological times, resulting in a brighter sun. The solar constant 4.5 billion years ago (when the earth formed) was about 70% of today's value. At the time of the dinosaurs (300 million years ago) it was about 2.5% less than today. Although this is a significant long-term change, the change per year is very small, about 13×10^{-9}% per year, an increase of 0.02 W/m^2 over the past 11,000 years. Changes to this forcing mechanism are probably too small to be significant over the span of human history. Shorter-term fluctuations in solar irradiance are discussed later in this chapter.

7.1.2 Milankovitch Cycles

It has long been known that the earth has periodically experienced ice ages, extended epochs lasting roughly 100,000 years when global temperatures were up to 10°C colder than at present. The periods of warmer climate in between ice ages, called interglacial periods, are thought to last between 12,000 and 30,000 years (see Section 7.3 for an explanation of the data behind these conclusions). The current interglacial period has been under way for about 11,000 years, approximately the duration of human history, and is expected to last perhaps another 20,000 years. The transition between ice age and interglacial period is abrupt, on the order of hundreds of years, and may be much shorter [7, 8, 9]. Geological data indicate that this periodic shift between ice age and interglacial periods has been occurring for at least the past 3 million years and probably much longer. Similar evidence suggests that even further in the past the earth may have been entirely covered with ice at some times (e.g., 500 million years ago) and totally free of ice with average temperatures as much as 15°C warmer at other times (e.g., 50 million years ago) [3].

A careful analysis of the data for the past 3 million years, first performed by Milutin Milanković, indicates a periodicity in the fluctuation of ice ages that matches known variations in the earth's orbit. The three variations in the earth's orbit are periodic changes in the earth's obliquity, eccentricity, and precession. None of these changes affect the total amount of energy reaching the earth averaged over the year, but all of them affect the seasonal variations in solar radiation reaching a particular location on the earth, thus affecting the energy actually absorbed. It is thought that these changes, coupled with various feedback mechanisms (discussed later in this chapter) are responsible for the conditions needed for the nearly periodic cycle of ice ages seen over the past 3 million years.

Obliquity is the tilt of the earth's axis with respect to the plane of its orbit. During the earth's annual trip around the sun the axis of the earth remains pointed in roughly the same direction, much like a top or gyroscope. This means the northern hemisphere is tilted toward the sun for approximately 6 months and away from the sun for the other 6 months when it moves to the other side of the sun. When the North Pole is tilted toward the sun, the sun's energy strikes the northern hemisphere at an angle closer to 90°, causing summer conditions. The other half of the year, when the North Pole is tilted away from the sun, the southern hemisphere experiences summer. Currently the earth's axis is tilted by an angle of about 23.4°, resulting in a seasonal variation of 400 W/m^2 at middle northern latitudes, but this angle varies slowly between 22.1° and 24.5° over a 42,000-year period. The periodicity of the orbits of the other planets, primarily Jupiter and Saturn, effects this change because their orbits periodically bring them closer to the earth, thus exerting a stronger gravitational force on earth's equatorial bulge (the earth is not perfectly spherical). The total energy reaching the earth does not change with changes in obliquity, but climate could be affected by different absorption rates in the northern and southern hemispheres because summers will be hotter and winters cooler for larger angles of tilt. The difference in forcing between the maximum and minimum angle of obliquity is thought to be about 100 W/m^2 for northern latitudes, but averaged over the entire globe the effect is thought to be only about 6 W/m^2 because the effect tends to cancel between the northern and southern hemispheres [5].

Eccentricity is a measure of how elliptic the earth's orbit is. The general formula for the eccentricity of an ellipse is

$$(7.5)\quad e = \frac{r_a - r_p}{r_a + r_p},$$

where r_a is the distance to the sun at closest approach and r_p is the farthest distance from the sun. The earth's eccentricity varies from near zero to about 0.06 over a 96,600-year period and is currently about 0.017. The current eccentricity results in about a 3% change in the solar constant (40 W/m^2) during the year for northern latitudes as we move farther and closer to the sun. Notice that this is about one tenth the seasonal change due to orbital tilt; summer and winter are caused by the obliquity of the earth, not the eccentricity. Again it should be pointed out that the total energy received by the earth averaged over the year will be the same, regardless of the eccentricity. Variations in the eccentricity also result from gravitational interactions with other planets, mainly Jupiter, which slightly changes the shape of the earth's orbit when the two planets are on the same side of the sun together.

Precession is a wobble of the earth's axis of rotation that occurs in addition to the slower changes in obliquity. Precession is caused by the gravity of the sun and moon acting on the earth's equatorial bulge and occurs over a 21,000-year period. Because it occurs somewhat more rapidly than changes in obliquity and eccentricity, this effect was first noted by the ancient Greeks. Both precession and obliquity change the direction in which the axis of the earth points; the North Star is the pole star for only part of this 21,000-year period. As is the case for the other orbital variations, precession affects the seasonal variation of solar radiation arriving at a particular location on the earth but does not affect the total annual average energy reaching the earth.

The maximum total change in climate forcing due to changes in the earth's orbit over a 100,000-year cycle including the annual seasonal change is less than 100 W/m^2, boosting the seasonal change to 500 W/m^2 for a northern latitude of 60° during times when all these effects add together [6]. However, it should be kept in mind that these are changes experienced at a particular latitude; the total annual average energy reaching the earth does not change as a result of these orbital variations. The precise mechanism of how orbital variations that affect only seasonal variations in the distribution of insolation could result in the observed glacial periods is evidently connected to complex feedback interactions between the incoming solar radiation and atmosphere, ocean, ice, and land masses. These feedback mechanisms, which include changes in the atmospheric carbon dioxide level, will be discussed in more detail later in this chapter.

The fact that small differences in insolation between northern and southern hemispheres and a complex set of feedback mechanisms eventually result in significant and rapid changes in climate from ice age to interglacial indicate that the climate is very sensitive to small perturbations. Elevated sensitivity of a system to small changes in input parameters is a signature of chaotic behavior. The final state of a chaotic system can be radically different depending on slight variations in initial conditions, making the outcome extremely difficult to predict. It is possible that the earth's climate is a chaotic system that has two (or more) semistable states. We know that the shorter-term changes of weather are the result of chaotic behavior of atmospheric systems, which is why weather is so hard to predict more than a week ahead. It appears that climate could also be chaotic but on a much longer time scale: tens of thousands of years, compared with the weekly chaotic changes in weather. The possible chaotic nature of longer-term climate change and the fact that the switch between ice ages and interglacial periods can be very brief, sometimes decades, has caused a growing concern among climate scientists. Humans are changing the makeup of the earth's atmosphere in a significant way, and the long-term effect on climate could be small or large, rapid or slow. It is possible that small, human-caused changes will be enough to perturb the earth's climate into some other final climate state, significantly different from the present.

7.1.3 Sun Spots

In addition to the longer-term change in the solar irradiance mentioned in Section 7.1.1, shorter periodic variations in the solar constant have been measured [10]. The sun's magnetic field reverses every 22 years, resulting in an 11-year sunspot cycle. Although darker, cooler regions on the sun's surface decrease the radiant energy leaving the sun when they are present, other regions on the sun, called faculae, are more active, resulting in an increase in total solar radiation when there are large numbers of sunspots. Sunspot activity also appears to be connected to other solar phenomena such as solar flares, which produce short-term bursts of electromagnetic radiation that can reach the earth. These bursts typically originate from sunspot regions and often interfere with earth-orbiting communication satellites. The 22-year periodicity has been linked to drought cycles on the earth using tree ring data. Data for the past 1,000 years indicate that solar intensity changes by less than 1 W/m^2 during these cycles [6]. The change over the past 250 years is estimated to be 0.12 W/m^2 [5], although some scientists estimate a slightly larger effect [10].

7.1.4 Ocean and Atmospheric Circulation Effects

The sun heats the surface of the earth unequally, reaching the earth's equator at an angle of approximately 90° year-round but striking the northern and southern hemispheres at a larger angle that changes significantly during the year. This uneven heating between the equator and the poles drives ocean and atmospheric circulation patterns, which affect the global climate as a feedback mechanism. Although none of the mechanisms in this subsection constitute a forcing of the climate, they appear to be instrumental in explaining short-term climate variations and may well be part of the feedback mechanism that amplifies the small orbital variations of the earth into the extreme changes between ice age and interglacial climates.

Figure 7.1 shows the atmospheric circulation patterns. Surface winds in the northern hemisphere between the equator and 30° are called the Trade Winds, whereas those above 30° are called Westerlies. This part of the atmospheric circulation pattern made trade possible between Europe and the Americas with the use of sailing vessels. Ships could sail to the Americas at lower latitudes and return at higher latitudes, facilitating the transport of slaves from Africa to the Americas and raw materials back to Europe. The upper atmospheric patterns, known as the Polar, Ferrel, and Hadley cells, which drive the surface winds, are also shown. Notice that warm air rises from the equator because of direct heating of the sun and sinks at about 30° latitude. The location of these cells is determined by the rotation rate of the earth; Venus, which rotates much more slowly, has only a single upper atmospheric circulation cell between the equator and poles. A stationary planet would have a single cell of circulation with warm air rising at the equator of the side facing the sun, traveling at high altitude directly north and south over the poles, and sinking at the equator on the dark side. Surface air traveling in the opposite direction would complete the circuit. Rotation twists the direct northward and southward surface motion into easterly and westerly winds, respectively, as a result of the Coriolis effect. The jet stream, a quasistable stream of air circulating the globe in the upper atmosphere at about 60° latitude, is formed from the turbulence at the upper boundary between the Polar and Ferrel cells.

FIGURE 7.1

Atmospheric circulation patterns for the earth. Warm, moist air rises at the boundary between the Ferrel and Polar cells at latitude 60°; cool, dry air descends at the boundary between the Hadley and Ferrel cells at latitude 30°. The rotation of the earth twists the various cells to form surface winds going east and west, depending on latitude.

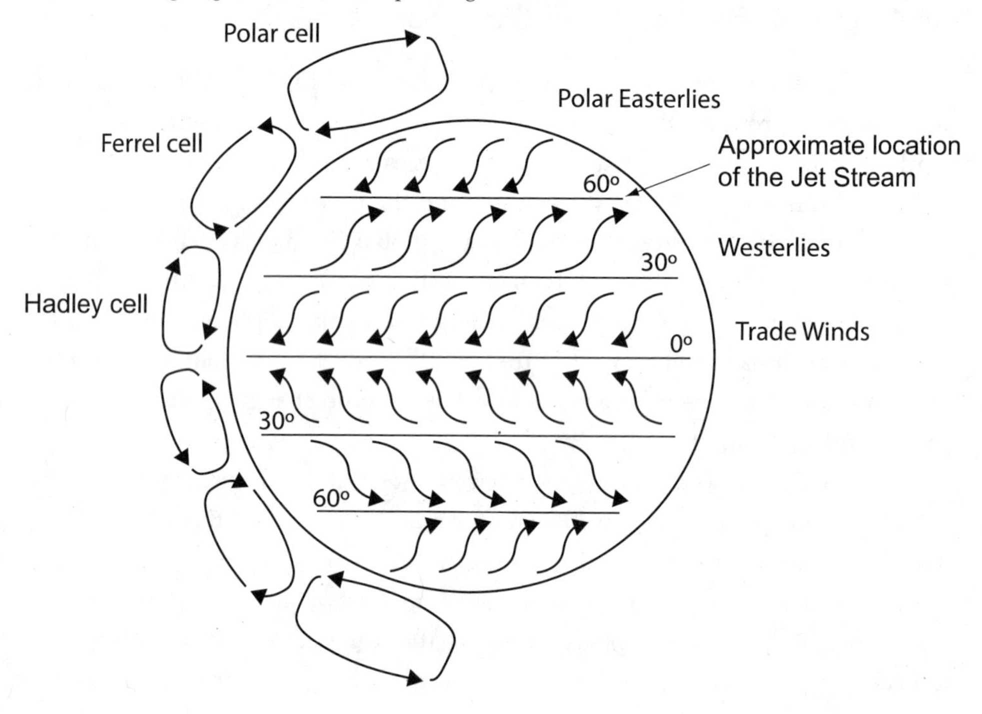

Current flow in the ocean moves warm water toward the poles at the surface of the ocean and cooler, saltier water toward the equator at a deeper level. This flow, called the global thermohaline circulation, is actually quite complex; a simplified representation is shown in Figure 7.2. As the warm water flows north it evaporates, making the remaining water cooler and saltier and thus denser. Near the poles this cooler, denser water sinks and returns to the equator along the bottom of the ocean. The entire trip takes about 1,000 years, and the rotation of the earth, surface wind patterns, and the location and shape of the continents strongly affect the circulation pattern. One important consequence of this circulation is that heat released in northern part of the North Atlantic Ocean, where the ocean is 5°C warmer than it would be otherwise, keeps the climate of northern Europe significantly warmer. Without this circulation pattern and the large heat capacity of water, which inhibits rapid temperature changes, northern

FIGURE 7.2

Major ocean circulation patterns [11].

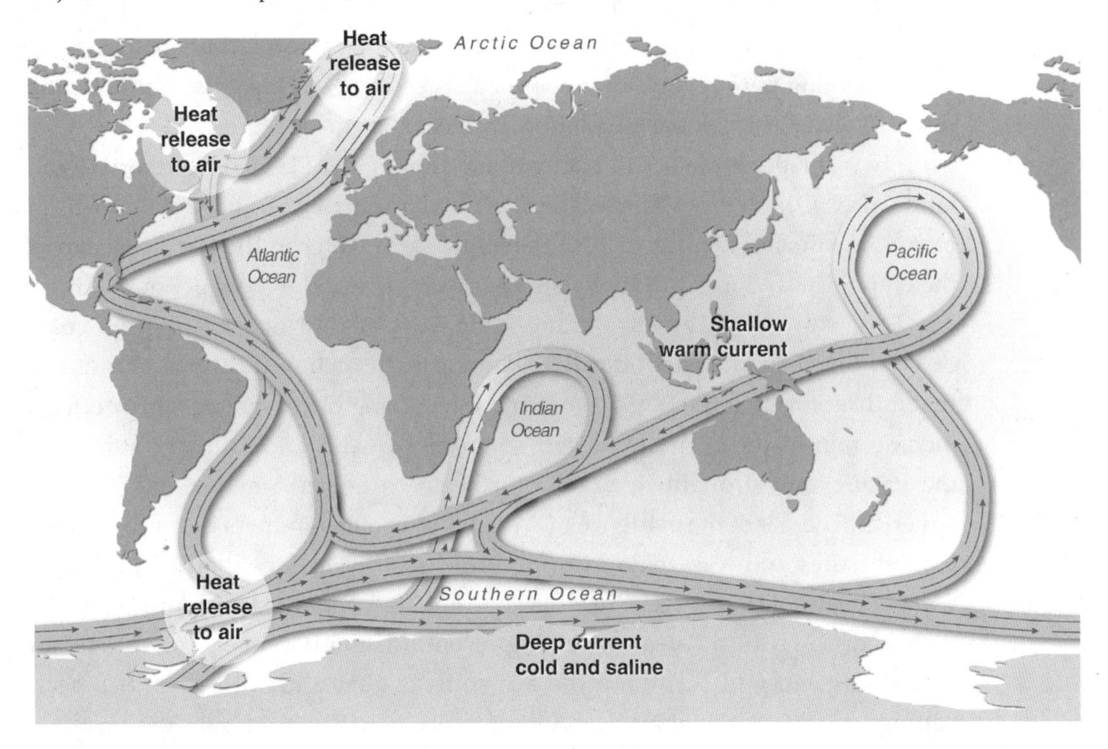

Europe, Great Britain in particular, would have a climate more similar to that of northern Canada,which is at the same latitude.

Climate models show that the world climate would be significantly different if ocean circulation were to change. Without the redistribution of solar energy provided by the thermohaline circulation, the north and south poles would, on average, be much colder. Changes in the circulation pattern affect the duration of ice coverage, thus changing the albedo of the earth, which is an important feedback mechanism. Some scientists believe that the thermohaline circulation plays a major role in the change between an ice age climate and interglacial periods, abruptly changing from one stable circulation pattern to another [7, 8, 9]. The Younger Dryas period was a short episode of global cooling that occurred about 12,000 years ago as the earth was coming out of the last ice age, right before current interglacial epoch. At this time the thermohaline circulation was beginning to carry warm water to the poles, melting the large glaciers covering the northern and southern polar regions. Data indicate that the circulation can be temporarily interrupted or slowed when larger amounts of freshwater flow into the North Atlantic [7, 8]. During the Younger Dryas a slowdown of the circulation resulted in a temporary, 1,300-year reversal of the temperature increase associated with the end of the last ice age. A similar but smaller-magnitude cooling period occurring about 8,200 years ago is also thought to be the result of changes in global circulation patterns.

Two smaller-scale ocean circulation patterns, the El Niño–Southern Oscillation (ENSO) and the North Atlantic Oscillation (NAO), have shorter-term effects on climate. During an ENSO event, warm water in the Pacific Ocean at the equator near Tahiti gradually spreads east toward the coast of South America. This warm surface water reduces the flow of cool water along the coast of Peru and Ecuador, periodically reducing fish catch, which was one of the first indications that the ENSO pattern is a recurring phenomenon. Other effects include rain in the Peruvian desert, changes in wind and rainfall patterns in the Pacific, and drought in Brazil, northeastern South America, and Central America. There is some evidence of wetter and cooler summers in southwestern United States and a lessening of hurricane activity in the Atlantic during an ENSO occurrence. The event itself lasts from a few months to a couple of years and occurs on a 2- to 7-year cycle. At present it is not possible to predict when an ENSO event will occur, but there is some evidence that the cycle has been getting shorter as global average temperature has increased over the past 200 years.

The NAO is a fluctuation between the nearly permanent low-pressure atmospheric system that exists over Iceland and a permanent high-pressure zone in the Atlantic Ocean over the Azores, off the coast of Portugal. When the low pressure over Iceland is lower than normal and the high in the Atlantic is higher than normal (a positive NAO), stronger storms are seen in the Atlantic, winters in Europe and North America are milder and wetter, and winters in northern Canada and Greenland are colder and drier. As is the case for the ENSO, NAO events last a few years.

Other ocean–atmosphere oscillations are known to exist; for example, the Arctic Oscillation and Pacific Decadal Oscillation occur with periods of 10 to 40 years. When we study climate change it is important to be able to identify these temporary fluctuations and to distinguish them from longer-term changes. For this reason a great deal of effort has been spent in trying to understand these short-term fluctuations in climate. It is especially important to understand factors affecting fluctuations in arctic climate because, as will be explained later in this chapter, the arctic is expected to react more rapidly than lower latitudes to global changes in climate and thus is a more sensitive indicator of climate change.

7.1.5 Aerosols

An aerosol is a collection of particles or droplets suspended in the air that are so small that they remain airborne for long periods of time. They may be solid or liquid and range in size from 1.0×10^{-5} m to 1.0×10^{-7} m in diameter. When they occur in the troposphere, the lowest layers of the atmosphere, rain eventually removes them from circulation. If they reach the layer above the troposphere, called the stratosphere, they may remain in circulation for much longer periods of time, up to a decade, because rain clouds do not form at that height in the atmosphere. Natural sources of aerosols include volcanoes, dust blown into the air from deserts, forest and grass fires, and droplets from sea spray. Through fossil fuel burning and land use changes (e.g., cutting or burning forests), humans contribute about 10% of the aerosols found in the atmosphere today. These aerosols can travel great distances. For example, satellite pictures from space and atmospheric sampling from airplanes and balloons show that prevailing winds carry significant amounts of dust across the Atlantic to South America from the Sahara desert in Africa.

Aerosols generally reflect solar energy (although some kinds of aerosols such as soot particles absorb energy) and thus have a negative effect on surface temperature. Changes in the earth's albedo over the past few hundred years due to human-caused changes in the amount of atmospheric aerosols are estimated to cause a negative change in forcing of about -0.5 W/m^2 [5]. Aerosols also affect cloud formation because tiny particles are needed as seeds for condensation of water into clouds. This is the least understood aspect of climate modeling, but changes in cloud condensation caused by increases in aerosols are thought to provide an additional negative feedback, with a forcing change of about -0.7 W/m^2 over the past 250 years [5]. This is an additional forcing added to the effect of normal cloud formation, which will be discussed later in this chapter.

Volcanoes add large amounts of particulate matter and greenhouse gases to the atmosphere over short periods of time. For large volcanoes the aerosols may reach the stratosphere, where they will stay in circulation for several years. Because of the aerosols released during a large eruption, more solar radiation is reflected, leading to a temporary cooling of the climate. These effects can clearly be seen in climate records over the past 100 years, and in fact the study of these events has led to greater confidence in computer models of the atmosphere that can accurately predict the result of the sudden addition of large amounts of particulates to the atmosphere. Based on global measurements, scientists estimate that the Mount Pinatubo eruption in 1991 resulted in a global decrease in average temperature of about half a degree the year after the eruption. Volcanism, coupled with variations caused by the sunspot activity mentioned earlier, are thought to be the reason for the "Little Ice Age" in Europe from approximately 1500 to 1850 (exact dates are not well established), when temperatures were below normal by about 1°C. Although the temperature variation was small on average, the lower temperatures during this period caused many social problems, including crop failures and famine.

Meteors colliding with the earth have effects similar to those of volcanoes, releasing large quantities of aerosols into the upper atmosphere, resulting in a temporary cooling of the climate for up to 10 years. Several major mass extinctions of flora and fauna found in the fossil record are now thought to be a result of meteor impacts that resulted in global cooling trends. The best-documented event was the Cretaceous extinction of many land and sea species, including dinosaurs, and was the result of a meteor striking the earth just off the coast of the Yucatan peninsula about 65 million years ago. Calculations of the environmental effects of global nuclear war show that a similar cooling effect, lasting about a decade, would result from soot launched into the upper atmosphere by nuclear explosions [12].

7.1.6 Albedo Changes

Albedo changes to the earth's surface can be both feedback and forcing mechanisms. During the winter and during glacial periods when the earth is partially covered with ice and snow, the reflection of solar radiation is greater, causing further cooling, the condition needed for more ice and snow to form. This is the classic definition of a positive feedback loop; once started, the mechanism pushes the system further in the same direction. The ice and snow albedo change feedback mechanism also works in reverse: Melting of ice and snow exposes more land, which absorbs more energy, leading to higher temperatures and more melted ice. This is perhaps the clearest example of the difference between feedback and forcing mechanisms. Positive feedback mechanisms increase the effect of, but do not act independently from, forcing mechanisms. The snow and ice feedback mechanism is thought to play a major role in the switch from an interglacial to glacial climate. Orbital variations cause cooler winters in the northern hemisphere, which allows ice to remain in regions where it normally melts, leading to an increase in albedo and further cooling.

Cloud cover is a main contributor to the net albedo of the earth, with about 20% of the incoming radiation from the sun reflected by clouds. Reflection from the atmosphere (6%) and the earth's surface (4%) make up the other 10% of the average albedo of 30% ($a = 0.3$). Scientists initially thought that a rise in the surface temperature of the earth might result in more ocean evaporation and this would act as a negative feedback, increasing the cloud coverage and thus returning surface temperatures to their original value. However, as we will see later in this chapter, the water vapor in clouds also increases trapping of infrared radiation from the earth, which makes the outcome of increased cloud coverage more complicated. During the global grounding of airplanes after the terrorist attack on the World Trade Center in 2001, global daytime temperatures rose slightly, but nighttime temperatures dropped. This was the result of the disappearance of jet contrails for 3 days, which allowed more sun to reach the earth's surface during the day but also allowed more energy to escape at night. The type of cloud also plays a role in whether a given cloud has a net warming or cooling effect, and it is not clear what kind of cloud will form if evaporation increases because of higher surface temperatures or increased aerosols in the atmosphere. The current scientific consensus is that the net effect of increased evaporation has a neutral overall impact on surface temperature, but a concerted effort is under way to better understand these effects.

Land use changes such as forest clearing and planting of irrigated crops in locations where there formerly was only desert are additional examples of albedo changes that constitute climate forcing mechanisms. In these cases changes occur that act as inputs rather than reactions to the climate system. On average, changes in land use have acted as negative forcing mechanisms. It is estimated that land use change has contributed a $-0.2\ W/m^2$ forcing to the climate over the past 250 years [5]. On the other hand, black carbon soot deposited on snow and ice through human industrial activities acts as a positive forcing by reducing the albedo of the snow or ice. This effect is also thought to be small, about $+0.1\ W/m^2$ [5].

7.1.7 Geomagnetism

The earth's magnetic field reverses on an irregular period of just under a million years, possibly resulting in a larger influx of charged particles from the sun and cosmic rays into the earth's atmosphere. The reversals are well documented in the magnetic fields found in rock that has cooled after being exuded during sea floor spreading. When the rocks cool, the iron in them becomes magnetized in the direction of the earth's field. The farther from the location where the molten rock is being deposited, the longer ago the rock was formed, and a clear record of magnetic reversals can be seen by examining rock farther and farther from the spreading zone. At present these magnetic reversals are not thought to play any role in climate. However, there is a slight possibility that during these reversals more charged particles from the sun and cosmic rays penetrate the earth's atmosphere, heating the atmosphere and oceans. The exact amount of forcing is still under debate, but obviously this would occur only during a magnetic reversal and there are no indications of climate change during these events.

7.2 Atmospheric Forcing

The largest natural climate forcings and feedbacks are connected to the makeup of the atmosphere. Because of the complex interaction of solar radiation with various chemical compounds found in the atmosphere, the earth's surface is some 33°C warmer than would be expected if the atmosphere were perfectly transparent. Without this increase in surface temperature, the earth would be uninhabitable. Similar mechanisms make the surface of Venus an uninhabitable 400°C warmer than it would be without its atmosphere of mostly carbon dioxide. In this section we discuss the physics of these mechanisms and provide a few sample calculations to show how they are responsible for the climate we experience today. The connection between the chemical makeup of the atmosphere and the ocean is also treated.

7.2.1 The Atmosphere

As discussed in earlier chapters, the sun emits blackbody radiation, which peaks in the visible part of the spectrum. The atmosphere is transparent in the visible part of the spectrum, allowing most of this solar radiation to reach the earth's surface. The earth, also acting as a blackbody, re-radiates this energy but at a much lower temperature, so that the peak in the spectrum is in the infrared range. Because of various gases present, the atmosphere is not as transparent in the infrared range, so some of this energy is absorbed and heats the earth's atmosphere. It is this effect we want to investigate in this section. Example 7.2 gives a simple calculation of the effect of a hot atmosphere on the surface temperature of the earth.

EXAMPLE 7.2

Second Calculation of the Earth's Surface Temperature

Calculate the surface temperature of the earth assuming an atmosphere of a single layer, all at the same temperature. A schematic is shown in Figure 7.3.

The atmospheric layer allows visible light from the sun to pass through unimpeded but captures infrared emitted from the earth's surface. This causes the atmospheric layer to heat up until it's (approximately) blackbody emission, mostly in the infrared, is equal to the incoming infrared radiation from the earth. The earth's surface area, A, at temperature T_E is in contact with the atmosphere at temperature T_a. The atmosphere emits infrared in two directions, so the energy balance between emission and absorption of radiation in the atmosphere and the earth is

$$(7.6)\quad 2\sigma\varepsilon_a A T_a^4 = \sigma\varepsilon_a A T_E^4,$$

where the emissivity of the atmosphere is used on both sides of the equation because this equation is the energy balance for the atmosphere. This ignores convection, ocean evaporation, and the fact that the atmosphere has a temperature profile that varies with height, but it is a useful first approximation. Simplifying gives

$$(7.7)\quad T_a = 2^{-1/4} T_E.$$

The energy balance equation for the earth's surface, Equation (7.3), now has an additional term representing the extra infrared energy received from the radiating atmospheric layer. For an earth with emissivity of 1, we have

$$(7.8)\quad (1 - a)S/4 + \sigma\varepsilon_a T_a^4 = \sigma T_E^4.$$

Substituting $T_a = 2^{-1/4}T_E$ into this equation gives

$$(7.9) \quad T_E^4 = \frac{(1-a)S}{4\sigma(1-\varepsilon_a/2)}.$$

The emissivity of the atmosphere, ε_a, indicates how much energy it emits at a given temperature and depends on its exact chemical makeup. A value of $\varepsilon_a = 1$ gives a surface temperature for the earth of 303 K, which is higher than the current measured average of 288 K. A value of $\varepsilon_a = 0.75$ gives a surface temperature of 287 K, very close to the global average.

As noted in Example 7.2, the emissivity of the atmosphere is a key factor in determining the earth's surface temperature. Emissivity is defined as the ratio of the emitted radiation to that of a perfect blackbody at the same temperature. For objects in thermal equilibrium with their surroundings using blackbody radiation as the only means of heat transfer, the emissivity is equal to the absorptivity, which is the ratio of energy absorbed to that of a perfect blackbody at the same temperature. An object that did not have this property could possibly absorb energy more efficiently than it emitted at a given temperature. This would break

FIGURE 7.3

A simple radiative equilibrium model for the earth's surface and a single atmospheric layer of uniform temperature. IR = infrared.

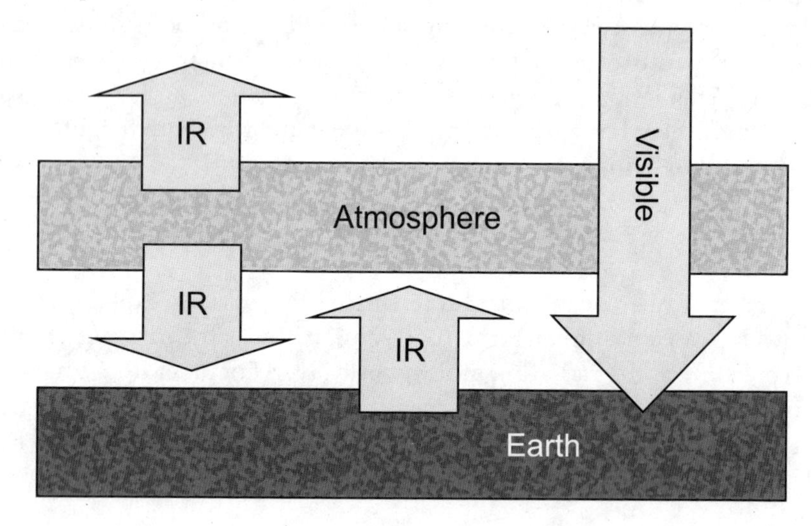

the second law of thermodynamics because the object could spontaneously heat itself at the expense of the surroundings without any input work. The absorption of radiation by the atmosphere is thus an important factor in determining the earth's surface temperature because it equals the emissivity. In the preceding two calculations we assumed the earth has an emissivity of 1, but in fact neither the earth nor the atmosphere is a perfect blackbody. The atmosphere's emissivity is further complicated because it is frequency dependent. Some wavelengths are completely transmitted, whereas others are partially or completely absorbed because of the precise chemical makeup of the atmosphere. Figure 7.4 shows the solar spectrum at the top of the earth's atmosphere and at the earth's surface. The visible spectrum ranges from 380 nm to 750 nm; higher wavelengths are in the infrared range, lower wavelengths are ultraviolet.

In Chapter 2 we found, using Equation (2.20), that for the sun with a surface temperature of 5,780 K, the peak in the solar spectrum is 520 nm, squarely in the middle of the visible spectrum. A similar calculation for the surface of the earth, at 288 K, gives a peak wavelength of 10,000 nm, which is in the far

FIGURE 7.4

The solar spectrum at the top of the atmosphere (upper curve) and at the earth's surface (lower curve). Parts of the spectrum that have been absorbed (blocked) by a specific atmospheric gas are marked by the respective gas (water vapor, carbon dioxide, oxygen, and ozone) [13].

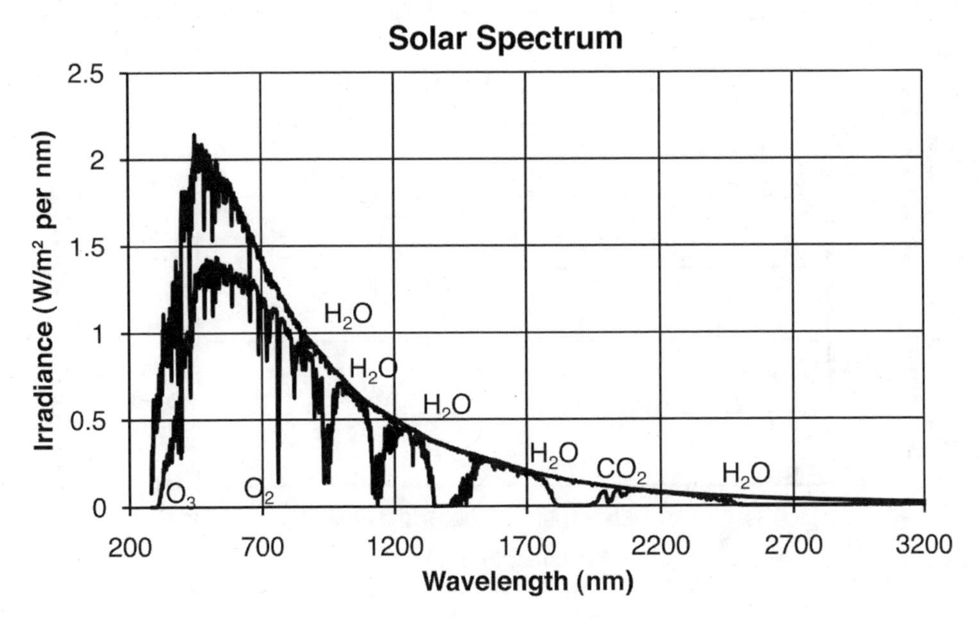

infrared. Figure 7.5 compares the spectrum of the sun to the radiation given off in the infrared by the earth. It is clear from the figure that most of the radiation coming from the sun has wavelengths below 5,000 nm, whereas much of the radiation leaving the earth has wavelengths above 5,000 nm. The trapping of radiative energy at wavelengths above 5,000 nm by atmospheric absorption is responsible for the warming of the atmosphere and the earth's surface.

Wavelengths that are absorbed by the atmosphere show up as dips in the solar spectrum as measured at the earth's surface. These dips are clearly evident in Figure 7.4. Nitrogen, constituting 78% of the atmosphere, does not block reradiated infrared energy from the earth. Oxygen, which makes up 21% of the atmosphere, blocks very little infrared, although oxygen together with ozone blocks short-wavelength ultraviolet (UV) radiation from reaching the earth. As mentioned in Chapter 1, the UV blocking property of O_2 and O_3 is important in protecting life on Earth from these harmful wavelengths. Radiation that is not absorbed is transmitted through the atmosphere. A more detailed picture of atmospheric transmission for other gases is shown in Figure 7.6 for the infrared range of wavelengths.

FIGURE 7.5

The emission spectra of the sun on the left (the same data as in Figure 7.4) and the approximate emission spectra of the earth (assumed to be a blackbody at 288 K) on the right. Notice that the scale for the radiance of the sun is 100,000 times larger than that of the earth.

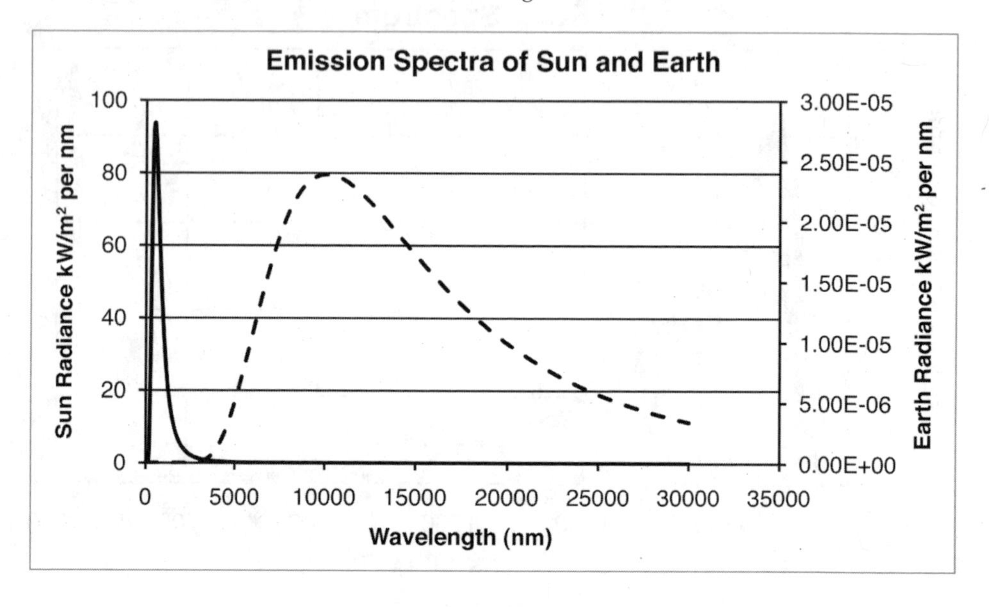

FIGURE 7.6

Transmission spectra of various gases in the atmosphere at room temperature for infrared wavelengths [14, 15]. A transmission rate of 1 means 100% of the radiation at that frequency passes through the gas. Oxygen and nitrogen are not included because they are largely transparent for this range of wavelengths. In this figure equal concentrations of each gas are compared at the same pressure and temperature; however, it should be kept in mind that the atmosphere has different concentrations of different gases.

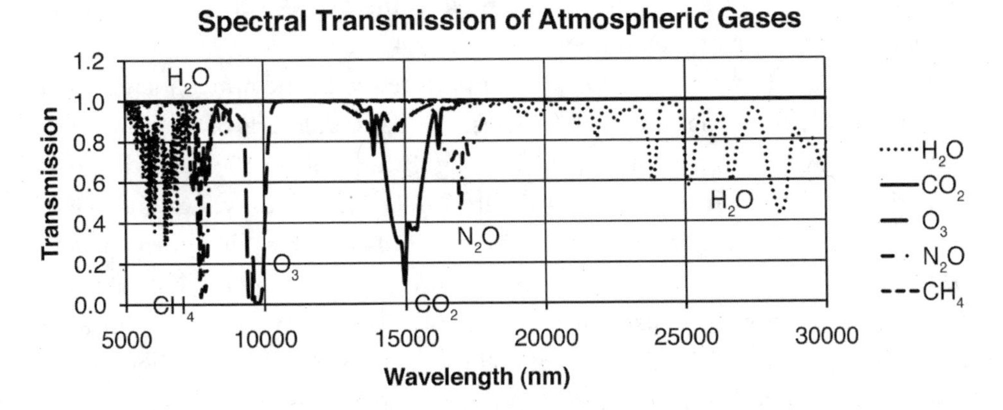

As explained in Chapter 2, the electrons associated with individual atoms and molecules have quantized energy levels. This means that only photons with precisely the right energy can be absorbed by changing an electron from one energy level to another. This is the basis for discrete absorption and emission spectra specific to a given element or molecule and is used by scientists to identify the particular elements or molecules in a sample of gas. In addition to the electron energy levels, complex molecules can have vibrational and rotational energies associated with the relative motion of their components. These energies are also quantized, which means that the molecule can change to a different frequency of vibration or rotation only by the absorption of a discrete quantity of energy.

The energy of a photon is given by $E = hf$, where h is Planck's constant and $f = c/\lambda$ is the photon frequency. For isolated molecules at low temperatures, the particular quantized energies of a given molecule can absorb only photons of certain wavelengths (there is actually a small range of wavelengths because of Heisenberg's uncertainty principle, which does not concern us here). For groups of molecules at temperatures above absolute zero this strict condition for absorption is relaxed a little because of the thermal motion of the molecules and collisions with other molecules. A molecule in motion already has a little extra kinetic energy, so when it meets a photon that doesn't quite have the right energy it may still be able to absorb that photon by using some of its own kinetic energy to make up the difference. A molecule may also absorb a photon with a little too much energy if the extra energy can be transferred into kinetic energy of the molecule. Collisions of molecules at higher pressure also have a similar effect on the emission and absorption of photons, allowing a wider range of wavelengths to be absorbed. For these reasons the discrete absorption spectra spread into broad bands, as shown in Figures 7.4 and 7.6.

Although it makes up a very small percentage of the atmosphere, water vapor clearly plays a significant role in absorbing several parts of the solar spectrum, as shown in Figure 7.4. It also has a significant role in absorbing infrared radiation from the earth, as seen in Figure 7.6. In fact, water vapor and water droplets in clouds together are thought to be responsible for approximately 65% of the 33° increase in surface temperature of the earth [16]. Other gases, now called greenhouse gases, which also play a role in absorbing infrared radiation, are CO_2, N_2O, CH_4, O_3, and the various chlorofluorocarbons mentioned in Chapter 1. Carbon dioxide, much of which occurs naturally in the atmosphere, contributes as much as 25% of the increase in the earth's surface temperature above the no-atmosphere prediction of Example 7.1. As seen in Figure 7.6, the main absorption band of carbon dioxide occurs very near the peak of the earth's blackbody spectra (10,000 nm), so it is not surprising that it is a key greenhouse gas. Other greenhouse gases contribute the remaining measured warming effect of 33°C. As mentioned previously, without this 33°C warming the earth would be too cold for most life as we know it. The concerns of climate scientists today are related to changes in the natural forcings and feedbacks which produce this surface temperature.

An important difference between water vapor and the other greenhouse gases such as carbon dioxide is that the amount of water vapor in the atmosphere adjusts very quickly according to the temperature. Any given water molecule spends only a few weeks in the atmosphere before falling as rain, so the balance between evaporation and rainfall, the hydrological cycle, quickly regulates the amount of water vapor in the air. Changes in global average temperature can change the equilibrium amount of water vapor in the atmosphere (the relative humidity), and the changes are rapid, but the amount remains steady for a given temperature. This makes water a positive feedback mechanism; a warmer climate means more water in the atmosphere, which increases the amount of global warming. In contrast, a carbon dioxide molecule, once emitted, stays in the atmosphere for as long as 100 years, where it can continue to have an effect on climate. This means that the addition of carbon dioxide to the atmosphere has a long-term effect as a forcing mechanism. Other greenhouse gases have similar behavior. As discussed in Chapter 1, chlorofluorocarbons remain in the upper atmosphere for dozens of years. It is for this reason that water vapor is treated as a feedback mechanism in climate modeling, but carbon dioxide and other gases are considered as forcing mechanisms.

The amount of carbon dioxide found in the atmosphere is determined by various external factors, including natural emissions (e.g., volcanoes) and human fossil fuel burning, whereas any natural or human-caused changes to the amount of water vapor quickly readjust to the equilibrium amount. This is also the reason most global warming literature focuses on changes in the amount of carbon dioxide and other greenhouse gases in the atmosphere and the forcings associated with these changes, even though water vapor plays the more significant role in earth surface temperature.

Table 7.1 shows the major components of the atmosphere, ranked in order of most abundant. Some of the gases listed in the table, such as argon, neon, helium, and krypton, absorb very little solar energy at longer wavelengths and are not considered greenhouse gases. Other gases result in a forcing of the climate, depending on how much of the gas is present, which in turn depends on temperature and pressure. The radiative efficiency is a measure of the forcing effect of that gas per part per billion in the atmosphere and is measured in $W/m^2/ppb$. The actual concentration in the atmosphere of a given gas is multiplied by the radiative efficiency to find the radiative forcing for that gas, and these figures are also shown in Table 7.1. From the table it is clear that chlorofluorocarbons, although they have a high radiative efficiency, are not a significant greenhouse gas at present because they do not constitute a significant portion of the atmosphere. Although radiative efficiencies of methane and nitrous oxide are much larger than carbon dioxide, CO_2 has a much larger radiative forcing because it is much more abundant in the atmosphere. Anthropogenic forcings listed in the table indicate the amount of forcing as the result of changes in the normal background levels of these gases caused by human activity.

In the following we will focus mainly on carbon dioxide and the fossil fuel source of this greenhouse gas. However, from Table 7.1 we can see that other gases that are also the result of human activity have a much stronger potential for climate forcing. For example, the radiative efficiency of methane is 10 times

TABLE 7.1

Major components of the earth's atmosphere by percentage of molecules present and their radiative efficiencies [5, 17, 18]. Figures listed as anthropogenic indicate forcings caused by changes in the average background amounts of these substances caused by human activities.

Gas	Percentage of the Earth's Atmosphere	Radiative Efficiency ($W/m^2/ppb$)	Radiative Forcing (W/m^2)
Nitrogen (N_2)	78.1		~0
Oxygen (O_2)	20.9		~0
Argon (Ar)	0.93		~0
Water vapor	0.48		~56
Carbon dioxide (CO_2)	0.035	1.4×10^{-5}	~23 (1.66 anthropogenic)
Liquid and solid water	0.002		A feedback mechanism (see text)
Neon (Ne)	0.0018		~0
Helium (He)	0.00052		~0
Methane (CH_4)	0.00017	3.7×10^{-4}	0.55 (anthropogenic)
Krypton (Kr)	0.00010		~0
Nitrous oxide (N_2O)	0.00003	3.0×10^{-3}	0.15 (anthropogenic)
Ozone (O_3)	0.000007		0.14 (anthropogenic)
Chlorofluorocarbons (CFCs)	0.00000014	0.01–1.4	0.30 (anthropogenic)
Aerosols	0.00000002		–0.5 (anthropogenic)

that of carbon dioxide, and chlorofluorocarbons have as much as 10,000 times more impact per molecule. As we saw in Chapter 1, the use of many chlorofluorocarbons is being phased out, but methane emission remains a concern. There has been a 150% increase in atmospheric methane since the emergence of human civilization. Anaerobic bacteria in rice fields, the digestive tracts of domesticated cattle, handling of fossil fuels, sewage treatment, landfills, and termite colonies appear to be the major sources of this recent increase in atmospheric methane levels.

7.2.2 The Carbon Cycle

The hydrological cycle describes the movement of water from oceans into the atmosphere by evaporation and transpiration of plants and then back to the ocean as rain, runoff, and ground water. The rate at which this cycle operates ensures that, for a given surface temperature, the ocean–atmosphere system remains approximately in equilibrium as far as water vapor is concerned. Similarly, the carbon cycle describes the movement of carbon compounds into the atmosphere and back to earth. However, this cycle takes much longer to complete. Although water is the most important greenhouse gas, the amount of water in the atmosphere does not change significantly over the time scale of human civilization. Conversely, the concentration of carbon dioxide, the second most important greenhouse gas, is rising rapidly because of changes in the carbon flux and the slow rate of the carbon cycle. A simplified diagram of the earth carbon cycle is shown in Figure 7.7.

Plants absorb carbon dioxide from the atmosphere and use energy from the sun to drive the photosynthetic process. Oxygen is the byproduct of this process, and in fact the atmosphere today is 20% oxygen as a result of plant photosynthesis; before about 3 billion years ago the atmosphere had no oxygen. The uptake of carbon dioxide by plant photosynthesis varies according to the season of the year, and although this flux is only about 1.4 Tkg/yr, the annual fluctuation can be easily measured (see Problem 7.22).

FIGURE 7.7

The carbon cycle, 2000. Reservoirs of carbon in terakilograms are shown as boxes, annual fluxes in terakilograms per year are shown as arrows. Oval sources are human generated [5, 18, 19].

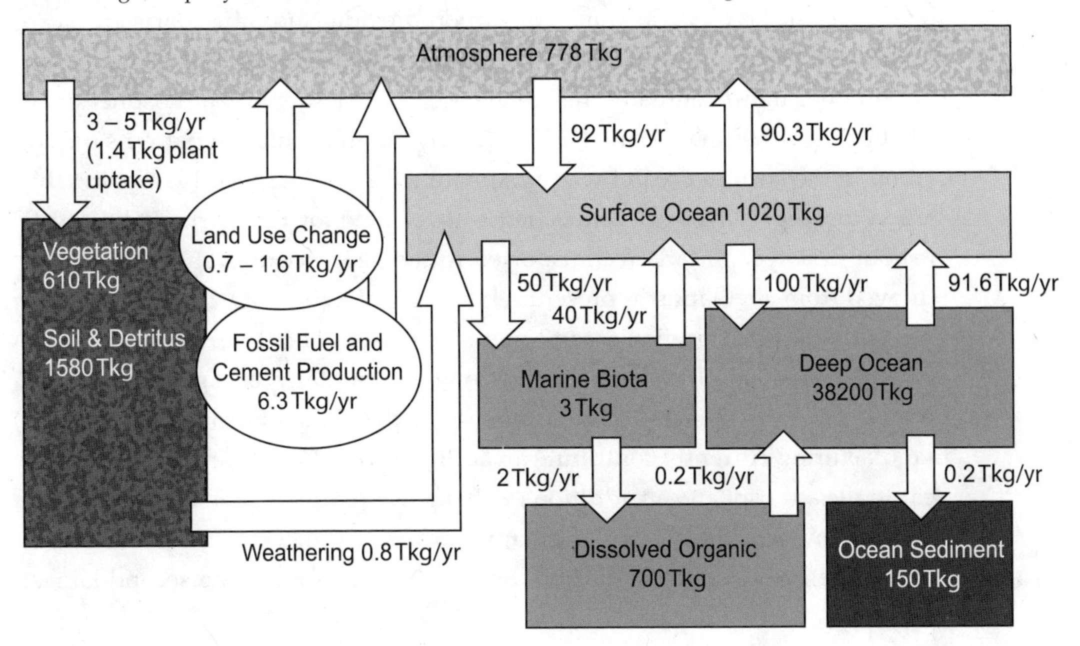

The quantity of carbon dioxide taken up by plant metabolism can be determined with some degree of accuracy, but additional carbon uptake or emissions caused by changes in land use or other factors are difficult to estimate. These carbon transfers vary by region; for example, a terrestrial carbon absorption of 2 Tkg to 4 Tkg per year occurs in the northern hemisphere beyond the known 1.4 Tkg/yr reduction caused by normal plant metabolism. The exact mechanism is not well understood, but the current consensus is that better land management, including reforestation, regrowth in farm lands, reduction of forest fires, and better planting methods together with longer growing seasons and changing rainfall patterns may account for most of this carbon sink [18, 19].

As can be seen in Figure 7.7, the ocean stores the bulk of the carbon in the carbon cycle. The figure also shows that the flow of carbon between the atmosphere and the ocean currently provides a net 2 Tkg per year absorption by the ocean surface. This is equivalent to a reduction by about 30% of the human contribution to atmospheric CO_2 caused by fossil fuel burning. It is not yet clear how long the ocean will continue to absorb this 2 Tkg per year or what climate conditions might affect the absorption rate. It is known that this additional carbon is gradually increasing the acidity of the oceans, as evidenced by changes in the variety and range of sea life, coral in particular. There are also regional variations in the ocean–atmosphere carbon flux that are not well understood; some parts of the ocean are sinks of carbon dioxide, and other parts emit carbon dioxide to the atmosphere. Some computer models show that higher global temperatures may eventually change the ocean into a source of atmospheric carbon dioxide rather than a sink [19]. Currently ocean circulation patterns reduce carbon in the surface ocean by approximately 8.4 Tkg per year by moving this carbon to deeper layers. Because the turnover time for the thermohaline cycle is about 1,000 years, carbon transported to the deep ocean is effectively being removed from circulation at present.

Although carbon dioxide and methane exist in only trace amounts in the atmosphere, as noted in Table 7.1, the total amount, at 778 Tkg, is quite large because of the large size of the atmosphere. Fossil fuel burning and cement manufacturing currently contribute an additional 6.3 Tkg of carbon dioxide per year to the atmosphere, an addition of about 2.95 ppm per year. Cement manufacturing involves the heating of calcium carbonate (limestone) to form calcium oxide (lime), which releases carbon dioxide, but this source, the second largest human-caused CO_2 flux, is less than 3% of the total global fuel use emission. It is the burning of carbon, $C + O_2 \rightarrow CO_2$ + heat, that contributes the bulk of human-related carbon dioxide in the atmosphere.

EXAMPLE 7.3

Carbon Flux Caused by Human Activity

In the year 2000, global primary energy use was about 400 quad, of which 85% was from fossil fuels. Estimate the average amount of carbon emitted per person.

Various fossil fuel sources such as coal, natural gas, and petroleum have varying carbon content, but we can make a simple estimate if we assume that the 400 quad was generated using oil with a 75% carbon content by weight.

If 85% of the energy used was from fossil fuel, then $400 \times 0.85 = 340$ quad of fossil fuel was used in 2000. A quad is the energy content of about 1.82×10^8 bbl of oil, so to provide this much energy using oil we need 6.19×10^{10} bbl of oil. Assuming petroleum is 75% carbon by weight, the total amount of carbon used in 2000 was $0.75 \times 6.19 \times 10^{10}$ bbl $\times$ 42 gal/bbl $\times$ 3.5 kg/gal = 6.82×10^{12} kg of carbon, where we have used 3.5 kg/gal as the weight of petroleum.

For a global population of 6.3 billion this comes out to 6.82×10^{12} kg/6.3×10^9 = 1,083 kg = 1.1 tons of carbon per person per year.

The figures of 6.82×10^{12} kg of carbon total or 1.1 tons of carbon per person per year are very close to the actual emitted amount shown in Figure 7.7. The ratio of carbon mass released to fuel burned for petroleum is about 0.76 kg C/1 kg fuel[1] and about 0.83 kg C/1 kg fuel for natural gas (however, this does not include energy in the form of hydrogen mixed in with these sources, which provides additional energy without additional carbon emission). Coal, which provides about a third of the fossil fuel energy used globally, can vary from 50% to 95% in carbon content. A more careful calculation of the carbon emitted from fossil fuel sources that takes into account the different carbon and energy contents of various fossil fuels currently being used yields the measured value of 6.3 Tkg/yr shown in Figure 7.7.

EXAMPLE 7.4

Converting Mass to Parts per Million

The amount of carbon in the atmosphere in 2000 was about 778 Tkg. What is this in parts per million?

The number of carbon atoms in the atmosphere is 778 Tkg × (6.02 × 10^{23} molecules/mole)/12 g/mole = 3.9 × 10^{40} molecules.

The atmosphere provides a pressure of about 10^5 Pa on the earth's surface. Pressure is force per area, and weight is mass times the acceleration of gravity, so the mass of the atmosphere is approximately $m = PA/g = 10^5 \text{ Pa} \times 4\pi \times (6.4 \times 10^6 \text{ m})^2/9.8 \text{ m/s}^2 = 5.25 \times 10^{18}$ kg.

For an atmosphere of mostly oxygen and nitrogen with average molecular weight 29 g/mole, the mass of air is 5.25 × 10^{18} kg × (6.02 × 10^{23} molecules/mole)/29g/mole = 1.10 × 10^{44} molecules.

Dividing the number of carbon molecules by the number of air molecules gives the number of carbon atoms per atom of air, which is 0.000355 or 355 ppm. This is approximately correct for the year 2000; the current measured amount is 380 ppm.

In these examples and figures we have looked at the total carbon emission and uptake. The carbon significant in climate change is found in the form of carbon dioxide, as mentioned previously. Carbon has a molecular weight of 12 g/mol, whereas carbon dioxide has a weight of 44 g/mol. A kilogram of carbon released by fossil fuel burning contributes 44/12 or 3.7 kg of carbon dioxide to the atmosphere. In the remainder of this chapter we will use *carbon* and *carbon dioxide* interchangeably because there is 1 atom of carbon involved for each, but the reader should keep in mind that the actual weights of the two molecules are different.

The amount of carbon dioxide in the atmosphere has fluctuated between 190 ppm during glacial periods and 280 ppm during interglacial periods over the past 650,000 years (see Section 7.3 for more details). The measured shifts in atmospheric carbon dioxide appear to lag behind average temperature changes by a few hundred years; cooler climates initiate a reduction in atmospheric carbon dioxide, whereas warmer climates cause an increase in atmospheric carbon dioxide. This means that in the geological past carbon dioxide has acted as a positive feedback; as orbital parameters changed to warm the climate, carbon dioxide levels increased, causing further warming [18]. This feedback mechanism, albedo, and ocean circulation changes are believed to be the main mechanisms by which the earth shifts into and out of a glacial period over short periods of time, even though changes in forcing caused by the Milankovitch cycle occur very gradually.

In the past 250 years, however, atmospheric carbon dioxide levels have increased dramatically to more than 380 ppm, 35% higher than at any time in the past 650,000 years (five ice age periods). The quantity of methane in the atmosphere shows a similar history and has increased over the past 250 years to a value 150% higher than it has been in the past 650,000 years [5]. Two very convincing lines of reasoning have led scientists to conclude that this increase can be attributed directly to human activities. First, carbon in atmospheric carbon dioxide can be radioactive carbon dated (see Example 4.3), and this tells us that at least three quarters of the extra 100 ppm of carbon in the atmosphere came from fossil fuels. Second, as a more refined version of the calculation in Example 7.3 shows, human activity contributes a net positive increase of about 6.3 Tkg per year of carbon to the global carbon cycle; this much additional atmospheric carbon per year for only 35 years would account for the measured 35% increase. As mentioned earlier, the ocean removes about 2 Tkg per year, and the use of fossil fuels was significantly less 35 years ago; a more careful calculation shows that the calculated increase in carbon dioxide over the past 250 years due to human activity does accurately match the measured 35% increase. Example 7.5 shows one way this type of calculation can be done and provides a prediction of future carbon dioxide levels for a modest increase in fossil fuel use.

EXAMPLE 7.5

Future Carbon Dioxide Levels

Estimate the amount of carbon dioxide in the atmosphere in 2050 assuming a global growth rate for fossil fuel consumption of 2% (recall from Chapter 4 that the growth rate for the past 30 years has been 5%, so 2% is a very conservative estimate).

We can use the calculations shown in Example 7.4 to find that the 6.3 Tkg of carbon that was added to the atmosphere in 2000 accounts for an addition of about 2.9 ppm. Recall from Chapter 1 that a percentage growth rate based on the current amount is an example of an exponential increase. In the present case we want to find the cumulative increase over a period of time as the result of this exponential growth rate. So the additional concentration of carbon dioxide 50 years in the future is given by the integral of the exponential over 50 years, or

$$C_{CO_2} = \int_0^{50} 2.9e^{0.02t}\,dt = 2.9(e^{0.02\times 50} - e^0)/0.02 = 249.2 \text{ ppm.} \quad (7.10)$$

Adding this to the present 380 ppm gives 630 ppm in 2050, which is similar to predictions found in the Intergovernmental Panel on Climate Change reports discussed in the next subsection. Possible changes in ocean absorption of carbon dioxide or other potential parameter changes have not been included.

7.2.3 Summary of Anthropogenic Climate Forcing

The Intergovernmental Panel on Climate Change (IPCC) is a politically neutral group of scientists responsible for reviewing and assessing climate change data [5]. It includes scientists from more than 130 nations and includes some 2,500 scientists acting as reviewers, more than 800 contributing scientific authors, and 650 lead authors. The IPCC does not do any research but rather reviews and summarizes the scientific data found in published peer-reviewed papers. Because the IPCC is evaluating material that has already been peer reviewed and uses a multiple-level approval process that involves dozens of scientists, all of whom must agree as to the content of a report, the resulting IPCC reports are exceedingly reliable and are the most accurate assessment of climate change data available. Figure 7.8 is an IPCC summary of the anthropogenic climate forcings discussed here. These figures represent estimated human-caused changes in climate forcing over the past 250 years; natural forcings and feedback mechanisms are not included. Carbon dioxide is clearly the most significant, accounting for a forcing change of 1.6 W/m^2 in excess of natural forcing.

7.2.4 Climate Modeling

The model in Example 7.2 shows what can be done with very simple calculations; a reasonable estimate of global surface temperature can be made using only a few lines of algebra. A first step in making a more sophisticated computer model of the climate is to include more than one atmospheric layer. The simultaneous energy balance between as many as 100 atmospheric layers, each at a different temperature, is often included in more comprehensive models, as is the movement of air, both vertically and horizontally. Analogous calculations can be made of the pressure and temperature variation as a function of altitude, humidity, cloud formation, and other parameters that quantify various

FIGURE 7.8

Anthropogenic changes in climate forcing occurring over the past 250 years [5]. A color version of this graph can be found on the author's Web site for the book or the IPCC Web site.

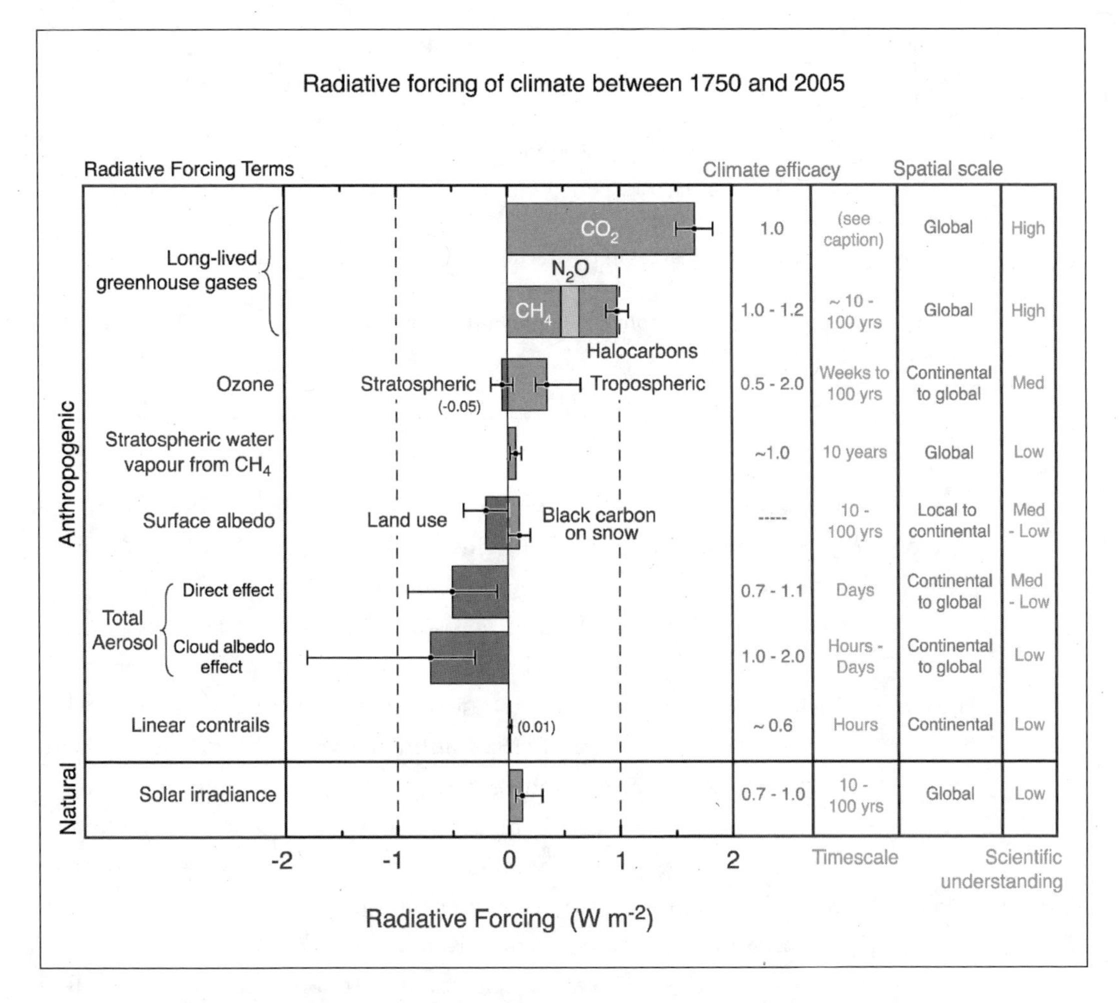

properties of the atmosphere [1]. Other important features of a complex climate model include different albedos for land and water, ocean circulation patterns, seasonal variations in energy flux for each hemisphere, ice and snow coverage, aerosols, the carbon cycle (Figure 7.7), and the hydrological cycle. Table 7.2 shows several typical parameters that are included in elementary climate models (references [1], [2], [20], and [21] show how these parameters are calculated and can be incorporated into the heat balance Equation (7.4) to increase the sophistication of a simple model).

TABLE 7.2

A few additional atmospheric parameters included in climate modeling and their current values.

Effect	Description	Parameter
Infrared loss	Infrared radiation escaping directly into space	40 W/m^2
Evaporation	Heat lost by surface oceans to atmosphere through evaporation	78 W/m^2
Convection	Heat transported from the surface to the atmosphere through convection	24 W/m^2
Albedo change	Changes in the fraction of cloud cover, f_c, versus fraction of exposed surface, f_s	$a = f_c a_c + f_s a_s$
Adiabatic lapse rate	Atmospheric temperature change with altitude	–6.5 K/km
Relative humidity	Concentration of water vapor	0.8
Concentration of CO_2	Current concentration	388 ppm

Initially, separate models analyzing each of the main components of climate were developed. There were individual computer models for the atmosphere (which included the parameters in Table 7.2), land surfaces, ocean and sea ice changes, the carbon cycle, and aerosols. Over the past 25 years these separate components were tested and gradually integrated into comprehensive models, which include all the forcings and feedback mechanisms mentioned earlier. As outputs the models predict variations in sea and land surface temperature, stratosphere and troposphere temperature, temperature variations between daily extremes, sea level changes, average precipitation, ice coverage, ocean circulation, changes in the frequency and intensity of extreme weather events, and deviations in atmospheric water vapor. There are at least 23 complete, major climate models, all of which require supercomputers to run. The interested reader can download and experiment with a real climate model that will run on a desktop computer from the Educational Global Climate Modeling Web site (http://edgcm.columbia.edu/; see Problem 7.43). This model uses the same set of programs used for scientific research but with lower resolution so that the results can be calculated with less computational power.

One important test of a sophisticated climate model is whether it can accurately account for climate change in the past. Figure 7.9 shows two graphs of global mean surface temperature for the past 100 years [5]. The graphs show the measured temperature and the predictions of 14 computer climate models and 58 simulation runs; the top graph includes anthropogenic and natural forcing, the bottom graph includes only natural forcing. The effect of three volcanoes,

FIGURE 7.9

Surface temperature anomaly observations, dark line, and computer models for the past 100 years. The top graph includes anthropogenic forcings; the bottom graph does not. The thick line is the model average of 58 simulations using 14 different models, and the thin lines are individual computer models [5].

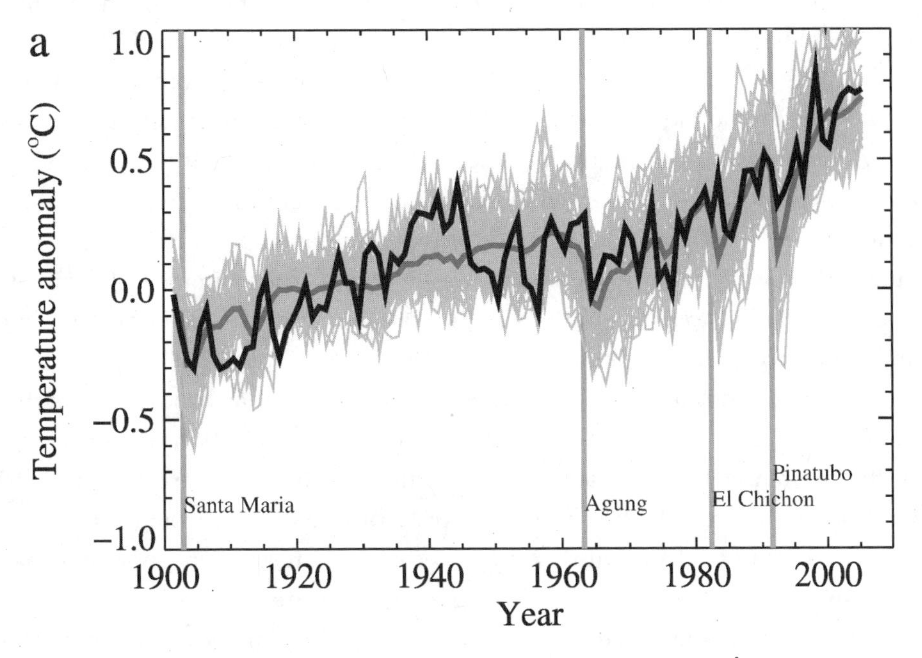

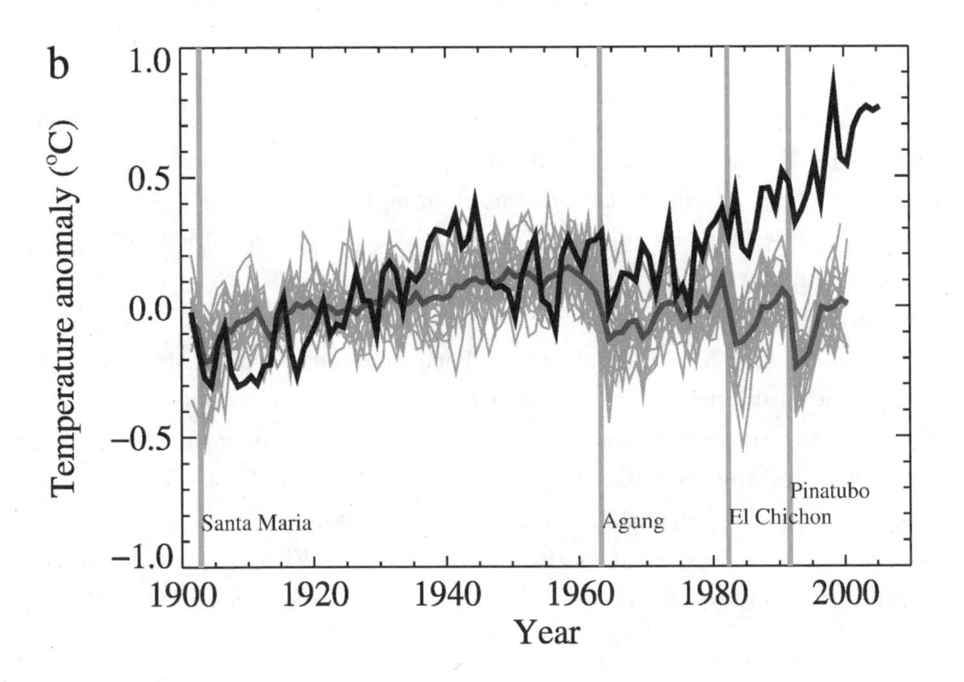

Agung, El Chichon, and Pinatubo, can clearly be seen in both the observations and the computer simulation.

As can be seen from the graphs, the computer models can accurately predict the past surface temperatures, but only if the anthropogenic forcings in Figure 7.8 are included. This is only one of many tests used to verify that computer models are accurate. Similar graphs have been made for regional rainfall, regional temperature fluctuations, sea ice, and many other climate phenomena [5]. Being able to explain the individual trends of several different parameters simultaneously demonstrates the strength of these models. The fact that the models can accurately explain a documented global surface temperature increase and at the same time account for a documented decrease in the temperature of the stratosphere indicate that they are robust representations of the climate. That climate models are found to be accurate is not surprising; they are based on the same simple physical principles such as those shown in Example 7.2. The computer is simply performing the same calculations but including many more details that are not easily done by hand. As we saw in the case of projections of energy use for the future in Chapter 4, accurate predictions of the future are complex and occasionally in error. However, the majority of scientists working on climate problems concur that present-day computer models are accurate enough to take seriously the predictions currently being made of future climate.

In an effort to minimize the error inherent in predicting the future, computer modelers use scenarios. Different scenarios take into account possible adjustments in human behavior that might occur as reactions to economic, population, and cultural changes. In the following we are focusing only on global surface temperature, but it should be kept in mind that the models also provide detailed predictions of other climate factors such as regional temperature changes, sea ice and sea level changes, and precipitation. Figure 7.10 shows the predictions found in the ICPP's last three reports and several scenarios for future global average temperature change. In the graph the black line labeled "Observed" is the running average over 10-year periods, and the dots are annual averages for years that are anomalies from the trend. Each solid curve represents the average of multiple simulations run using many different climate models. It should be clear from the graph that measured temperature changes over the past 15 years fall well within the error bars of predictions made in the First Annual Report, Second Annual Report, and Third Annual Report of the IPCC.

The "Commitment" scenario shown in Figure 7.10 is a prediction that assumes that carbon dioxide levels remain where they were in 2000; in other

FIGURE 7.10

Model predictions of average global surface temperature rises for various scenarios [5]. The curves represent climate model predictions of the First Annual Report (FAR), Second Annual Report (SAR), and Third Annual Report (TAR) of the IPCC. The shaded area up to 2005 represents the estimated error in the prediction for prior models. The curves extending into the future are current climate model predictions for various IPCC scenarios. See the text for a fuller explanation of the abbreviations for different scenarios (Special Report on Emission Scenarios [SRES]) used in the figure. A color version of this graph can be found on the author's Web site for the book or the IPCC Web site [5].

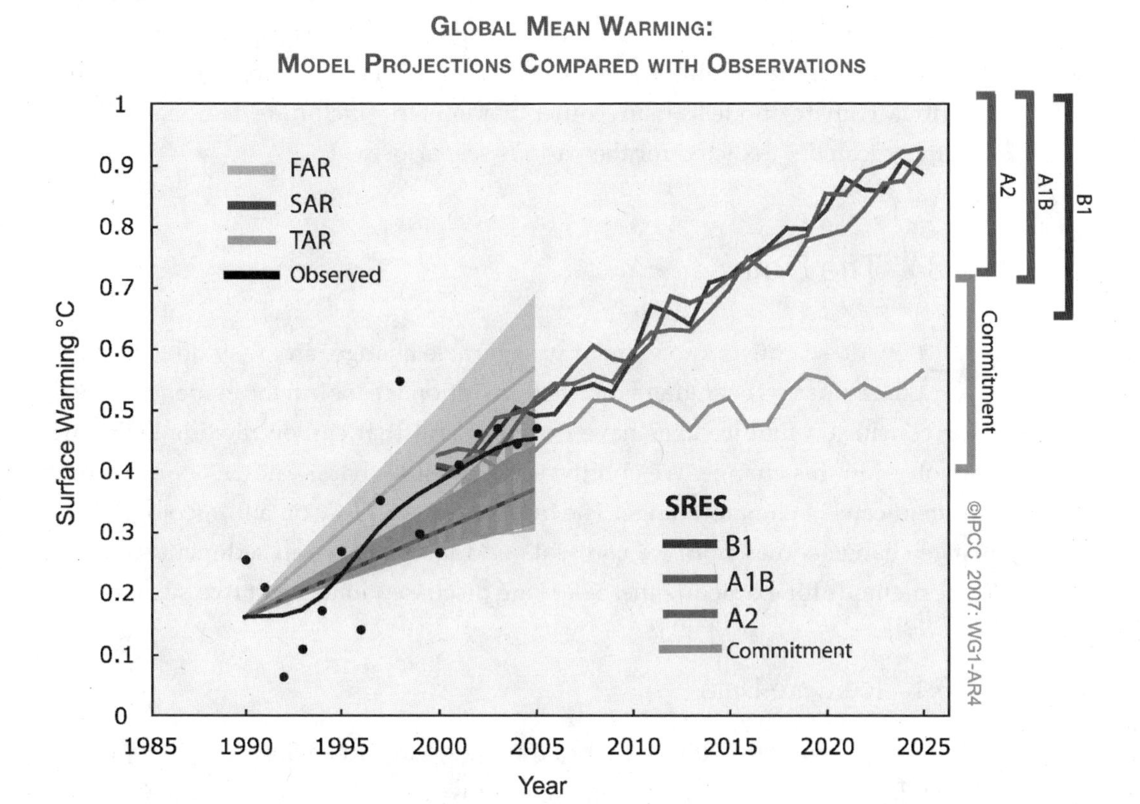

words, we are committed to that much change in surface temperatures even if no additional carbon dioxide was released by humans after the year 2000. The scenario labeled A1B corresponds to a world where there is rapid global economic growth, the global population peaks and begins to decline by 2050, and new technologies are introduced that improve energy efficiency. Fossil fuels are still used as a significant source of the world's energy. For the A2 scenario the population does not decrease as rapidly, and economic development is strongly regional rather than global. Economic development is slower overall. In the B1 scenario the global economy moves toward a service economy, with a reduction in material intensity, and there are no special incentives toward a reduction in carbon emissions. The bars to the right of the graph show the range of error expected in the predictions. Models have been run for many other scenarios that include different estimations of economic and population growth. The scenarios presented here represent little or no effort made on the part of governments or markets to reduce carbon emission. The models can also be run for various proposed carbon dioxide emission regulations. The lower limit to temperature change, zero growth in carbon emissions starting in 2000, is the "Committed" scenario. Again it should be emphasized that this is not the only output of the various climate models; many other parameters, including regional climate changes, can be predicted for the various scenarios.

7.3 ■ The Data

How do scientists know about past climate change, and why do they think the climate is changing now? In this section we look at the evidence behind the conclusion that ice ages have occurred and that carbon dioxide levels are involved in this change. We end the section with what is known about current human-caused climate change. The bulk of the scientific data for geological climate change comes from ice core data and lake and ocean sediment samples. The methods for extracting these data are discussed in the next two subsections.

7.3.1 Ice Core Data

As snow accumulates, it is at first porous, allowing air to flow in and out of the snow layer. As more snow accumulates, lower layers compress, and the snow becomes a type of ice called firn, with pockets of trapped air that are no longer in contact with the atmosphere. In regions where the ice does not melt in the summer, these trapped bubbles constitute a record of the makeup of the atmosphere at the time they were trapped. Additional snow falling the next year creates a year-by-year layer of trapped bubbles. By carefully examining the chemical composition of the gases found in these bubbles, we can extract a great deal of information. Several locations have layers of ice going back 800,000 years. Locations where ice core data have been examined include Greenland, Antarctica, the Himalayas, the Andes, and Mount Kilimanjaro in Africa. In all these locations cylinders of ice about 15 cm in diameter and sections totaling hundreds of meters in length are extracted and taken to laboratories for analysis.

In regions where the snow does not melt during the summer, there are seasonal differences in the quantity and quality of the deposited ice. These differences leave behind clearly defined layers in the ice, marking the seasonal changes. By counting the layers back from the present, we can determine the age of a particular layer in a manner analogous to determining the age of a tree from tree rings. Annual variations from year to year mean that any sequence of multiple layers is unique. This makes it possible to compare cores taken at different locations. If the time spans for two cores overlap, they will share a unique sequence of layers, and this allows the correlation of data from one core to that of another.

Some data, such as carbon dioxide and methane levels, can be measured directly from the air trapped in bubbles. Other information, such as global temperature, is extracted by the use of proxies, or data that correlate with temperature. The most significant proxy for temperature is the hydrogen isotope deuterium, which has an extra neutron compared with normal hydrogen. This extra neutron means that water formed with deuterium is slightly heavier than normal water. Ocean evaporation will favor the lighter form of water over water with deuterium (a process called fractionation), and this difference is more pronounced when ocean surface temperatures are lower. The deuterium levels in an ice layer are a proxy for ocean surface temperatures during the year the layer formed; more deuterium means a warmer ocean.

Several other proxies can be used to verify the temperature changes found using the deuterium isotope data. In warmer years more of the ocean is ice free and there is more wind. This means that aerosols, including dust and ocean salt from increased ocean spray, show up in the ice layer. These changes in chemical makeup change the acidity of the ice, which can be measured by determining the electrical conductivity of the layer. Oxygen isotopes found in water can also be used as a proxy for temperature. As in the case of deuterium, the heavier ${}^{18}_{8}O$ isotope is found in lower quantities than the lighter ${}^{16}_{8}O$ isotope in cooler years because of lower evaporation rates and lower wind speeds. Measured levels of hydrogen peroxide are higher in years that have more sunlight because sunlight acts to form hydrogen peroxide from water vapor. Some trapped gas, such as atmospheric nitrogen, tends to diffuse slowly out of the bubble through the ice. The rate of this diffusion depends on the temperature difference between the ice layer and the surface and on the isotope of nitrogen. The lighter $^{14}_{7}N$ isotope diffuses faster than the heavier $^{15}_{7}N$ isotope, which means that for any given layer, a larger separation difference between $^{14}_{7}N$ and $^{15}_{7}N$ indicates a warmer climate the previous year. The actual temperature of the ice at different levels is also a record of seasonal temperature changes. A warm climate sends a warm pulse through the ice, and these pulses remain as temperature variations in the ice, which can be measured for hundreds of years into the past.

Other information about climate besides surface temperature, amount of the ocean free of ice, and wind patterns can be extracted from ice cores. As mentioned previously, volcanoes eject large quantities of aerosols into the air, and these chemicals have signatures specific to volcanoes. The aerosols change the acidity of the ice layer, which is measured by electrical conductivity. A final piece of evidence involves the presence of the beryllium isotope $^{10}_{4}Be$ in ice core layers. More of this isotope is formed by changes in the levels of cosmic rays reaching the upper atmosphere. Changes in cosmic ray intensity in the upper atmosphere are caused by changes in solar activity (sunspots) and changes in the earth's magnetic field (which deflects incoming charged particles).

Temperature, carbon dioxide levels, nitrous oxide, and methane levels are shown in Figure 7.11 for the past 650,000 years, based on ice core data from Antarctica. The temperatures are determined from deuterium data, but as noted earlier, this is corroborated by other methods. This data are also supported by dozens of other ice cores drilled in other locations in both hemispheres, indicating that the recorded changes were global.

Several points should be noted in Figure 7.11. One is that methane and carbon dioxide levels change in step with temperature changes. As mentioned earlier, in the past CO_2 and CH_4 levels have acted as feedback mechanisms, pushing the climate into warmer interglacial periods and also back into ice ages. It should be mentioned that the temperature shift from glacial to interglacial periods is about 7°C to 10°C and occurs very rapidly, in as little as a few decades. Shifts from interglacial periods back to ice ages are not as rapid but are still fast. The figure also shows the very recent sharp increase in CO_2, CH_4, and N_2O levels, as measured directly from the atmosphere. Approximately 40% of the N_2O, another greenhouse gas, in today's atmosphere is the result of human activity including crop cultivation (especially where nitrogen is used as a fertilizer), biomass burning, and industrial processes such as nylon production. Recent global temperature changes are not shown in Figure 7.11 but are discussed later in this chapter. As a final note, the current interglacial period of the past 11,000 years

FIGURE 7.11

Ice core data from Antarctica [5]. The top curve is measured nitrous oxide levels, the next curve is carbon dioxide levels, the third curve is methane, and the last curve is the measured amount of deuterium, which is a proxy for temperature. Interglacial periods are shown in gray.

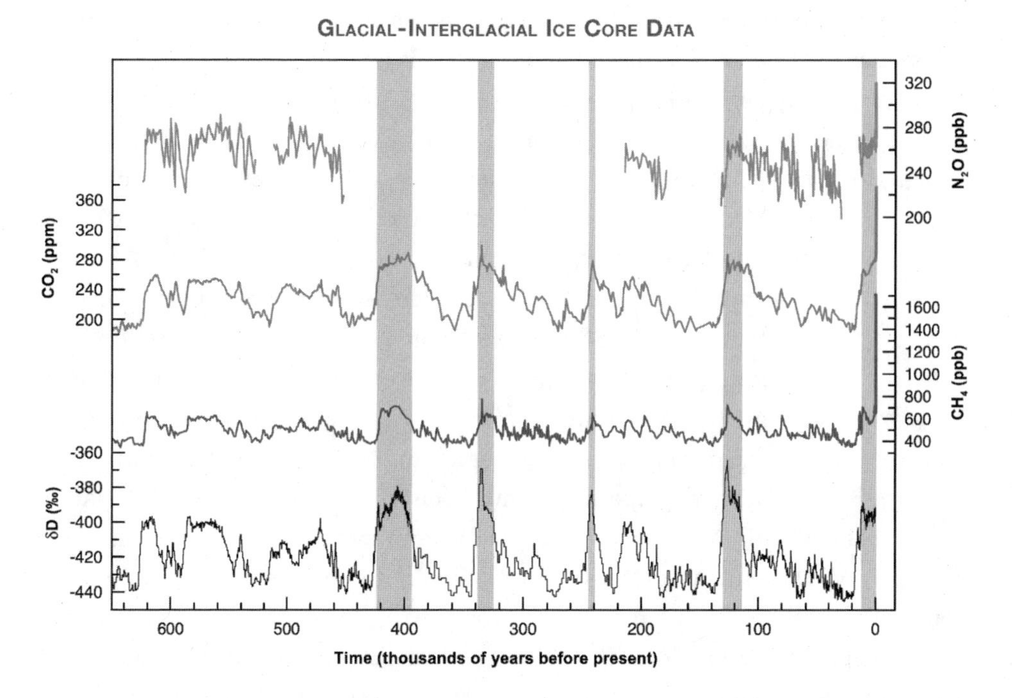

has seen more constant temperatures compared with other interglacial periods, when temperatures varied quite a bit. It is this long period of stable climate that allowed human civilization to develop.

7.3.2 Ocean Sediment Core Data (Varved Sediment)

In ocean and lake locations where the rate of sedimentation has been heavy, there is often enough data to extract climate information. Two main data sources are available, one based on chemical analysis and the other on the investigation of microfossil species. Seasonal changes leave an annual record in the sediment, called a varve, which shows up as a different-colored layer. Much like ice core data, cores of the sediment in the ocean floor are extracted and analyzed for climate information. The sediment layers record variations that match ice core data in locations where sedimentation is heavy enough so that there is sufficient detail. Carbon-14 can be used to radioactively date the age of the layers that contain organic material, and the thorium to uranium ratio can be used for non-organic material.

Plankton and algae that live at the surface of the ocean produce an organic compound called alkenone, which comes in two different forms. When ocean surface temperatures are higher, these organisms produce more tri-saturated alkenones relative to di-saturated alkenone. This adaptation makes these organisms less susceptible to freezing because the di-saturated alkenone acts as an antifreeze agent, leaving the organism more flexible in cooler water. These molecules do not degrade when the organism dies, and so they can be found in ocean sediment. Measuring the ratio of these two types of alkenones in an ocean sediment layer indicates sea surface temperatures at the time the plankton was alive. Also, as in the case of ice layers, an increase in the heavier isotopes of oxygen (the ${}^{18}_{8}O$ to ${}^{16}_{8}O$ ratio) found in shells of surface-dwelling plankton indicate warmer temperatures because there is more evaporation of water and more winds, hence higher heavy isotope concentrations. Heavy metal uptake in corals is also temperature influenced, appearing as rings in the annual growth, much like tree rings. Warmer temperatures mean faster growth, which means larger amounts of strontium, uranium, and magnesium in calcium compounds found in coral. Measurements of these heavy metals in sediment containing coral remains can be used as a temperature proxy.

Oxygen depletion, indicating slower ocean currents and hence a cooler climate, can be statistically determined by the relative abundance of organisms adapted to low-oxygen conditions. Some bottom-dwelling marine bacteria in particular adapt to low-oxygen conditions by preferentially using ${}^{14}_{7}N$. Changes in the ${}^{15}_{7}N$ to ${}^{14}_{7}N$ ratio in ocean sediment layers map ocean floor oxygen levels and therefore ocean circulation. Marine and lake sediments also show variations in the amounts of pollen grain blown into the water from land plants, and this indicates climate changes occurring over land. The ratio of cold-adapted plant pollen to warm-adapted plant pollen indicates long-term seasonal variations in land surface temperature.

The amount and type of inorganic material found in ocean sediment are also clues to past climate. Floating ice that originated from glaciers and polar ice caps has debris embedded on the underside from sliding along land surfaces before being deposited in the sea. If there is an increase in ice flow due to a warmer climate, the ice carries more and heavier debris farther into the ocean, and this shows up in ocean sediment layers. Differences in the inorganic mineral makeup of ocean sediment layers can be determined by measuring the magnetic susceptibility of the layer.

Tree ring data from both living and petrified trees have long been used to gather information about past climate. By overlapping rings in fossil trees with more recent trees, a record of climate change can be established going back 10,000 years or more in some locations. These data correlate well with ice core and ocean sediment data and have been used to establish temperature changes occurring over the past 1,000 years.

7.3.3 Climate Changes of the Past 1,300 Years

Figure 7.11 includes atmospheric changes of carbon dioxide, methane, and nitrous oxide occurring over the past 250 years but does not show surface temperature changes for this recent time period. Figure 7.12 shows average surface temperature changes for the northern hemisphere as estimated in eight different studies of temperatures over the past 1,300 years. Ice core, ocean sediment, and tree ring data are included.

FIGURE 7.12

Eight measurements of changes in northern hemisphere temperatures over the past 1,300 years using ice core, sea sediment, tree ring, and other data [5]. Abbreviations indicate the study name (see the IPCC report for details) followed by the year the study was made. The black line ("Instrumental") represents the average of direct measurements. The gray line (PS2004) is a numerical fit of data that existed before 2004. The vertical axis represents the deviation from the 1961 to 1999 mean temperature in Celsius. A color version of this graph can be found on the author's Web site for the book or on the IPCC Web site [5].

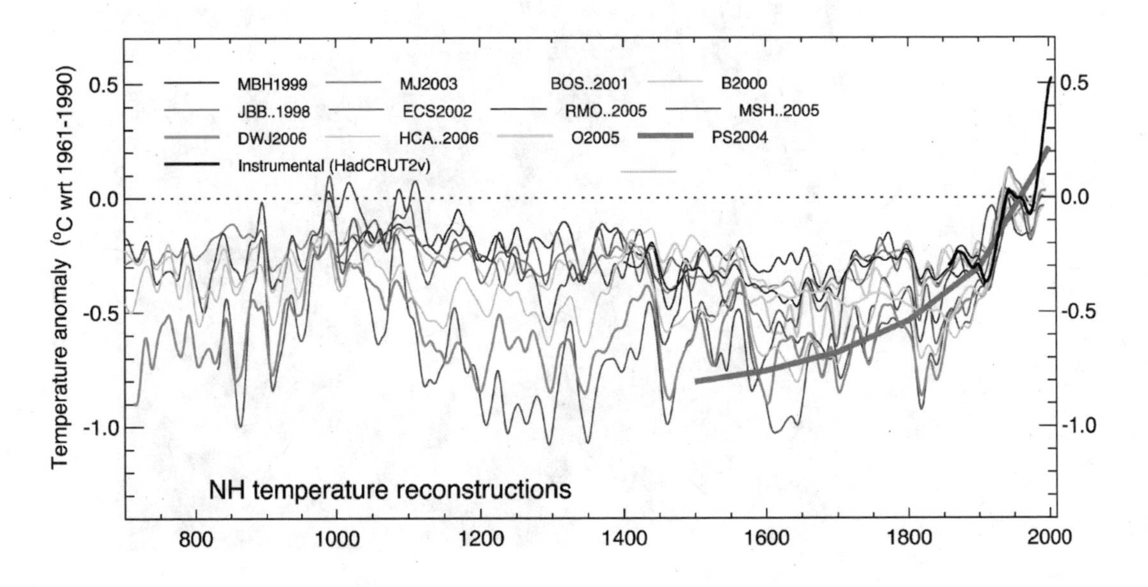

The data shown in Figure 7.12 are only a small part of the scientific data supporting the conclusion that global temperatures have risen by about 0.6°C over the past 250 years. Glaciers around the world are shrinking (Figure 7.13), permafrost is melting, and there are shifts in peak river flow caused by earlier

FIGURE 7.13

The sign marks the location of the face of this Alaska glacier in 1951. The current face has retreated to 2 km in the distance.

spring snow melting, effects found in lakes and rivers, and increased coastal erosion [22]. Because the earth radiates blackbody radiation almost equally in all directions but absorbs energy from the sun preferentially at the equator, the heat flux away from the earth will be larger than the incoming heat in polar regions. This means the effects of a warmer earth will be more observable in arctic regions, and this does appear to be the case. Changes in ice coverage over Greenland and other polar regions have been documented both on the ground and with satellite data and are significantly larger than for middle latitudes. Global sea levels have risen by approximately 23 cm in the past 100 years. Other supporting data include the more than 28,000 cases of changes in biological marine and land systems that have been documented, including shifts in migration patterns, reproduction times, leaf maturation, and plant blooming times [22].

7.4 ■ Summary

Most scientists reacted with deep incredulity to early reports from the 1980s that the global climate was warming and was caused by human activity [3]. Scientists are some of the most skeptical people on Earth, demanding proof from multiple sources before committing to a consensus on a topic, and even then the commitment is always contingent in the sense that they reserve the right to change their minds if better data are provided. Consensus on the main points of human-caused climate change has continued to grow as more data are collected. Nearly all scientists are convinced that carbon dioxide levels have increased by 35% over the past 250 years because of human activities, that the earth's global average temperature has increased by about 0.6°C in the past century, that the global sea level has risen by about 6.7 cm over the past 40 years, and that data from satellite and surface measurements show that the extent of arctic sea ice has decreased by 15% since the late 1970s [23]. Most scientists are reasonably confident that the rise in global mean temperature over the past 30 years is due to increases in greenhouse gases, that the continued emission of greenhouse gases will lead to an increase in global temperature of 1.4°C to 5°C over the next 100 years, that sea levels will rise by 40 cm over the next century because of thermal expansion of the ocean alone, that global rainfall patterns will shift, and that hurricane and typhoon intensity is likely to increase with higher ocean surface temperatures [3].

Many of the effects of a global temperature increase will not be benign. They may include food and water shortages, species extinctions, permanent submersion of some islands and coastal areas, higher rates of infectious diseases, and increased death during heat waves. Temperature changes are not projected to be uniform; some areas will be hotter and drier, others cooler and wetter. The predicted climate change means each region will have to make changes in farming and food production. Pests, including infectious diseases, are expected to migrate because of changes in temperature and rainfall. Sea levels will continue to rise because of thermal expansion of the ocean and melting of ice located over polar land masses. This will make coastal areas, where a large fraction of the world's population lives, more vulnerable to flooding, especially as the intensity of tropical storms increases with sea surface temperature increases. The additional question of the chaotic nature of climate and the possibility of changes that are much larger and much more rapid than anticipated make human-caused climate change a highly inadvisable experiment to undertake.

As mentioned in Chapter 1, when motivated with enough conclusive evidence and economically viable alternatives, humans can act positively to solve challenges facing them. Several proposals have already been suggested for the reduction of carbon dioxide and other greenhouse gases. More efficient vehicles; more efficient buildings; more efficient coal plants; replacement of coal by natural gas; capture of CO_2 at power plants; increased use of nuclear, wind, and photovoltaic power; reforestation; and improved farming techniques have all been proposed as means to reduce carbon dioxide emissions [24, 25]. Economic incentives such as the type of cap-and-trade system that was successful for acid rain (see Chapter 1) have also been proposed (see [26] for a discussion of how this might work). Some of these potential solutions will be discussed in Chapter 8. Most experts agree that it will take a combination of all of these strategies to significantly reduce global carbon emissions. One positive note is that nearly all these solutions are also solutions to the problem of finite fossil fuel resources, discussed in Chapter 4. Solving the world's pending energy crisis will go hand in hand with addressing human-caused global climate change.

Questions, Problems, and Projects

For all questions,

- Cite all sources you use (e.g., book, article, Wikipedia, Web site). Before using Wikipedia as a source, please read the policies of research and citing in Wikipedia: http://en.wikipedia.org/wiki/Wikipedia:Researching_with_Wikipedia and http://en.wikipedia.org/wiki/Wikipedia:Citing_Wikipedia. For Wikipedia you should verify the information in the article by reading the sources listed.
- For all sources, briefly indicate why you think the source you have used is a reliable source, especially Web resources.
- Write using your own words (except for direct quotes, which should be enclosed in quotation marks and properly cited).

1. It has been said that climate is what you expect, but weather is what you get. Explain this statement.
2. In Example 7.1 the earth's albedo was given as 0.3. Dry sand has an albedo of 0.4, leafy crops or forests have albedos of about 0.2, and calm seawater has an albedo of 0.05. Repeat the calculation in the example for an earth covered entirely by sand, forests, and water. How sensitive is the surface temperature calculation to albedo?
3. (a) What albedo should a spherical satellite of 1 m diameter have in order to maintain a surface temperature of 300 K? Assume the satellite is always in the sun but rotates so that it radiates in all directions, as the earth does. (b) If you paint the satellite with stripes that are alternately black ($a = 1$) and silver ($a = 0$), what portion of the satellite should be black?
4. Repeat Problem 2.9 in Chapter 1 to find the solar constant at the distance of Venus's orbit. Use this value and follow Example 7.1 to find the predicted surface temperature of Venus, assuming no atmosphere and an albedo of 0.76. How does this compare with the known surface temperature?
5. Find the ratio of r_p to r_e in terms of eccentricity using Equation (7.5). What is this ratio for an eccentricity of 0.017 (the current value)? Using a value of $r_p = 1.5 \times 10^{11}$ m, what is r_e? Follow the calculation in Problem

2.9 to find the effect on the solar constant due to this annual change in the distance to the sun.

6. Using reliable sources, investigate and report on the effects on human civilization of the 1°C temperature change during the Little Ice Age, which occurred between 1600 and 1900 in Europe.
7. Using reliable sources, write a summary of the Arctic Oscillation, the Pacific Decadal Oscillation, and the Madden and Julian Oscillation. What effects do each of these have on the local climate?
8. Go to the Ocean Surface Current Analysis Real-Time Web page (http://www.oscar.noaa.gov/). Select an interesting region (e.g., the El Niño region off the west coast of South America) and look at the surface current patterns over several months. Report on the seasonal variations in surface currents and any other interesting changes you find.
9. Knowledge of the thermohaline circulation comes in part from radioactive tracer elements that were released into the atmosphere during nuclear weapon testing in the 1950s and 1960s. Various radioactive isotopes were deposited on the sea surface and began to circulate with water in the ocean. Because the time and location of the deposit of these isotopes are known, their present location gives some indication of circulation patterns. Using reliable sources, write a summary of this accidental scientific tool and how it has been used to map ocean circulation patterns.
10. Some current data show a slowdown of the thermohaline circulation. Using reliable sources, report on the current state of knowledge about the recent slowdown and its possible effects (you may want to start with references [6], [7], [8], and [9]).
11. Using reliable sources, find a discussion of other large volcano eruptions besides those mentioned in the text and their effect on climate. Summarize the discussion, listing the amount and duration of the effect, and include a summary of the evidence used to make these conclusions.
12. Using reliable sources, explain the connection between studies of ENSO, NAO, Arctic Oscillation, Pacific Decadal Oscillation, and longer-term climate change studies. (Search the NASA surface solar radiation data page http://data.giss.nasa.gov and the British Atmospheric Data Centre http://badc.nerc.ac.uk/browse for the effects of El Niño and solar cycle on the warming trend.)
13. Using reliable sources, write a summary of the Cretaceous extinction event. How did this meteor change global climate?

14. Using reliable sources, report on the transport by wind of dust, pollutants, and land-based bacteria over large distances. The case of sand from the Sahara reaching South America is particularly interesting.
15. Using reliable sources, write a summary of what is known about geomagnetic reversals. How often do they occur, and what is the mechanism for the reversal? What effects are thought to result from such a reversal?
16. Verify the results of Example 7.2 for $\varepsilon_a = 1$ and $\varepsilon_a = .75$. Try a few other values (remember emissivity is between 0 and 1). How sensitive is the surface temperature to emissivity in this calculation?
17. Assume a solar constant at the orbit of Venus of 2,626 W/m^2 and an albedo of $a = 0.76$, follow Example 7.2, and find the surface temperature of Venus, assuming a single atmospheric layer with emissivity equal to 1. How does this compare to the known surface temperature of 730 K (see the next problem)?
18. Your answer to the previous problem will be about half the observed surface temperature because in the case of Venus the atmosphere is much thicker and has a larger temperature gradient. Extend Example 7.2 to include two atmospheric layers, making reasonable assumptions. How close does the two-layer atmosphere come to correctly predicting the surface temperature of Venus?
19. The solar constant at the orbit of Mars is 593 W/m^2, and the Mars albedo is $a = 0.25$. Repeat Examples 7.1 and 7.2 for Mars and compare your answers with the actual surface temperature of 220 K. Make reasonable estimates of parameters you do not have at hand (or find them).
20. Use Equation (2.20) from Chapter 2 to find the peak of the blackbody radiation of the earth, assuming a surface temperature of 288 K. Locate this wavelength in Figure 7.5. What part of the spectrum is this?
21. Using reliable sources, summarize the steps of the earth's hydrological cycle.
22. Using reliable sources, investigate the Keeling curve and summarize what is known about the annual fluctuation of carbon dioxide related to plant photosynthesis.
23. Repeat Example 7.3 using the same assumptions for the typical North American assuming a national energy use of 100 quad. How does this per capita carbon dioxide emission compare with the world average?

24. Update Example 7.3 using figures for the present year (or the most recently available data). How different is this result from that of Example 7.3?

25. Repeat Example 7.3 for the following cases: (a) a 30% improvement in energy efficiency; (b) using the coal equivalent of energy instead of oil; (c) using a 95% carbon content (anthracite coal) instead of 75%. How sensitive are your results to the parameter choices used in Example 7.3?

26. The conversion between parts per million (ppm) in the atmosphere and kilograms of carbon is approximately 1 ppm = 2.13 Tkg. The global average of carbon dioxide has fluctuated between 260 ppm and 280 ppm over the past 650,000 years but has grown to more than 380 ppm in the past 250 years. How many metric tons of carbon do these figures represent?

27. Calculate the carbon you would emit by driving a car 10,000 miles a year, assuming it gets 30 mpg. Assume a carbon to fuel ratio of 0.8 kg carbon/1 kg fuel.

28. The amount of lead emitted per year into the atmosphere averaged about 0.4 Gkg per year in the 1980s. Repeat Example 7.4 to find the total amount of lead emitted into the atmosphere in parts per million during this same time period.

29. Find a reliable Web site that helps you calculate your carbon footprint. What is your family's carbon footprint? What are the biggest factors in your carbon footprint?

30. Go to the Carbon Disclosure Project Web site (http://www.cdproject.net/) and look at the Carbon Disclosure Leadership Index (CDLI) list. The list reports the direct carbon emissions (Scope 1) and indirect emissions from power generation and heat (Scope 2) in 1,000 mt of CO_2 emitted for various large companies. The emissions per year are reported as "Intensity." The CDLI score ranks the company's performance on a number of issues related to carbon emissions. Compare a few companies and report on any surprises or interesting findings.

31. Using reliable sources, find out the percentage of carbon emission for several major sectors of the economy (e.g., transportation, agriculture, manufacturing, energy providers, housing, construction, waste management). Find these figures for the United States or any other individual country and for the world.

32. Using reliable sources, find and report on the carbon emissions associated with the production of beef, chicken, pork, fish, and an assortment of fruits and vegetables. Which food sources result in the largest carbon dioxide emissions? These may be reported as CO_2 per food calorie or total CO_2 production.

33. Repeat Example 7.5 with fossil fuel consumption growth rates of 1% and 3%. Also find the estimated atmospheric carbon dioxide for the year 2100. Repeat these calculations but include the assumption that the ocean will continue to absorb 2 Tkg of carbon per year.

34. Equation (7.10) can also be used to estimate the carbon added to the atmosphere from preindustrial time until now. Repeat the calculation using an upper limit of zero (present day) and a lower limit of minus infinity (the distant past). Subtract this number from the present value of 380 ppm to find the preindustrial atmospheric carbon level. How close is this to the known value of 280 ppm for interglacial periods (the answer is not exactly correct because global growth rate has not always been 2%)?

35. Using reliable sources, summarize how the IPCC functions. Find any legitimate criticisms of the IPCC and evaluate them.

36. Go to the IPCC Web page and download and read the executive summary of the 2007 report. Report on any interesting things you learn.

37. Other than the IPCC Web page, probably the most informative Web site on global warming is the RealClimate page (http://www.realclimate.org/). Evaluate this page in terms of the information available and write a few summary paragraphs about it. Who are the authors of this page, and what is their intent?

38. Find and summarize the official positions on climate change of several scientific organizations. Examples of prominent scientific organizations include the National Academy of Sciences, Goddard Institute of Space Studies at NASA, State of the Canadian Cryosphere, Oxford University, Union of Concerned Scientists, Environmental Protection Agency, National Oceanic and Atmospheric Administration, Royal Society of the United Kingdom, American Geophysical Union, American Institute of Physics, American Meteorological Society, National Center for Atmospheric Research, and Canadian Meteorological and Oceanic Society.

39. In a couple of paragraphs, summarize a section of Coby Beck's "How to Talk to a Global Warming Skeptic" Web page (http://gristmill.grist.org/skeptics). Make a list of any questions you have about climate change that are not answered on this page.
40. Find a Web site of a climate change skeptic (someone who does not believe in global warming). Evaluate and summarize their arguments. Find out as much as you can about the person or organization presenting the information. We will discuss these pages in class.
41. Using reliable sources, trace the history of the response of Exxon (now ExxonMobil) to the evidence in support of human-caused climate change. How has their position changed over time?
42. Computer climate models can be broken down into several broad categories. Using reliable sources, write an explanation of each of the following types of models: energy balance models, one-dimensional radiative–convective models, two-dimensional statistical–dynamical models, and three-dimensional general circulation models.
43. The Educational Global Climate Model can be found at http://edgcm.columbia.edu/. Unfortunately, at this time the program can be downloaded for free for only a limited trial. Download the software and run the snowball earth simulation (or a simulation of your choosing). Report on your findings.
44. Using reliable sources, find and report on the growing regions (the plant hardiness zones) used by farmers and horticulturists to determine which plants will grow where. Speculate on changes in the locations of these zones in the event that the earth goes into an ice age and the average temperature decreases by 10°C. Pick one or two locations and report on how far north the equivalent zone is for a 10°C change.
45. Using reliable sources, summarize the controversy surrounding the early versions of the so-called hockey stick graph (Figure 7.12). What can you conclude about the current version of the graph?
46. A common climate myth is that the urban heating effect accounts for the measured increase in temperature over the past century (i.e., locations where temperatures are measured are near cities, which are becoming larger and hotter). Using reliable sources, write a brief rebuttal to this myth.

47. Using reliable sources, investigate and report on the illegal leaking of e-mails from the University of East Anglia ("Climategate") in 2009. From a scientific point of view, have these e-mails changed the basic conclusion that humans are causing climate change?
48. In early 2010 the IPCC admitted making a mistake that involved using non–peer reviewed data to estimate the rate of melting of the Himalayan glaciers. Using reliable sources, investigate and report about this error. From a scientific point of view, have these problems changed the basic conclusion of the IPCC that humans are causing climate change? What have been the political consequences of this problem?
49. A common climate myth is that satellite data measuring lower atmospheric temperatures do not match earthbound measurements. Investigate and report on this myth and how the temperature measurement discrepancies were eventually resolved.
50. What difference does a few degrees' increase in global temperature make? Using reliable resources, explore the following: (a) increase in sea level and its effect on population centers; (b) increase in temperature and the connection with the spread of infectious diseases such as malaria, schistosomiasis, dengue, and Lyme disease; (c) changes in rainfall patterns and farming; and (d) changes in heat stress problems for humans and other organisms.
51. Find a reliable Web site that shows where sea level rise caused by global warming will have the largest effect. Report on your findings. Include a discussion of two different effects: thermal expansion of the ocean and sea ice melting.
52. Using a reliable source, find out the current status of the polar ice caps and the rate of tundra melting.
53. Go to the Climate Hot Map page (http://www.climatehotmap.org/) and in half a page, summarize some interesting facts found on this site.
54. A proposed solution to human carbon emissions is to plant trees, which absorb carbon. Suppose a hectare of trees can sequester 10 mt of carbon per year. How many hectares of trees would be needed to absorb the 6.3 Tkg emitted in 2000? Is this a feasible solution for carbon emissions? Why or why not?

55. In addition to the suggestions mentioned in the text for solutions to human-caused carbon dioxide emissions, a number of radical solutions have been proposed, such as injecting aerosols into the atmosphere to block sunlight. Find and report on a few of these potential solutions. Evaluate the advisability of having to rely on these solutions (compared with the solutions listed in the text).

56. Find reference [25] ("Stabilization Wedges," by S. Pacala and R. Socolow) and summarize the authors' main points (there are also several versions of this article on the Web). What ideas are they proposing to solve the carbon dioxide emission problem?

REFERENCES AND SUGGESTED READING

1. F. W. Taylor, *Elementary Climate Physics* (2005, Oxford University Press, Oxford).

2. J. N. Rayner, *Dynamic Climatology: Basis in Mathematics and Physics* (2001, Blackwell, Boston).

3. K. Emanuel, *What We Know about Climate Change* (2007, MIT Press, Cambridge, MA).

4. J. Houghton, *Global Warming: The Complete Briefing*, 3rd ed. (2004, Cambridge University Press, Cambridge).

5. Reports of the IPCC on global warming (http://ipcc-wg1.ucar.edu/wg1/wg1-report.html).

6. Climate forcing data, NOAA Paleoclimatology Program (http://www.ncdc.noaa.gov/paleo/forcing.html).

7. E. Bard, "Climate Shock: Abrupt Changes over Millennial Time Scales," *Physics Today* 55(12) (Dec. 2002):32.

8. S. Weart, "The Discovery of Rapid Climate Change," *Physics Today* 56(8) (Aug. 2003):30.

9. R. B. Alley et al., "Abrupt Climate Change," *Science* 299 (March 2003):2005.

10. J. Lean, "Living with a Variable Sun," *Physics Today* 58(6) (Aug. 2005):32.

11. The Government of Norway's GRID-Arendal program in collaboration with the United Nations Environment Programme (http://maps.grida.no/go/graphic/world-ocean-thermohaline-circulation1).

12. O. B. Toon, A. Robock, and R. P. Turco, "Environmental Consequences of Nuclear War," *Physics Today* 61(12) (Dec. 2008):37.

13. Renewable Resource Data Center, National Renewable Energy Laboratory, "Reference Solar Spectral Irradiance" (http://rredc.nrel.gov/solar/spectra/am1.5/).

14. L. S. Rothman et al., "The HITRAN 2008 Molecular Spectroscopic Database," *Journal of Quantitative Spectroscopy and Radiative Transfer* 110 (2009):533–572 (http://www.cfa.harvard.edu/hitran/).

15. The GATS, Inc. online spectral calculator (http://www.spectralcalc.com/).

16. G. A. Schmidt, "Water Vapour: Feedback or Forcing?" (http://www.realclimate.org/index.php?p=142).

17. U.S. Climate Change Science Program, "Strategic Plan for the Climate Change Program, Final Report 2003," (http://www.climatescience.gov/default.php).

18. U.S. Climate Change Science Program, "Frequently Asked Questions" (http://www.gcrio.org/ipcc/ar4/wg1/faq/index.htm).

19. J. L. Sarmiento and N. Gruber, "Sinks for Anthropogenic Carbon," *Physics Today* 55(6) (Aug. 2002):30.

20. D. Hafemeister, *Physics of Societal Issues* (2007, Springer Science, New York).

21. J. R. Barker and M. H. Ross, "An Introduction to Global Warming," *American Journal of Physics* 67(12) (Dec. 1999):1216.

22. C. Rosenzweig et al., "Attributing Physical and Biological Impacts to Anthropogenic Climate Change," *Nature* 453 (May 2008):353.

23. J. C. Comiso and C. L. Parkinson, "Satellite-Observed Changes in the Arctic," *Physics Today* 57(8) (Aug. 2004):38.

24. A. H. Rosenfeld, T. M. Kaarsberg, and J. Romm, "Technologies to Reduce Carbon Dioxide Emissions in the Next Decade," *Physics Today* 53(11) (Nov. 2000):29.

25. S. Pacala and R. Socolow, "Stabilization Wedges: Solving the Climate Problem for the Next 50 Years with Current Technologies," *Science* 305(8) (Aug. 2004):968.

26. J. M. Deutch and R. K. Lester, *Making Technology Work* (2004, Cambridge University Press, Cambridge).

Notes

[1] This is about 2.2 kg C/gal. For gasoline the value is about 2.4 kg C/gal, and it is 2.8 kg C/gal for diesel.

CHAPTER 8

Risk and Economics

In Chapter 7 we touched on some of the risks associated with continued use of fossil fuels and the resulting changes to the earth's climate. Yet alternative energy sources are not without their own particular consequences. For example, widespread use of corn-based ethanol could result in higher food prices, water shortages, or increased pollution from fertilizer. To properly evaluate an alternative source of energy or a potential solution to any environmental problem, it is necessary to perform a cost–benefit analysis. The first step in this process is to scientifically evaluate the risk involved.

The subject of scientific risk assessment has its foundations in the calculation of actuarial tables for insurance purposes and the evaluation of risk for accidents in nuclear reactors and chemical manufacturing plants. There are several alternative approaches to risk assessment; for example, we may start with probabilities based on historical data or probabilities extrapolated from laboratory studies on animals. For a new technology the risk can sometimes be estimated by looking at the effect of possible failure of individual components of a more complicated system.

A scientific risk assessment seldom is definitive in decisions about the application of given technology. Often economic and political factors play a significant role. Deciding to ban chlorofluorocarbons was relatively easy because the economic impact was small. Banning carbon dioxide emissions is much more problematic because 85% of the energy supply for modern society is implicated. In the final section of this chapter we extend the discussion begun in Section 2.7.2 of the economic factors involved in making decisions about new technologies.

8.1 ■ Risk

Risk as evaluated scientifically can be loosely defined as the probability of an undesired event occurring times the severity of that event [1, 2]. Compare the risk of a bicycle accident for experienced adult riders against that for inexperienced riders. The severity of an accident is approximately the same for both, but the probability, and therefore the risk, is higher for the inexperienced rider. By contrast, the risk of head injury in a bicycle accident involving experienced riders is higher for someone not wearing a helmet; the probability of an accident is approximately the same, but the severity when an accident occurs is higher without the helmet.

Often a risk analysis answers the question, "What is the probability that this event (e.g., exposure to radiation) will cause this result (e.g., cancer)?" [1]. The answer may be expressed in terms of the risk per mile traveled, per dose received, or some other parameter that indicates the amount of exposure. Alternatively, the analysis may provide the risk for some occurrence from unknown causes either as a risk rate per year or as risk accumulated over time. For example, the average annual cancer rate in the United States from 2002 to 2006 for cancers from all causes was 542 per 100,000 for men and 409 per 100,000 for women. Over an average lifespan of 78 years the cancer risk is 45,500 for men and 36,900 for women per 100,000.[1] In other words, the chance of developing some kind of cancer in your lifetime if you are male is 45.5% (36.9% if you are female), but your chance of developing cancer in 2005 was only 0.54% if you are male and 0.41% if female.

Two other useful concepts related to risk are the relative risk (RR) and the probability of causation (POC). RR compares the risk for some group or population that has been exposed to a substance compared with an equivalent group that was not exposed. A relative risk of 1 indicates that there is no additional risk due to exposure. RR is sometimes reported as excess relative risk (ERR), which is the RR minus 1. A POC analysis answers the question, "Was this outcome (cancer) caused by this event (exposure to radiation)?" [2]. Although the risk for certain types of cancer may be low due to exposure to some chemical agent, a person with that type of cancer may have a high POC, indicating a high likelihood that the cancer was caused by the chemical exposure compared with other possible causes. The POC for exposure to a given agent is calculated as the risk from that source divided by the risk due to all causes, whereas RR compares an exposed population with an unexposed group. Although the two have slightly different meanings, they are often used interchangeably, and when small exposure rates to large populations are evaluated the two measures yield nearly the same result.

Often we are interested in the risk of death due to a particular activity, technology, or exposure to a chemical substance. In these cases we are concerned with *premature* death; everyone eventually dies from some cause, but usually we want to know the risk of early death due to a particular cause. This risk can be reported in a number of different ways, such as loss of life expectancy, person days lost, working days lost, nonworking days lost, quality-adjusted life years, or disability-adjusted life years. The last two figures can also be used to report risk of unexpected experiences or diseases that are debilitating but not immediately fatal. In this chapter we will look at loss of life expectancy, the most commonly reported risk figure. The reader is referred to the literature for a discussion of these other measures of risk [1–3]. In making decisions that involve data from risk assessment, it also is important to clearly distinguish which groups are at risk. There is some risk to the general public due to coal mining, such as traffic accidents with coal trucks and water and air pollution, but these are much lower than the risks faced by coal miners.

8.2 ■ Probability

The first step in a risk assessment for a particular event is a calculation of the probability of the event regardless of the severity. For some assessments the probability is sufficient for making a decision; the severity is not evaluated. In cases involving multiple unrelated probabilities, the total probability of an event occurring can be found by multiplying the probabilities together. Example 8.1 demonstrates the calculation of the probability of an event occurring where the individual probabilities are unrelated. Notice that in this example we are calculating probability rather than risk; the severity of the event is not included. Risks that involve probabilities that are dependent on each other are considered in the next section.

EXAMPLE 8.1

Probability of an Event on a Given Day

Calculate the probability that if your tire goes flat, it will occur on a rainy Thursday in October.

There is a 1 in 7 chance of the day being Thursday and a 1 in 12 chance of the month being October (if we assume all months have equal numbers of days). The chance of it being a rainy day varies depending on where you live. Los Angeles, California gets about 35 rainy days per year, whereas Recife, Pernambuco State, Brazil typically has about 210 rainy days per year.

The total probability for the flat tire to occur on a rainy day in Los Angeles is $1/7 \times 1/12 \times 35/365 = 0.00114$, or a chance of roughly 1 in 880. For Recife the probability is 0.00685, or 1 in 150.

If the probability of an individual event is known, for example from historical data, the risk, assuming equal severity, of the event occurring in some future time frame can be calculated as shown in Example 8.2. Here again the severity of the event is ignored, so what is actually being calculated is probability. In the example we also assume that the risk does not change with time, an assumption that is not always the case.

EXAMPLE 8.2

Risk of Fire

Suppose the probability of a fire at a paper mill is 11×10^{-6} per year of operation. There are approximately 2,400 paper mills in the United States. What is the chance of a fire in a paper mill in the United States in the next 30 years?

We have $11 \times 10^{-6}/\text{yr} \times 2400 \times 30\ \text{yr} = 0.79$ or a 79% chance for a fire occurring in one of the mills in the next 30 years. Notice that this treats all paper mills as identical in risk, regardless of the age or size of the plant or the possibility of improvements in safety over time. Also note that what appears to be a very small probability yields a high chance of occurring if enough time passes or the probability applies to a large number of cases.

Risk assessments are often used to compare the risk between two different actions. We want to know whether performing a certain activity with known risks will be a wiser choice than the risks associated with a different course of action. For example, we may want to compare administering a vaccine that has a known (but very small) fatality rate due to allergic reactions in order to protect a population from a serious disease, as opposed to administering the vaccine to only those who have a high probability of exposure to the disease. There are several ways to attempt to make this kind of risk comparison, and they potentially give different results. The absolute risk for each action can be compared, or the incremental risk, the increase of risk in each case, may be compared for only the affected populations. If we decide the risk should be shared by the total population, then the total risk should be calculated. These analyses do not necessarily give the same outcome, as seen in Example 8.3 (modified from an example in [2]).

EXAMPLE 8.3

Compared Risks

Suppose the risk for illness related to air pollution is X per person in a city. Two different types of incinerator plants are proposed. Plant A will cause an incremental increase in risk of pollution-related illness of 15%, but the plant is small enough that it can be located so that only a third of the population is affected. The incremental increase in risk for plant B is only 5%, but it must be located in a site that will affect half the population. Which plant should be chosen?

The risk before the plant is built is NX, where N is the total population. The absolute risk for the people affected by building plant A is $N/3 \times 1.15X = 0.38\ NX$, and the absolute risk of the people affected by plant B is $N/2 \times 1.05X = 0.52\ NX$. This suggests that plant A is the better choice because the absolute risk is lower.

However, we get a different result if we calculate the incremental increase in risk for the at-risk portion of the population. The incremental risk for plant A is $N/3 \times 0.15X = 0.05\ NX$, and the incremental risk for plant B is $N/2 \times 0.05X = 0.02\ NX$. Here we see that the incremental increase in risk due to plant B would be lower.

In the previous two calculations we have not included the unaffected population. Perhaps a better evaluation is the shared or total risk to the population due to either course of action. The total risk for plant A is $N/3 \times 1.15X + 2/3 \times 1X = 1.05NX$, and for plant B we have $N/2 \times 1.05X + N/2 \times 1X = 1.02NX$. Our conclusion is that plant B is the lower risk.

Example 8.3 should make it clear that a scientific risk assessment that looks at probability is only the first stage in the process of making political or economic decisions. Obviously both groups affected by the choice of plant location in Example 8.3 would feel differently about which plant should be built. The decision of which plant to build may be affected if one group has more political influence. The severity of a particular risk must also be evaluated. If the pollution-related illnesses in the example are life threatening or debilitating, the decision becomes more difficult than if the effects are slight. Cost may also be an important factor, in which case a cost–benefit assessment may be performed after the scientific risk assessment. Suppose plant *B* in the example costs 10 times plant *A*, and the severity of the health problems is known to be very mild. How do we decide whether the additional cost warrants the reduction in total risk by 3%?

8.3 ■ Historical Risk

Historical risk projections are calculated based on the assumption that the death or accident rate from some activity or event will remain about the same over time. In other words, the probability of a future event is assumed to be the same as the rate at which the event occurred in the past. In these cases the probability and severity are treated as a single number. The risk predictions for airplane accidents, car accidents, and employment accidents for a particular industry (e.g., mining fatalities) are generally made using historical data. A prediction based on historical data is the simplest type of risk model to calculate and, given enough data to ensure a constant trend, is generally accurate.

When looking at tabulated tables of risk based on historical data, it is important to keep in mind how risk is actually calculated for the figures given. For example, the risk of death in a particular occupation can be listed by age, cause, number per year, number per normal lifetime, or number per quantity of output product. Slightly different conclusions can be drawn about risk in the coal mining industry, for example, depending on whether deaths are measured per tons of coal mined, per person-hour, or per employee-year. Historically the number of deaths per ton of coal mined has decreased much more rapidly than the number of deaths per person-hour [2]. A safety evaluation based on deaths per ton of coal mined might reach a different conclusion than an evaluation made on the number of deaths per person-hour. Table 8.1 compares vehicle risk using several different parameters.

TABLE 8.1

Comparison of fatality rates and risks for different types of transportation. For car, truck, pedestrian, SUV, motorcycle, van, bicycle, and bus the annual average fatality figures for the 1999–2005 were divided by the number of miles traveled, time traveled, and number of trips, as reported in the National Household Travel Survey by the U.S. Department of Transportation [4]. Aviation data are from [5], boat data from [6], and rail data from [7]; railroad workers and other nonpassengers are not included.

Vehicle	Fatalities per Year, 1999–2005	Fatalities per 10^6 Miles Traveled	Fatalities per 10^6 Minutes of Travel	Fatalities per 10^6 Trips per Person
Car	19,979	102	51	964
Light truck	6,056	99	56	1,239
Pedestrian	4,815	19	0.9	14
SUV	4,000	85	46	868
Motorcycle	3,449	3,382	1,761	61,076
Van	2,082	44	23	426
Bicycle	704	1,117	91	2,133
Recreational boat	692		17	
Bus	45	5	1	39
Air	106	0.01	5	9
Rail	3	0.42		

It is clear from Table 8.1 that the number of deaths from a given cause is not sufficient to evaluate risk. For example, the total number of deaths involving cars is much higher than for light trucks, but fatality rates per minute traveled and per person trip are lower. By all measures traveling by bus, air, or train is safer than traveling in any personal vehicle. Clearly motorcycles are by far the most risky means of transportation, and traveling by bicycle is also surprisingly risky per mile traveled. Several details are hidden in the rough figures given in Table 8.1. For example, SUVs based on truck bodies have a fatality rate twice that of crossover SUVs and 25% higher than that of midsized cars. Likewise, 1-ton pickup trucks have a 30% higher fatality rate than compact pickups, as do sports cars compared to midsized cars [8].

As mentioned previously, there are several ways to report historical risk. In many instances (particularly legal cases) the number of years lost due to premature death for a given cause is used. This is generally calculated from the number of years left in life (relative to some average life expectancy, currently taken to be 78 yr) times the probability of dying from the given risk, averaged over the population. Table 8.2 lists the loss of life expectancy for a few interesting risks.

TABLE 8.2

Loss of life expectancy for selected risks [2].

Risk	Loss of Life Expectancy (days)
Alcoholic	4,000
Smoker	2,500
Cancer	1,300
20% overweight	1,100
Grade school dropout	950
Motor vehicle accidents	210
Air pollution	75
AIDS	55
Spouse smoking	52
Radon	45
Fire, burning	25
Poison	25
Nuclear power	2

8.3.1 Fatal Traffic Risks

In order to examine some of the complications involved in using historical data to make risk predictions, we will examine the risk of fatal traffic accidents in detail. Example 8.4 predicts traffic deaths in 2001 based on deaths in 2000 as an example of historical risk prediction.

EXAMPLE 8.4

Traffic Deaths in the United States

The number of people who died in car accidents in the year 2000 was 41,945 out of a population of 281,421,906. Predict the number of people who died in 2001 due to car accidents.

The proportion of automobile deaths was $41{,}945/281 \times 10^6 = 1.49 \times 10^{-4}$ in 2000. This is a rate of 15 per 100,000 people (150 in a million). In making a prediction of the risk of dying in an automobile accident in the year 2001, we would estimate the risk to remain at 15 per 100,000. The population in 2001 was 285,102,075, so the number of deaths would be predicted to be about 42,500. The actual number of traffic deaths in 2001 was 42,116, slightly less than predicted.

As mentioned earlier, one problem with using historical death rates to make predictions is that there usually is more than one way to report these figures. In Example 8.4 we used the number of accidents and the total population to find the fatality rate per 100,000 people in the United States. But not everyone drives or drives the same number of miles per year, and not everyone owns a vehicle. A second way to calculate the traffic fatality rate is the number of deaths per vehicle. This figure is easier to calculate and is more commonly used in countries outside the United States. In the United States official traffic fatality rates are supplied by the National Highway Traffic Safety Administration and are given as number of fatalities per 100 million miles traveled. In the case of traffic fatalities, both these measurements show the same trends over time. The number of miles driven, the total number of vehicles, and the total population all influence the fatality rate, and it is not clear which figure is more pertinent for a given risk prediction. Comparing these different measures sometimes leads to interesting conclusions. Figure 8.1 shows the number of highway fatalities, the vehicle miles traveled, and the death rate per 100 million miles traveled. We can make the paradoxical conclusion from these figures that driving has gotten safer over time, although the number of fatalities has not decreased significantly.

FIGURE 8.1

Traffic-related fatalities in thousands of people, vehicle miles traveled (VMT) times 10^{11}, and fatal accident rate per vehicle mile traveled in the United States [8]. Pedestrians killed by vehicles are included, as are truck and bus accidents.

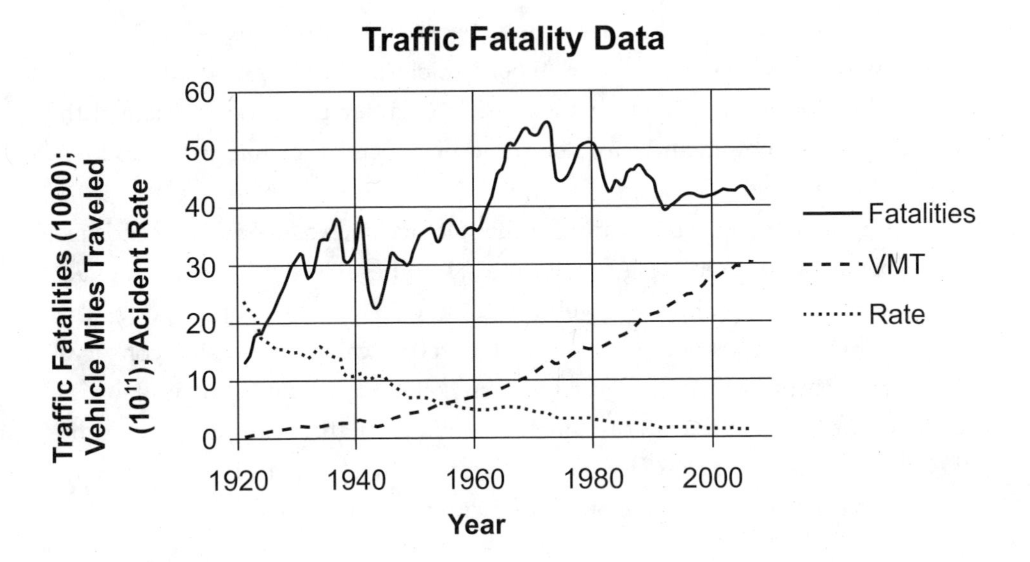

From Figure 8.1 we see that in 1921 the fatality rate was 24.08 deaths per 100 million vehicle miles traveled, but in 2007 the rate had decreased to 1.36. A similar trend can be seen in other countries under very different driving conditions, which indicates that as the population of a nation gains experience in driving and managing traffic, fatality rates decline. Although developing countries, such as those in Africa, have many fewer cars per person, they have much higher fatality rates per vehicle than developed countries. As can be seen in the graph for the United States, however, the change in the death rate is not linear, which means that trying to use past fatality rates to predict future fatality is problematic. However, the year-to-year change is fairly steady, so using trends to make short-term predictions is often of use.

Many factors affect car accident rates, and these change over time. In Figure 8.1 it can be seen that the total number of fatalities dropped significantly in 1974. This was largely the result of a reduction in the national speed limit to 55 mph in an attempt to reduce gasoline consumption after the oil shortages of 1973. However, changes in the trend of vehicle miles driven and the accident rate show less change in this time span. The age of the driver also drastically affects the accident rate. Male drivers between 18 and 21 years old have three times more fatal accidents than women of the same age and twelve times more fatal accidents than female drivers in their early 70s per vehicle. As drivers gain experience, they have fewer accidents; the rate of automobile deaths drops sharply as drivers enter their mid-20s and continues to gradually decrease until their mid-70s for men and women. The fatality rate per mile driven increases for people in their 80s, with the result that women in their late 80s have about the same fatality rate per mile traveled as men between 18 and 21 years old. Because older drivers drive less, the death rate per vehicle is still much lower for drivers over 80 than for young drivers. As we saw in Chapter 1, the average age of the population is changing, and this also could be expected to change the accident rate over time.

It might be thought that various other factors, such as vehicle age and size, would be factors in predicting fatality rates, but these do not have much effect. There is no correlation between vehicle age and fatality rate for countries that have significant numbers of cars [10]. A theoretical calculation using conservation of momentum shows that in a head-on collision of a car with an identical-sized vehicle with twice the mass, the person in the lighter car is 12 times more likely to suffer fatal injuries. However, this is not what happens in real accidents; car size does not directly correlate with death rates [10]. For example, the Ford

Explorer, with a weight of 4,255 lb, has approximately twice the fatal accident rate for drivers of both vehicles as a Volkswagen Beetle, with a weight of 2,762 lb [8]. This is in part because larger vehicles tend to be higher off the ground and so are more likely to roll in an accident. Sporty cars have higher accident and fatality rates than family sedans, probably because of the type of driver each attracts rather than the physical properties of the car itself.

The construction of the car body also has a large effect on survival, as do road conditions. As an extreme example of construction differences, race car drivers have survived collisions with solid barriers at 180 mph with only minor injuries in Formula One racing cars that weigh as little as 1,300 lb. These vehicles are built with a "survival cell" that protects the driver, even when the rest of the car is destroyed (see Problem 8.6). It is well known that driving at night and in the rain increases the chance of a fatal accident. It is also the case that urban roads are more than twice as safe as rural roads, but interstates in rural areas are safest of all [10].

In Example 8.4 we found the fatal automobile accident risk in 2000 to be 1.49×10^{-4} deaths per person per year in the United States. This does not sound like a large risk. Multiplying by 1 million gives a 150 in a million chance of a fatal car accident for residents in the United States in 2000, which also does not sound very risky. However, as noted, the total number of resulting deaths per year is approximately 42,000 because of the large population of the United States. This points out an important aspect of risk analysis: A seemingly small fatality rate applied to a large population over a long time has a devastating effect. Fourteen times as many people die on the highway each year as died in the terrorist attacks on the World Trade Center in 2001. A year and a half worth of traffic fatalities in the United States is approximately equal to the number of U.S. soldiers who died during the entire Vietnam War. Five times as many people have died on the highway in the United States since the end of World War II as U.S. soldiers who died during the war.

From this discussion it is clear that human factors play the biggest role in determining changes in traffic fatality rates. Although a huge amount of effort and money has been spent in the United States to make safer cars (e.g., seatbelts, airbags, antilock brakes) and safer roads, the most dramatic changes in safety have come because of decreased speed limits and a lower incidence of drunk driving. Lower fatality rates due to changes in driver behavior have been documented in other countries as well, where cars are sometimes much older and not always equipped with the latest safety devices. Predicting risk from historical data has the inherent difficulty that trends are subject to changes in human behavior. As an extreme example, a prediction of the death rate due to using horses as a means transportation made before the advent of the automobile would be grossly in error today.

8.3.2 Risk of Dying by Meteor Impact

As a further example of using historical risks, we examine the risk of dying due to a meteor impact. The statistics in this case are very well known, so a risk assessment is very accurate. However, the result is surprising.

After the formation of the planets, a large amount of debris remained in the solar system, traveling around the sun in various orbits, some of which were noncircular. Before about 3.8 billion years ago these asteroids produced large impact craters on all the inner planets, including the earth and the earth's moon. Because of their large size, the gravitational pull of Jupiter and Saturn gradually swept up many of the small remnants left over after planet formation. The remaining asteroids have been striking the earth, moon, and other planets at a lower but roughly constant rate for the past 3.8×10^9 years [11]. Although much of the evidence of crater formation on the earth has weathered away, more than 140 impact craters have been identified, and the mass extinction of flora and fauna occurring at the end of the Cretaceous period has been linked to a meteor impact.

The rate of impact and the size of the meteors currently striking the earth can be estimated from surveying cratering on the moon, where there is no erosion, and from direct observation of the earth's upper atmosphere. A smaller number of craters on Venus, which has a thicker atmosphere, and a larger number on Mars, which has a thinner atmosphere than earth, also yields information about the rate of cratering for the inner planets and the interaction of meteors with atmospheres. Some 25 tons of dust-size particles vaporize in the earth's upper atmosphere each day, leaving the characteristic traces in the night sky known as meteors [11]. About once a year an asteroid the size of an automobile enters the earth's atmosphere, and once every few hundred years an asteroid between 50 m and 300 m in diameter strikes the earth. The Tunguska event, an explosion occurring in 1908 over a remote region in Siberia, is now believed to have been caused by a large meteor impact. An explosion due to a meteor impact with energy roughly equivalent to the atomic bomb that exploded over Hiroshima occurs in the upper atmosphere of the earth about once a year but at altitudes too high to cause any surface damage [11]. Typically meteors reach the earth with speeds of about 20 km/s and have densities ranging from 1 g/cm^3 to 8 g/cm^3. The energy of a typical meteor is calculated in Example 8.5.

EXAMPLE 8.5

Energy of a Meteor

Estimate the kinetic energy of a meteor with a radius of 100 m and density $\rho = 4$ g/m^3 traveling at 20 km/s.

Kinetic energy is $\frac{1}{2}mv^2$, and mass is density times volume, ρV, so for a spherical meteor we have

$$(8.1) \quad \frac{1}{2}\rho V v^2 = \frac{1}{2}\rho \frac{4}{3}\pi r^3 v^2 = 3.35 \times 10^{18} \text{ J}.$$

The energy of an explosion is often given in terms of the equivalent tons of TNT. One ton of TNT gives off approximately 4.18×10^9 J of thermal energy. Dividing this into Equation (8.1) gives 8.02×10^8, or 800 Mt of TNT equivalent.

To date there have been no documented human fatalities due to meteor impacts. There are several known cases of automobiles and houses being struck by very small meteors, however. From calculations based on nuclear bomb testing and estimations of the result of the Tunguska event, the area of devastation from a large meteor that penetrates to a distance of less than 25 km from the surface is given in square kilometers by

$$(8.2) \quad A = 100Y^{2/3},$$

where Y is the yield in megatons (Mt) equivalent of TNT. Most of the buildings within such an area of impact would be flattened, and the population in this zone would probably be killed instantly.

A calculation of the yield of a meteor colliding with the earth can be estimated from kinetic energy, as in Example 8.5, including the angle of collision. Meteors entering the atmosphere at very shallow angles tend to be deflected back into space. However, a solid meteoroid with a diameter of 50 m traveling at 20 km/s has a yield of about 10 Mt and can penetrate into the lower atmosphere if it strikes at an angle of 90° to the surface. A solid meteor with a diameter of 250 m traveling at the same speed would have a yield of about 1,000 Mt and would probably leave a 5-km diameter crater on impact. Table 8.3 shows the estimated fatality rate for different-size meteor impacts [11]. The estimations in the table do not include deaths by tsunamis caused by impacts occurring over oceans. The global events listed, if they were to occur, would kill a significant portion of the present earth population (6.7×10^9).

Given the data in Table 8.3, a calculation of the risk of dying from a meteor can easily be performed, as shown in the following example. Because the historical data are well known, the risk calculation is surprisingly accurate.

EXAMPLE 8.6

Risk of Death by Meteor

Using the figures in Table 8.3, estimate the risk of death by a global-scale meteor impact for a person of lifespan 70 years.

Using 1.5×10^9 deaths for an event that occurs once every 7×10^4 years gives a fatality rate of $1.5 \times 10^9/7.0 \times 10^4 = 2.1 \times 10^4$ deaths per year (shown in the final column of Table 8.3), about half the number of U.S. highway deaths per year. Dividing this by the present global population of 6.7×10^9 gives a risk of 1 in 3.1×10^{-6} per year (a three in a million risk). If you live for 70 years, this would be a lifetime risk of 70 yrs $\times$ 3.1 $\times 10^{-6} = 2.2 \times 10^{-4}$, or 220 in a million, similar to the lifetime risk of being hit by a car as a pedestrian.

As we can see from Example 8.6, the risk of death from a meteor impact is significant; in fact, it is much greater than the risk of dying in a tornado or hurricane or being struck by lightning. Why then, has no one died of a meteor strike? The risk from meteors is an interesting example of a low-probability risk that has a very high consequence or severity. Tunguska type events occur only every 250 years, and an event that is large enough to kill off a significant fraction (20%) of the global population occurs only once in 70,000 years. Our perception of the danger of a meteor impact is low because we have no direct experience with such a disaster, unlike tornadoes or hurricanes. Statistically, however, the risk of death from a meteor impact is roughly the same as dying in an airplane crash or flood.

The known periodic meteor showers can be predicted because the orbits of the asteroids responsible for those showers have been identified. These objects

TABLE 8.3

Fatality rates for different types of meteor impacts [11]. The global deaths per year decrease for large-scale events because these events are rare.

Event	Diameter (m)	Energy (Mt)	Average Interval (years)	Local Deaths	Global Deaths per Year
Upper atmosphere	<50	<9	Daily	0	0
Tunguska type	50–300	9–2,000	250	5×10^3	20
Medium–large	300–600	2,000–1.5×10^4	35×10^3	3×10^5	8
	300–1,500	2,000–2.5×10^5	25×10^3	5×10^5	20
	300–5,000	2,000–10^7	25×10^3	1.2×10^6	45
Global	>600	1.5×10^4	7×10^4	1.5×10^9	2×10^4
	>1,500	2×10^5	5×10^5	1.5×10^9	3×10^3
	>5,000	10^7	6×10^6	1.5×10^9	250
Rare (Cretaceous type)	>10,000	10^8	10^8	5×10^9	50

are generally quite small and are not thought to be a threat. NASA has a program for larger objects that have orbits close enough to the earth that there is a risk of collision [12]. The threat is very real; in 1989 asteroid 1989FC, with kinetic energy of 1,000 Mt, passed inside the moon's orbit. Although comets are less dense than most asteroids and have orbits that cause them to strike the earth much less often, they typically have speeds two or three times as high as asteroids, making them a significant threat as well. Pictures of comet Shoemaker–Levy colliding with Jupiter in 1994 show spectacular, Earth-sized holes in the surface layer of clouds on Jupiter.

8.4 ■ Risk Due to New Technology

Risks due to new technology cannot be assessed directly from historical data. For example, there have not been enough nuclear reactor failures to yield statistically significant death rate data. In these cases the probability of failure of a complex system can sometimes be calculated based on the likelihood of failure of the individual components. The probability of a critical water supply pipe breaking in a chemical plant can be made from historical experience with similar pipes. The probabilities of failure for each component in the process are then multiplied to get the probability of a failure of the entire device. This procedure is called a fault tree or event tree analysis and is applicable when probabilities of failure are not independent. In the following we apply these ideas to nuclear power plants, but it should be kept in mind that the same types of calculations are routinely made for other technological applications, such as coal power plants, chemical manufacturing plants, incinerators, and other industrial processes.

As explained in Chapter 5, nuclear reactors cannot explode as atomic bombs. However, it is possible for them to release radioactive gas to the atmosphere in the event of a serious disaster, such as a core meltdown or graphite fire (Chernobyl). For this reason nuclear reactors are built with multiple levels of safety mechanisms designed to prevent a release of radioactive material in case of mechanical or electrical failure in the reactor. Each stage in a long chain of backup systems must fail in order for a serious reactor failure to occur. This is an example where, unlike Example 8.1, the probabilities do affect one another. The first stage in the sequence of safety features for a light water reactor (the most common reactor design) is the cooling system. An extremely simplified fault tree for a loss of cooling accident in a nuclear reactor is shown in Figure 8.2.[2] The possible stages for emission due to a loss of coolant shown in the figure include (a) rupture of a supply pipe (probability P_r), (b) loss of electric power (probability P_e), (c) failure of the emergency cooling system (probability P_c), and (d) failure of containment vessel (probability P_v). The probability of each of these not occurring is 1 minus the probability of it happening.

FIGURE 8.2

Extremely simplistic fault tree analysis of an emission of radioactive gas from a nuclear reactor due to a loss of coolant.

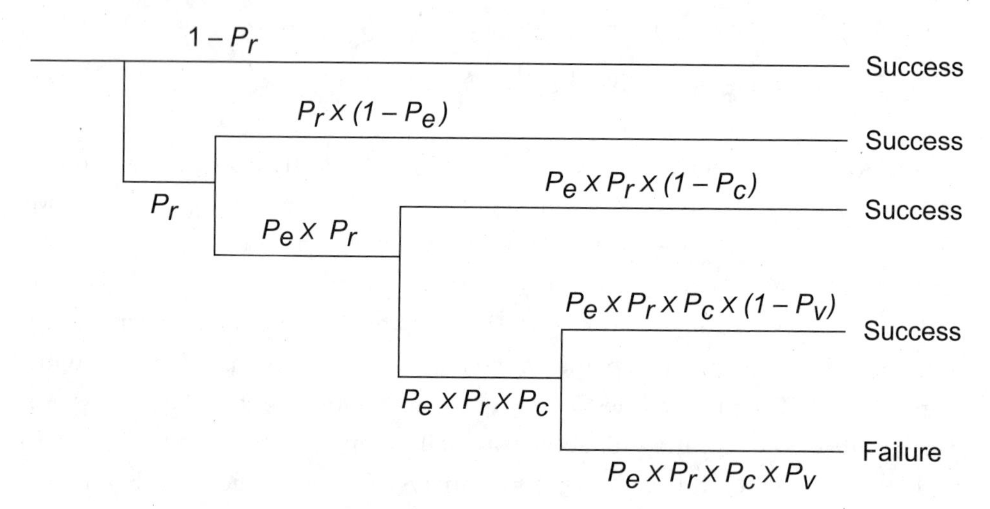

If the probabilities for each component in a fault tree are known, the probability or risk of an event occurring can be calculated. Example 8.7 is an example of how to perform such a calculation.

EXAMPLE 8.7

Simplified Fault Tree Analysis for a Nuclear Reactor

Suppose the probability for each of the stages (rupture of a supply pipe, loss of electric power, failure of the emergency cooling system, and failure of containment vessel) is known to be 0.1 (10%) per year based on historical failure rates of similar components. What is the reactor-year probability of failure of the backup systems of the simplified model in Figure 8.2?

The bottom-most branch of the fault tree gives the probability that all of the four backup systems fail. We have $P_e \times P_r \times P_c \times P_v = 0.0001$ per reactor-year or a 1 in 10,000 chance per reactor per year.

The first step in an actual calculation of a fault tree analysis of a nuclear power plant is to identify all possible events that might lead to plant shutdown, core meltdown, or a release of radioactivity. There are several problems inherent in making an estimate of such events. The failure rate for a particular water valve might be well known because of historical data of its use in other applications, but unexpected changes over time might occur. For example, the amount of deterioration of steel water supply pipes in nuclear reactors as a result of radiation damage was not anticipated when the plants were initially constructed (fortunately, this has not resulted in any serious reactor failures). The Chernobyl disaster was primarily the result of human actions; the operators turned off many safety features, changing the fault tree and making any earlier risk analysis useless [13]. A real evaluation of the risk of a reactor leaking radiation into the environment would be much more complicated than Figure 8.2 because of the large number of components in a reactor. The cooling system is only one of several lines of defense against a total reactor failure resulting in the release of radioactivity to the atmosphere. The current estimate of the probability of death due to failure of the safety features of a reactor is 3×10^{-5} deaths per reactor year of operation; a single reactor would have to operate for 100,000 years for three accident-related deaths to occur [2].

It should also be pointed out that the risk of failure of reactor components is not the same thing as the risk of exposure of the population to radioactive material. In the Three Mile Island incident in 1979 in Pennsylvania, the most serious reactor accident in the United States to date, there was a coolant system failure, which led to a core meltdown, but there was no hazardous exposure to the public. Once the risk of exposure has been calculated, the probability of it leading to disease or death must be calculated using other methods (discussed in the following sections).

As a final point, a fault tree analysis can sometimes be corroborated by calculating an upper bound of the likelihood of failure of the system based on an absence of historical failure. For example, there have been more than 8,000 reactor-years of operation worldwide for light water reactors without a loss-of-cooling accident (neither the Three Mile Island reactor nor the Chernobyl reactor was a light water reactor; see Chapter 4). This puts an upper bound on the likelihood of a loss-of-cooling accident of $1/8{,}000 = 1.25 \times 10^{-4}$ per reactor-year for light water reactors.

8.5 ■ Risk Calculations from Epidemiology

In most of the preceding examples the cause of the risk is clearly and directly related to the risk; it is the car accident that causes the death, and the death occurs very soon after the accident most of the time. In many cases, however, the cause-and-effect relationship is not clear, usually because it is delayed. In these cases the severity of the risk is in question. It is still possible to make a risk assessment connecting effects with their probable cause, but the procedure is more complicated. For example, the effects of smoking were not obvious initially because there is about a 20-year lag, on average, between when a person starts to smoke and when he or she shows signs of lung cancer. The exact mechanism of how this occurs is still not completely understood, but the data are strong enough to conclude that smoking causes cancer. Comparisons of large population groups with different exposure rates are called epidemiological studies, the topic of this section.

Most epidemiological studies also involve an estimate of the risk associated with exposure to different dosages of a given contaminant. As we will see later in this chapter, the question of quantity of exposure is very complex. The complexity of the effects of different dosages is one reason the much lower dosage of carcinogens in second-hand smoke has only recently been identified as a cancer agent. Often animal studies are used to validate and extend data first revealed by epidemiological studies. Ideally there should be information from both sources to make a decision about limiting exposure. In reality epidemiological data can never establish cause, only a strong cause-and-effect association.

There are two parts to a calculation of risk based on epidemiological studies or on animal studies. First, the amount of exposure must be calculated to determine the dosage received. Once the amount of exposure is determined, the effect of that dosage on the health of a person must be determined.

8.5.1 Stock and Flow Models

Before the risk to life or health of a particular chemical substance can be determined, the rate of exposure must be either measured or calculated. The exposure may be airborne, liquids in contact with the skin, or ingested substances such as food additives. In many cases it is not possible to measure the actual exposure because it happened in the past. However, in some instances it is possible to calculate the likely exposure from known data. Such a calculation is called a stock–flow model, of which there are several varieties depending on possible scenarios for exposure. A person might be exposed to a substance that is continuously being emitted into the air or a substance that is emitted in a single event and eventually decays or is diluted over time. Examples of continuous exposure are ambient pollutants from nearby factories or power generation plants. Single events might include a radioactive leak from a nuclear power plant or a one-time exposure to a chemical spill. Exposure might also be periodic, such as food additives that are ingested on a daily basis. Occupational contact with toxic fumes or liquids would also be an example of periodic exposure. The exposure to airborne toxins is often a function of the rate of air exchange in a building, as seen in the following example.

EXAMPLE 8.8

Exposure Concentration

Suppose a particular carpet adhesive continues to emit vapor at a constant rate of 12 $\mu g/m^2$ per minute for several months after the carpet is laid. What is the equilibrium concentration in a 20-m^2 room, 8 ft high, if the air is exchanged once every 4 hours?

The flow of vapor into the room from the carpet is 12 $\mu g/m^2 \times 20\ m^2 = 240\ \mu g/min$. At equilibrium concentration the flow into the room from the carpet equals the flow out of the room due to ventilation. The flow out of the room is Flow = $V \times E \times C$, where V is the volume of the room, E is the air exchange rate, and C is the concentration. Solving for the equilibrium concentration, we have $C = Flow/VE = [240\ \mu g/min \times 60\ min/hr]/[20\ m^2 \times 8\ ft \times (1\ m/3.28\ ft) \times (1/4\ hr)] = 1180\ \mu g/m^3$.

Example 8.8 assumed that the rate of flow into and out of the system of the toxic chemical was constant. For many substances, however, the flow rates are not constant. An example of variation in flow rate out of the system is radon, a naturally occurring radioactive gas that enters houses from the ground and decays over time. Radon, $^{222}_{86}Rn$, is a radioactive daughter product of naturally occurring radium found in some kinds of rock formations. Because radon is a gas and decays via alpha decay (see Chapter 4) with a half life of 3.8 days, it poses a potential lung cancer risk if the indoor concentration is sufficiently high (in fact, it is the second leading cause of lung cancer after smoking in the United States). The series of daughter products of radon decay have very short half lives that quickly result in a stable isotope of lead that is not a health threat. The following example treats a constant incoming flow, as in Example 8.8, but includes the possibility of the substance being removed from circulation by radioactive decay [1]. In this example the decay is not very effective in removing the toxin, but the model is applicable to other situations in which a decay or chemical reaction results in a significant reduction of the toxin over time. Additional aspects of radiation risk are treated in Section 8.5.4.

EXAMPLE 8.9

Home Radon Concentration

Suppose the room in Example 8.8 has a radon flow into the room from the ground of 3.5×10^5 pCi/h (1 pCi = 0.037 decays per second). The half life of radon is 3.8 days, so the radon is being removed by radioactive decay as well as by ventilation. How much is the equilibrium concentration affected by the decay rate?

Recall from Chapter 4 that the rate of decay for a radioactive substance is $\lambda = \ln(2)/t_{1/2}$, where $t_{1/2}$ is the half life. So the flow out due to decay is $\text{Flow}_{decay} = C \times \ln(2)/(3.8 \text{ days})$. Once again, the flow into the room equals the flow out, so we have $\text{Flow}_{in} = V \times E \times C + C\ln(2)/(3.8 \text{ days})$. Solving for C, the equilibrium concentration, gives $C = \text{Flow}_{in}/\left(EV + \frac{\ln 2\,\text{m}^3}{3.8\,\text{day}}\right) = (3.5 \times 10^5\,\text{pCi/h})/(12.2\,\text{m}^3/\text{h} + 0.008\,\text{m}^3/\text{h}) = 2.87 \times 10^4\,\text{pCi/m}^3 = 28.7\,\text{pCi/l}$. Notice that the radon decay rate is slow enough that it does not affect the concentration very much.

The concentration of radon in a house can easily be measured with a home test kit [14]. A carbon collector that absorbs radon is left exposed in the house for a week and then sent off for analysis. The average house in the United States has a concentration of about 1.5 pCi/l, but the concentration can vary by a factor of 20, depending on location and house construction. Because of the risk of lung cancer associated with radon, the Environmental Protection Agency in the United States recommends radon remediation if concentration levels are found to be greater than 4 pCi/l [15]. A remediation procedure starts with the known concentration, and the flow rate into the house is calculated by reversing the calculation in Example 8.9. Remediation measures to reduce the flow of radon into a house include sealing the basement, pressurizing the basement, increasing the ventilation rate, and introducing ventilation underneath the foundations of the house to remove the radon before it reaches the basement floor or walls.

The concentration of a substance after a one-time exposure may also involve exponential decay, this time without a constant flow rate into the system. Often the rate at which a toxin disappears is proportional to the concentration. This is true for substances introduced directly into the blood by injection and also by ingestion if the substance can easily cross the stomach lining. As the body removes the substance, the rate of removal is proportional to the amount left. If $C(t)$ is the concentration at time t, then we have

$$(8.3) \quad dC(t)/dt = -kC(t),$$

where $dC(t)/dt$ is the rate of change and k is the rate constant. The negative sign indicates that the concentration is decreasing. Equation (8.3) is the differential equation for exponential decay and has solutions

$$(8.4) \quad C(t) = C(0)e^{-kt},$$

where $C(0)$ is the initial concentration. We have seen this equation previously in the context of radioactive decay in Chapter 4 and with a positive exponent for population increase in Chapter 1. In Example 8.9 we also used this result for radioactive decay where $k = \ln(2)/t_{1/2}$.

EXAMPLE 8.10

Blood Level of a Drug

The treatment for various diseases requires that a drug remain in the bloodstream at a minimum concentration. The timing of the next dose depends on the rate of elimination of the drug from the body. Suppose a patient is injected with 1.2 g of a substance that is eliminated from the body at a decay rate of $k = -0.6/\text{h}$. If the minimum blood level for an effective dosage is 0.04 g/l in the blood, when should the next dose be administered? Assume an adult human body has about 5.0 l of blood.

We want the time when $C(t) = 0.04$ g/l. The initial concentration is $C(0) = 1.2$ g/5.0 l in which case, from equation (8.4) we have 0.04 g/l = (1.2 g/5.0 l) $e^{-0.6t}$. Solving for t gives $t = 3.0$ h.

It should be noted that the calculation in Example 8.10 can be used to find the blood level of a toxic substance due to periodic exposure by ingestion, inhalation, or skin exposure if the rate of transfer from the stomach, lungs, or skin to the blood is known. The amount of nicotine in the blood due to the periodic smoking of cigarettes can be calculated using this method. In order to find the risk of a chemical agent once the amount of exposure is known, the effect of a dose of a known amount must be determined, which is the subject of the following subsection.

8.5.2 Effective Dosage

The effect of a particular dosage of a chemical substance is usually not known precisely, although in some cases there are enough epidemiological data to determine the effect. We now have enough data comparing smokers with nonsmokers to conclude that there is a correlation between smoking and lung cancer and heart disease. The data for smoking include sufficient information about the degree of exposure (e.g., number of cigarettes per day or amount of secondhand smoke) and death rates to have a very good idea of the effect of tobacco smoke on health. Data from miners, who have a much higher exposure to radon, eventually led to studies that established a fairly complete understanding of the effects of radon on health. In another example, the risk of colon cancer from a diet high in fat came from studies involving immigrants coming to the United States. Before changing their diets to match those of people in the United States these immigrants had lower rates of colon cancer, but the rates changed to match those of U.S. citizens as their diets changed to match those of North Americans. In many other cases only partial or inconclusive data are available.

When the effect of a given dosage is known, a comparison to other dosages can sometimes be made. Lung cancer rates of miners exposed to radon are known. We might then ask what is the equivalent exposure or risk for the average nonminer or for a house with an elevated level of radon. The following example uses the concept of working-level months, which in the case of radon is defined as exposure to 100 pCi/l of radon in air for 173 h a month. The death rate due to lung cancer among miners is known to be between 200 and 600 deaths per million working-level months [1].

EXAMPLE 8.11

Radon Death Rates

The average home level of radon in the United States is 1.5 pCi/l. What is the number of working-level months (WLmo) of radon exposure for a person in the United States? What are the chances that the average person will have lung cancer at this rate of exposure? How many additional lung cancer deaths are expected from the average U.S. exposure?

If we assume the average person spends three fourths of the day at home, we have 0.75×1.5 pCi/l $\times$ 365 days/yr $\times$ 24 h/day $\times$ 1 WLmo/(173 h/mo $\times$ 100 pCi/l) = 0.57 WLmo/yr for the average exposure in the United States.

For an average lifespan of 72 yr the total exposure is 0.57 WLmo/yr $\times$ 72 yr = 41 WLmo. The cancer rate is between 200 and 600 deaths per million working-level months, so 200×41 WLmo/10^6 WLmo = 0.0082, or a 0.82% chance of lung cancer. For the higher estimate of deaths per million, there is a 2.4% chance.

For an estimated population of 306 million we have $306 \times 10^6 \times 0.57$ WLmo $\times$ 200/10^6 WLmo = 35,000 deaths per year. For the higher cancer rate we get about 104,000 deaths per year. Notice that the high estimate is more than twice the number of traffic accident deaths per year. The actual figure is estimated by the U.S. Environmental Protection Agency to be about 21,000 deaths per year. This is because the average level of 1.5 pCi/l does not account for the fact that many houses have no radon, whereas a small number of houses have a disproportionately high concentration of radon, so the at-risk portion of the population is actually much smaller than the entire population.

For many substances, such as new chemical products or food additives, the health effects of the substance are entirely unknown. In these cases often animal studies are required by the Food and Drug Administration or the Environmental Protection Agency, particularly if the substance will be administered as a medical treatment. Animal studies are quite expensive; a typical animal study with 300 rats might cost as much as $5 million [1]. For this reason, newly produced chemicals not used for medicinal purposes are sometimes not tested if there is little reason to think they might be toxic or if the exposure is thought to be slight. If, after the substance has been on the market for a while, the epidemiological data begin to indicate a problem, then a regulatory agency such as the EPA or FDA may require the manufacturer to perform animal tests.

Animal testing is generally performed with specially bred laboratory animals, typically rodents. These animals are bred to be genetically identical, as much as possible, and are usually selected to be sensitive to carcinogens, with the idea that greater sensitivity will reflect the worst-case scenario for human exposure. If the substance is being tested for carcinogenic properties, identical groups of animals are given varying amounts of the toxin under study, and after some period of time the animals are killed and examined for tumors to see whether exposure correlates to increases in cancer rates. A second kind of test is to give different groups of animals different amounts of a toxin and count how many die prematurely during some set period of time. Results from this kind of study are given as the LD_{50},[3] the dosage that kills 50% of the test animals in some standard time interval. LD_{10} values, which give the dosage that kills 10% of the animals, are also often reported. In some cases toxicity studies can be performed on biological tissue or cell samples rather than live animals; however, because of the complex interaction of various systems in a living organism, animal studies are generally considered to be superior to tissue studies. A drawback of animal study results is the fact that there is no guarantee that they will extrapolate well to humans, particularly for chemicals that target specific tissues or organs that may not react similarly in humans and animals.

In many risk analyses we want to evaluate whether the results of two different data sets are significantly different. In the simplest examples there are only two possible outcomes; for example, either the animal in a test trial dies during the test or it does not. A full discussion of the statistics involved in making these kinds of decisions is beyond the scope of this text, but a simple method of hypothesis testing in the case of only two possible outcomes is the *Z* test, used in the following example. Other tests include the chi-squared test, Student's

t test, and the f test. A very brief introduction to statistics is given in Appendix B; however, the reader should consult more advanced texts for further explanation of statistical methods.

EXAMPLE 8.12

Does Drug X Cause Cancer?

Suppose an animal exposure test involves 50 rats that are administered a dose of the chemical X and 50 control animals. In the control group (no exposure), two animals end up with tumors at the end of the trial period, and in the other group 10 have tumors. Can we conclude that the substance causes cancer and, if so, with what confidence?

For the control group, the probability of a tumor is 2/50 = 0.04, and for the other group the probability is 10/50 = 0.20. In the case in which there are two possible outcomes (tumor or no tumor), the data are expected to match the statistical model known as the binomial distribution. For the binomial distribution, the standard deviation is given by

$$(8.5)\quad \sigma = \sqrt{Q(1-Q)\left(\frac{1}{n_1}+\frac{1}{n_2}\right)},$$

where $Q = (p_1 n_1 + p_2 n_2)/(n_1 + n_2)$ (see reference [1] for a derivation). Here the p_1 is the probability of a tumor in the control group (0.04), p_2 is the probability of having a tumor in the second group (0.20), and n_1 and n_2 are the total numbers of subjects in each case. We calculate $Q = 0.12$, and Equation (8.5) gives $\sigma = 0.065$. The z score for comparing two binomial distributions is given by

$$(8.6)\quad z = \frac{|p_1 - p_2|}{\sigma} = 2.46.$$

Now we look up this value in a table of z scores or use a z score calculator [16], which tells how many standard deviations away from a perfect match ($z = 0$) these results are. The value in the table for $+2.46\sigma$ is 0.9862, or 98.6% of the data are expected to lie closer (within $\pm 2.46\sigma$) to the mean than these results. This means there is only a 1.4% chance that the two outcomes are statistically equivalent. This is a strong indication that the tumors were caused by the chemical administered during the test and did not occur randomly, as was the case for the two tumors in the control group.

8.5.3 Dose Models

The assumption that the risk (or effect) is directly proportional to the dosage (or cause) is known as the linear default model. For example, it is known from epidemiological data that exposure of one to five parts per trillion of arsenic in drinking water causes cancer at a rate of one in a million (of the lung, kidney, and bladder; combined risk). Background levels of arsenic in drinking water often exceed this by a factor of 1,000, in which case for a linear dose model we would expect the risk to be 1,000 in 1 million (1 in 1,000) for people so exposed. Although it is a useful first approximation, there are several problems with the linear default model.

From clinical trials we know that the linear default model does not apply to information from animal testing. The effects on small animals (usually mice or rats) cannot be assumed to extrapolate in a linear way to humans with larger body mass. A similar problem is encountered in determining the appropriate dosage of medicine for children when the effective dosage for an adult is known. The average daily dosage, d_h for a human that has the same effect for a given animal dosage, d_a, is

$$(8.7) \quad d_h = sf \times d_a,$$

where the scale factor, *sf*, is given by

$$(8.8) \quad sf = (W_h/W_a)^{1-b}.$$

Here W_h is the weight of the human, W_a is the weight of the animal, and b is usually taken to be approximately 0.75, a value based on historical data and used by the U.S. Environmental Protection Agency. Notice that the scale factor is not linear; a human weighing 20 times as much as a test animal does not have a scale factor of 20.

EXAMPLE 8.13

Extrapolation of Mouse Data

Suppose the rats in the exposure trial of Example 8.12 averaged 0.5 kg and had a normal lifetime of about 2.5 yr. Further suppose that the dosage for the group of 50 rats that resulted in 10 tumors (a probability of 0.20) was 30.0 mg/kg/day. What is the expected dosage that would produce the same probability of tumors in a population of humans of mass 65.0 kg?

Body weight is proportional to mass, so we may combine Equations (8.7) and (8.8) to get $d_h = (W_h/W_a)^{1-b} \times d_h = (65.0/0.5)^{0.25} \times 30.0$ mg/kg = 101.3 mg/kg per day. Were a group of humans to be exposed to this daily dose of the drug, we would expect a 20% (0.20 probability) incidence of tumors.

The total dosage per day needed to produce this effect in a 65-kg person would be 101.3 mg/kg × 65 kg = 6.6 g, and the lifetime exposure would be about 1,692 kg.

One difficulty implied by the linear default model is the political decision of how low a dosage is acceptable. If we assume the linear default model (at least for adults), we might try to reduce the exposure risk to zero for anything known to be hazardous at any level. There are at least two reasons not to do this. For most substances there is a threshold dosage below which the effects cannot be distinguished from other causes of death in the general population. It is probably impossible and not necessary to insist that public drinking water be completely free of arsenic [2]. Instead, in such cases we usually try to reduce the risk to an acceptable level, such as 1 in 100 million. At this rate there would be fewer than three cases of cancer caused by arsenic in the entire United States per year, which would be very difficult to distinguish from other cancer causes. In effect, lowering the arsenic exposure risk to 1 in 100 million lowers it below the background of other causes of death, and we can no longer be sure that someone with cancer got it from exposure to arsenic. In other words, the probable cause of death (POC) for arsenic would be lower (much lower, in fact) than the average for all other causes.

Another reason to be cautious in applying the linear default model is that for some substances there is a dose level below which the substance is actually beneficial. The general term for such an effect is *hormesis*. Figure 8.3 shows the linear dose model, a typical hormesis curve, and two other possible dose–response models. In the figure a positive response may indicate a beneficial or a detrimental effect depending on whether the substance is a medical drug or toxin. Alcohol consumption is an example of hormesis, where a positive dose response is considered to be detrimental. Drinking alcohol has adverse risks (cancer, accidental death) if consumed at dosages of more than three drinks per day. However, people drinking less than one drink per day have higher death rates than those drinking one to two drinks per day. The death rate associated with alcohol is lowest for men drinking one and a half drinks per day and for women taking one drink a day, where one drink is defined as one 12-oz beer, one 5-oz glass of wine, or one 1.5-oz shot of whisky. The probable reason for this

FIGURE 8.3

Various possible dose–response curves (arbitrary units). The response may be detrimental or beneficial, depending on the substance. The linear model assumes a direct relationship between dose and response. The hockey stick or threshold model assumes no response until some threshold dose is reacted (about 15 units in this graph). For the hormesis case there is a negative reaction (assumed to be beneficial in this case) up to some threshold dose (about 22 in the figure).

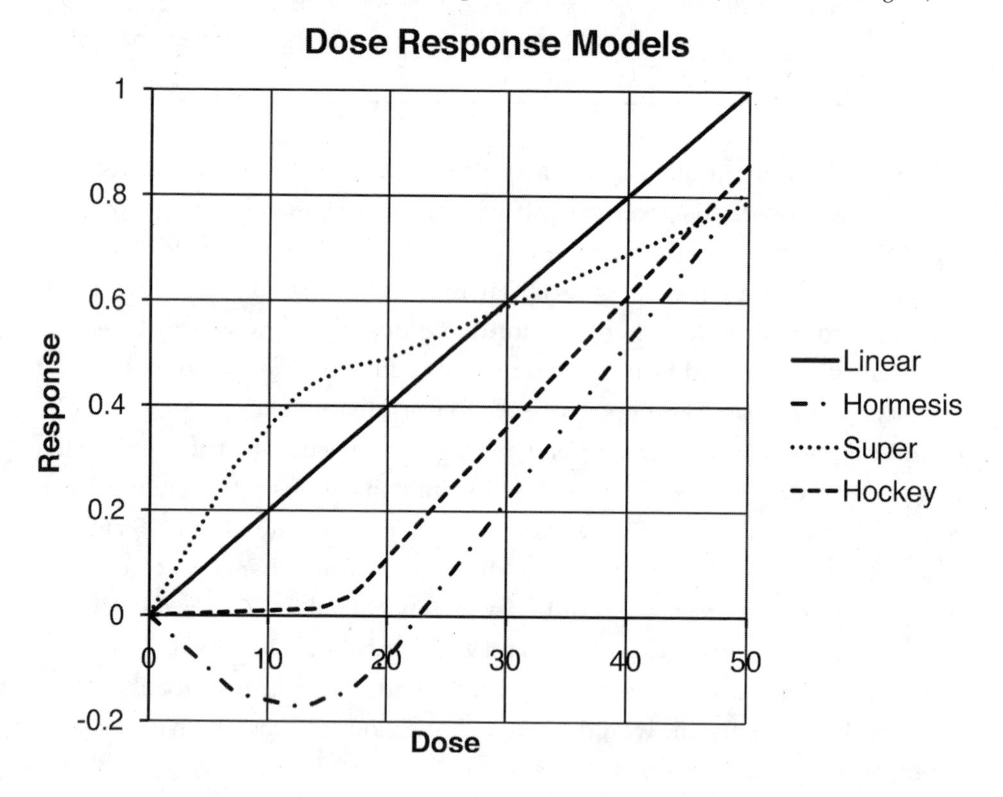

effect is that alcohol increases cancer risk and risk of accidents but decreases risk from heart disease, and these two effects intersect at about one drink per day.

As can be seen in Figure 8.3, there are other possible reactions to exposure to a chemical agent. In the threshold or hockey stick model there is little or no reaction until a threshold dosage, sometimes called the maximum tolerated dosage, is reached. Some types of carcinogens seem to fit this model; there must be repeated exposure before any effect is seen, probably because repair mechanisms in the body can recover from low exposure rates. For chemical agents obeying the super model the risk or effect of high dosages may be overestimated because of a low dose response that is initially stronger than linear. Drug tolerance (good or bad drugs) due to constant exposure fits the super dose–response curve; initially the drug causes a large response, but taking more of the drug does not produce proportional increases in response. Differentiating between the various possible response curves requires sufficient data, from animal studies or historical data. For example, if only a few data points at dosages between 45 and 50 in the figure were known, it would not be possible to distinguish between the four dose curves.

Because of the time delay between exposure and illness (as is typical for cancer), it is sometimes difficult to assign values to the dose level that causes a given health problem. Humans are also exposed to many potentially harmful agents simultaneously. This makes it very hard to distinguish between different causes and also raises the problem of synergistic effects. The interaction between radon and cigarette smoking, for example, required a great deal of research because both cause lung cancer, and until recently the amount of radon exposure for various groups was not well established. It is now known that smoking and radon together cause a greater risk than simply multiplying the risk of each.

As a final example of epidemiological studies, we consider the effects of air pollution on health. An early example of the health risks of air pollution occurred in London in 1952. Because of manufacturing and power generation, smoke, soot, and sulfur dioxide levels were already quite high at this time. In the first week of December of that year, the level of smoke and soot rose from about 500 $\mu g/m^3$ to 1,500 $\mu g/m^3$, and the level of SO_2 rose from about 0.15 ppm to 0.75 ppm because of a temperature inversion that trapped pollutants in the London area. In lockstep with this tripling in emissions, the number of deaths increased from about 250 per day to 900 per day during this 1-week period. It is estimated that approximately 12,000 people died as a result of the sudden increase in pollutants. Similar but less severe events occurred in London in 1956, 1957, and 1960. This and comparable episodes occurring in other cities have made a compelling argument for the connection between air pollution and health. We now have a reasonable amount of epidemiological data related to air pollutants, and consequently air standards are now mandated in many countries, including the United States.

One source of air pollution is in the energy production sector. As we saw in Chapters 4 and 5, many alternatives are being examined as potential replacements for dwindling supplies of fossil fuels. One part of this decision involves the amount of pollutant for each energy source. In evaluating the pollutants given off by an energy source, it is important to look at not only the pollutants given off during the operation of the plant but also the pollutants involved in the fuel supply cycle and the construction of the plant. Table 8.4 gives the life cycle emissions of five key atmospheric pollutants for most of the energy sources being considered for future energy supplies.

TABLE 8.4

Pollutants emitted per kilowatt-hour electricity production, including fuel cycle and construction of the plant [17]. The ranges reflect the variability of fuel sources (e.g., low-sulfur vs. high-sulfur coal) and the variability of manufacturing processes (in the case of wind and solar).

Fuel Source	Greenhouse Gas (grams CO_2 equivalent) per kWh	SO_2 (mg/kWh)	NO (mg/kWh)	NMVOCs (mg/kWh)	Particulates (mg/kWh)
Hydropower	2–48	5–60	3–42	0	5
Nuclear	2–59	3–50	2–100	0	2
Wind	7–124	21–87	14–50	0	5–35
Solar	13–731	24–490	16–340	70	12–190
Biomass	15–101	12–140	701–1,950	0	217–320
Natural gas (combined cycle)	385–511	4–15,000	13–1,500	72–164	1–10
Coal (modern)	790–1,182	700–32,321	700–5,273	18–29	30–663

NMVOCs = nonmethane volatile organic compounds.

In the following example we calculate the number of excess deaths due to sulfur dioxide emissions from a typical coal plant. Similar calculations can be performed for other energy sources and other pollutants.

EXAMPLE 8.14

Sulfur Dioxide from Coal

Suppose a coal plant uses 5.5×10^6 kg of coal that contains 2% sulfur to generate 18,000 MWh of electricity per day. Assume the plant also uses sulfur emission controls to remove 90% of the sulfur dioxide from the plant exhaust. The exact exposure rate depends on the location of the population relative to the plant, but a reasonable estimation is an exposure of 5 to 100 $\mu g/m^3$ per person per ton of sulfur dioxide emitted. The estimation of the effect of SO_2 also ranges from 2 excess deaths (National Academy of Sciences estimate) to 37 excess deaths (Brookhaven National Laboratory) per million people exposed at 1 $\mu g/m^3$ per person. The loss of life expectancy (see Section 8.5) is estimated at 0.2 to 370 years. How much SO_2 is emitted by this plant, how many excess deaths will occur, and what is the loss of life expectancy as a result?

The amount of SO_2 produced per year is 5.5×10^6 kg $\times \frac{0.02\,\text{kg}\,S}{\text{kg coal}} \times \frac{1\,\text{ton}}{1000\,\text{kg}} \times \frac{32\,\text{kg}\,SO_2}{16\,\text{kg}\,S} \times 0.1 \times 365$ day $= 5.0 \times 10^4$ tons/yr.

The number of excess deaths per year (using the lower estimate) is 0.5 excess deaths per year. The upper estimate is 185 deaths per year.

The loss of life expectancy is 5.0 × 10^4 tons/yr × $\frac{5\,\mu g / m^3}{ton\ SO_2 / yr}$ × $\frac{0.2\ years}{10^6\,\mu g / m^3}$ = 0.05 years for the lower estimate and 1,850 yr for the higher estimate.

The total per capita electrical use in North America is about 13,000 kWh a year, so the plant would supply the energy needs of a population of about 5.1 × 10^5 people.

Example 8.14 highlights the two aspects of risk analysis discussed in this section: exposure and dosage. In this example both exposure and dosage are poorly known. Exposure is highly dependent on the location of the power plant and depends on local features of terrain and weather patterns. The U.S. Environmental Protection Agency sets a limit of 0.5 ppm for sulfur dioxide, and it is clear that this level of exposure has severe health effects. However, the precise relationship (the dose–response curve) between exposure and death risk is poorly known. The result in Example 8.14 shows a very wide estimate of the effects of the power plant and demonstrates some of the difficulty in making a useful risk assessment.

8.5.4 Radiation Dose

Historically, several different units have been used to measure radiation and its effects. Table 8.5 gives several definitions useful for discussing radiation risk. Initially the curie was used to define the activity of a radioactive substance (the decays per second); this has been replaced by the becquerel. The roentgen, used until recently, is defined as the amount of ionizing radiation that produces an electric charge of 3.33 × 10^{-10} C of charge in 1 cm^3 of air under standard conditions. The rad and gray are units that measure the amount of energy per kilogram absorbed, regardless of the effects. The sievert has replaced the rem as a unit that measures the biological effects of ionizing radiation, as compared to the energy absorbed. Currently the gray and the sievert are in use, but the other quantities listed can be found in much of the older literature.

The relative biological effectiveness (RBE) measures the effect of a particular type of radiation and depends on the type and energy of the decay products and how the body is exposed. The RBE multiplied by the energy absorbed gives

TABLE 8.5

Radiation units. The gray and the sievert are the accepted international units.

Quantity	Units	Definition
Rate of decay	Curie Becquerel	1 Ci = 3.7×10^{10} decay/s 1 Bq = 1 decay/s
Energy intensity	Roentgen	1 R = 87 erg/g = 0.0087 J/kg
Absorbed energy	Rad Gray	1 rad = 100 erg/g = 0.01 J/kg 1 Gy = 1 J/kg = 100 rad
RBE	RBE	RBE = 1 for X-rays, γ-rays, and β RBE = 5–20 for neutrons RBE = 20 for α and fission fragments
Effective absorbed dose	Rem Sievert	rem = rad × RBE 1 Sv = 1 J/kg = 100 rem

RBE = relative biological effectiveness; rem = roentgen equivalent man.

the effective absorbed dose, measured in rems or sieverts. Radiation that does not have enough energy to cause damage to cells (e.g., visible light) is called nonionizing radiation. In contrast, ionizing radiation disrupts normal biological processes in the cell, and in general the more massive the decay product the more damage it can do. In fact, it is the property of ionization that makes radiation therapy useful. Sometimes it is possible to target a fast-growing tumor with a particular radioactive decay product, either by beaming the radiation at a specific target in the body or by administering a radioactive chemical compound that will be taken up preferentially by the cancer cells. Although alpha particles are easily blocked by as little as a piece of paper, if these particles do strike a living cell they are much more effective at causing damage than X-rays or electrons (β) because of their large size. If the alpha source is inhaled or otherwise ingested, it can be much more lethal than a beta or X-ray source of the same energy. On the other hand, gamma rays, X-rays, and even neutrons require significant shielding to prevent unwanted exposure, whereas beta particles are blocked by a thin sheet of aluminum.

The typical exposure to naturally occurring radiation in the world is 1.5 mSv to 3.5 mSv, although some locations have levels as high as 200 mSv. Because of a higher incidence of radon, North Americans receive an average radiation dosage of about 3.6 mSv per year [18]. Approximately 3 mSv of this exposure comes from natural causes such as radon and cosmic rays, the rest from human-made sources such as medical X-rays. Table 8.6 gives exposure rates for several natural and human-made sources.

TABLE 8.6

Typical radiation exposure [18].

Source	Exposure (mSv/year)
Single whole-body medical X-ray	14
Annual exposure, airline flight crew	2–9
Indoor radon	2.0
Cosmic radiation	0.31
Background radiation, sea level	0.28
Background radiation, Denver	0.81
Natural radiation in the body	0.39
Single mammogram	0.3
Single dental X-ray	0.4
Single chest X-ray	0.1
U.S. annual average X-ray exposure	0.53
One round-trip transcontinental flight	0.05

High dosages of ionizing radioactivity cause changes in body chemistry that lead to death quickly. A single dose of 4–5 Sv kills 50% of the victims within a month, and more than 6 Sv is lethal to 100% of the victims in the same time period. Fortunately, except for victims of the bombs dropped on Hiroshima and Nagasaki during World War II, most exposure is much less than this. The lifetime effects of large single doses of radiation are known from Hiroshima and Nagasaki survivors and workers exposed during the Chernobyl disaster. Table 8.7 shows data from Hiroshima and Nagasaki survivors. A plot of dose versus cancer rate (excess cancers divided by number of victims) shows an approximate linear dose model for a one-time, large exposure with a slope of 0.45 cancers per sievert exposure. It should be made clear that these figures are for a single dose over the normal background exposure, and the excess cancers occurred over the lifetime of the victims. The cancer rate of 0.45/Sv is then a lifetime excess cancer risk.

Effects of smaller doses of radiation (less than 100 mSv) are extrapolated from animal studies, uranium miners, and populations with varying amounts of natural exposure. A single low dose of radiation above the natural background shows an increased lifetime cancer risk of 0.08 cancers per sievert exposure for men and 0.13/Sv for women [19]. This suggests that radiation exposure follows something like the threshold dose model shown in Figure 8.3; low-level doses have a different dose–response curve slope than high-level doses.[4] It is also the case that different tissues in the body have different susceptibility to radiation. The thyroid gland and breast tissue are much more sensitive to radiation than other parts of the body, and the rate of leukemia is much higher for radiation exposure than for other known cancer-causing agents [20]. A final factor affecting cancer rates among exposed populations is age. Children exposed to radiation have a higher lifetime cancer risk than adults. For these reasons, predicting the exact cancer rate for a given exposure of ionizing radiation is difficult.

As noted in Table 8.7, a substantial number of cancers are expected to result from other causes, even if there were no exposure to radiation. A number that tries to evaluate the additional risk due to the exposure of a particular substance is the relative risk, which is the disease rate among the exposed population divided by the disease rate among the unexposed [18]. For example, the relative risk for the first group in Table 8.7 is (4,286/42,702)/(4,267/42,702) = 1.02. A relative risk of 1 is an indication that the exposure has not caused any additional cancers at all. A relative risk of 1.02 means an increased risk (or excess risk) of 2% in the exposed population compared to the unexposed group. The following example finds the relative risk for cancer rates due to average radiation exposure in the United States.

EXAMPLE 8.15

Relative Risk due to Average Radiation Exposure

The National Academy Board on Radiation Effects Research reports that for men we should expect 800 extra cancers per 100,000, averaged over a lifetime, due to a one-time 0.1-Sv exposure [19]. The rate for women is 1,300 cancers per 100,000 for women. The normal lifetime rate of cancer assuming no exposure is 45,500 for men and 36,900 for women per 100,000. Based on these rates, estimate the lifetime cancer rate expected for the national average radiation dosage of 3.6 mSv/year and calculate the probability of cancer caused by radiation compared with all types of cancer. Also calculate the number of people affected.

The lifetime risk for men for a single exposure of 3.6 mSv would be (800/0.1 Sv)/100,000 × 3.6 mSv = 0.00029 and 0.00047 for women. If you are exposed once every year for 78 years, the lifetime risk would be 0.00029 × 78 = 0.022 or 2.2% for men and 0.037 or 3.7% for women. The lifetime risk of having cancer from all sources, assuming no exposure, is 45,500/100,000 = 45.5% for men and 36.9% for women.

The relative risk for a lifetime exposure at the national average of 3.6 mSv would be 0.022/0.455 = 0.048 for men and 0.10 for women. So if you are a man and have cancer before you die, there is a 4.8% chance it was caused by radiation exposure. If you are a woman the chance is 10.0%. Note that the higher risk for women is partly because the radiation risk is higher and partly because the risk from other causes is lower.

If we assume that half the population of the United States is male and half female, there would be $152 \times 10^6 \times (800/0.1\ \text{Sv})/100{,}000 \times 3.6\ \text{mSv} = 4.4 \times 10^4$ radiation-caused cancers in men and 7.7×10^4 in women. These are the figures for new cancers occurring over a lifetime due to 1 year of exposure. The cancer rate of all types was 542 per 100,000 for men and 409 per 100,000 for women per year in the years 2002 to 2006. The total number of cancers from all causes was $152 \times 10^6 \times 542/100{,}000 = 8.2 \times 10^5$ for men and 6.2×10^5 cancers for women per year.

From Example 8.15 we see that, assuming the linear dose model for low-level radiation is correct, the cancer risk due to average annual radiation exposure from all causes is predicted to be low, accounting for about 2.2% of the lifetime cancer risk in the United States due to all causes for men and 3.7% for women. Only about 17% of the average radiation exposure is human caused, so the cancer risk due to X-rays, computed tomography scans, and the like is much lower. It should be noted that these results are for cancer rates, not the death rate by cancer. The number of excess cancer deaths caused by an exposure of 0.1 Sv is 410 for men and 610 for women per 100,000 [19].

TABLE 8.7

Exposure and excess cancer rates of Hiroshima and Nagasaki survivors [18]. Expected cancers are the number of cancers expected for a population of the same size but with no excess radiation exposure.

Dose (Sv)	Victims	Expected Cancers	Actual Cancers	Excess Cancers
<0.01	42,702	4,267	4,286	19
0.01–0.1	21,479	2,191	2,223	32
0.1–0.2	5,307	574	599	25
0.2–0.5	5,858	623	759	136
0.5–1.0	2,882	289	418	129
1.0–2.0	1,444	140	273	133
>2	300	23	55	32

As a final consideration of the risks of radiation, it should be noted that different radioactive wastes have different chemical properties in addition to different radioactive decay products. For example, iodine-131 undergoes alpha and gamma decay and has a short half life. Chemically, however, the body stores iodine in the thyroid gland instead of eliminating it. Likewise, strontium is more likely to be taken up and stored in the body, which makes these radioactive elements more dangerous than other radioactive materials with the same or even higher radioactivity. Nonradioactive iodine was given to people exposed to radiation in the region around Chernobyl in hopes that increasing the normal iodine in the body would prevent or slow the uptake of radioactive iodine by the thyroid. Other radioactive isotopes such as barium are used in medical testing because they pass through the body quickly. Radioactive cesium-137 is water soluble, so direct exposure is quickly eliminated from the body, but cows eating grass coated with cesium fallout concentrate this element in their milk. As a result of fallout from the Chernobyl disaster, 5,000 tons of radioactive powdered milk was destroyed in 1987 by the German government.

8.6 ■ Risk Perception

Our perception of risk is sometimes very different from the actual risk. For example, living 2 months in Denver, Colorado carries about the same risk (one in a million) for dying from radiation-induced cancer as living 150 years within 20 miles of a nuclear power plant [2]. At an altitude of 6,000 feet, Denver has a higher incidence of cosmic rays, resulting in an increased radiation risk compared with nuclear power plants. The risk of a single chest X-ray is also about one in a million, but many people would choose not to live next to a nuclear power plant but think nothing of having a chest X-ray or moving to Denver. Likewise, traveling 300 miles in a car carries about the same risk of an accident (one in a million) as traveling 1,000 miles in a plane, but many people would rather drive the 1,000 miles (and face three times the risk) than take a plane. The difference in risk of death between cars and planes is even greater than the accident risk. A short list of the risk of death for various actions or causes is found in Table 8.8.

The average person evaluates risk differently depending on several psychological factors. As should be obvious from the discussion in this chapter, driving or riding in an automobile is a very risky activity, yet most of us do it every day without thinking much about it. The following paragraphs examine

TABLE 8.8

Risk of death due to an action or cause [1, 2].

Cause	Annual per Capita Risk per 100,000
Heart disease	271
Diabetes	24
Pneumonia and influenza	35
Cigarette smoking (per smoker)	300
All fatal cancers	200
Motor vehicle accident (total)	15
Alcohol related	6
Using cell phone*	0.15
Home accidents (all ages)	11
Illegal drugs	5.6
Alcohol	6.3
Falls	6
Over age 70	43
Suicides	11
Accidental poisoning	4
Yearly coast-to-coast flight	1
Being hit by falling aircraft	0.004
Being hit by a meteorite	0.04
Tornado	0.015
Flood	0.045
Lightning	0.016
Heat wave	0.4
Bites or stings from venomous insect	0.017
Radon	
U.S. average of 1.5 pCi/l	0.015
U.S. high of 400 pCi/l	0.045

*The latest data shows driving while texting on a cell phone is approximately equivalent to driving with a blood alcohol level equal to the legal limit.

various psychological factors that affect the perception of risk but do not change the actual scientifically calculated risk [2].

Stereotyping biases the perceived risk of an activity, event, or product. For example, it may be the case that most of the models of tires of a tire manufacturer are very safe, but the perception of the risk of the entire brand may be heightened by a highly publicized failure of one model. Someone who has a bad personal experience with a product or activity will evaluate the risk of similar products or activities as higher.

Although a particular type or brand of equipment may have a low risk of failure, if a failure occurs early rather than later in the production lifetime, the perceived risk of the brand is much higher. For example, suppose a particular model of airplane has a landing gear failure rate of 1 in 100,000 landings. If the failure occurs in the first landing rather than in the 100,000th landing, the perceived risk of that airplane model will be much higher.

If we can easily imagine a way (or multiple ways) for something to fail, we tend to evaluate the risk as much higher than if we cannot think of a way for the failure to occur. The degree of complication of a fault tree analysis for a nuclear power plant can easily convince someone that there is a high likelihood of failure because of all the possible combinations of scenarios that might lead to failure, when in fact an increase in fault tree complexity indicates a lowering of the chance of failure.

Personal choice and control play a large role in our evaluation of risk. A situation in which we make the decision to take a risk may seem safer than if someone else or nature makes the decision. As can be seen in Table 8.8, the risk of dying in a car accident as a result of talking on a cell phone is 10 times the risk of dying in a tornado. However, because we can choose when we talk on the phone but cannot affect the timing of tornadoes, many people are willing to talk on the phone while driving, but very few of us would choose to experience a tornado close up. The risks for many popular recreational activities (e.g., all-terrain vehicles) are much higher than many occupational risks, but because we choose to do them we discount the amount of risk involved.

Perceived necessity also plays a role in risk perception. Some people feel compelled to take a particular risk (e.g., to overconsume alcohol or take recreational drugs) because of social pressure whereas others do not. Driving is fairly risky, but many people will make several trips to the store rather than wait and do most of their shopping in one trip because they feel it is necessary to have a particular item now instead of later.

Latency, or the delay between cause and effect, is often sufficient to decrease the perception of risk. Many smokers discount the risk of smoking because they are not yet ill. Illegal drug use is another example because being arrested for a single incident is unlikely, which may sway a decision to use the drug.

The subjective evaluation of dread plays a large role in risk perception. Cancer from radiation seems much worse to some people than cancer from smoking. Radiation cancer risks from nuclear power plants are much lower than cancer related to many other chemical exposures, yet most people perceive nuclear power plants to be more dangerous.

Rare but catastrophic risks are usually evaluated as worse, up to a point. Airplane crashes are rarer and have a much lower risk per mile than auto crashes, but plane crashes kill a large number of people all at once. For this reason airplane flights are often perceived as riskier even though the total number of deaths is much lower than that of automobiles (e.g., no one died in a commercial plane accident in 2002, compared to more than 42,000 traffic-related deaths in the United States). An exception to the idea that most people overestimate the danger of catastrophic disaster is the case of the risk of death by meteor collision with the earth. Because there has not been a single example in historical times, this risk is usually evaluated as very low although it constitutes a risk similar to flooding and higher than death from insect stings, tornadoes, or lightning.

Familiarity with the risk plays a role in risk perception. Unfamiliar, new, or unknown risks sometimes appear more threatening than known risks. Nuclear fission is less familiar to the general public than coal as a power source, and as a result it is often perceived as riskier, even though it is significantly safer in terms of number of people dying per kilowatt-hour generated (Section 4.6). More people die in car wrecks per month than died in the attacks on the World Trade Center in 2001, yet the perception is that terrorism is more threatening.

High risks to small groups typically are not perceived to be as serious as smaller risks to large numbers of people. Only recently has the seriousness of food allergies come to the attention of food manufacturers and the general public. The risk of dying due to exposure to peanuts, for example, is very high for those with an allergy to nuts but historically has been overlooked or ignored. Saccharine, on the other hand, has been the center of a long debate over its effects on health, even though there is still very little evidence to indicate that it presents a significant health threat. Because it has the potential to affect a large number of people, saccharine received more public attention than exposure to nuts, which affects a smaller number of people, although this smaller number is affected more severely.

The perception of risk is lower if people trust the people or agency in charge of that risk. Knowledge of this fact is the reason why the tobacco industry continues to donate large sums of money to charity in an attempt to sway the public perception of risk introduced by tobacco products. As a converse example of this effect, the perceived risk of nuclear power is much higher than the actual risk in part because of public distrust of officials in charge of the nuclear power industry.

As a final comment on perception of risk, we present Figure 8.4, which shows the death rate by age group for the past 100 years in the United States.

There are several interesting features of this figure. The mortality rate for all groups declined during this time period, but the rate of decrease for children under 1 year of age was much steeper. This is primarily because of the increase in the use of antibiotics, immunization, and better health care for children. There is a slight rise in death rates among the 15- to 24-year-old group starting in the late 1960s. This was a result of the rapid increase in the number of cars available and a sharp increase in the number of young drivers (the "baby boom"). The slight increase in death rates for 25- to 44-year-olds in the mid-1980s reflects the AIDS epidemic.

The most dramatic features of Figure 8.4 are peaks in mortality for age groups between 5 and 44 years old in the years 1918 to 1920. This sharp increase in death rate is the result of a flu pandemic, sometimes called the Spanish flu epidemic. Between 250,000 and 500,000 people die every year from seasonal flu, but during the Spanish flu epidemic between 50 and 100 million people died worldwide, and as much as half the world's population was infected. Unlike

FIGURE 8.4

Death rates for various age groups in the United States [21].

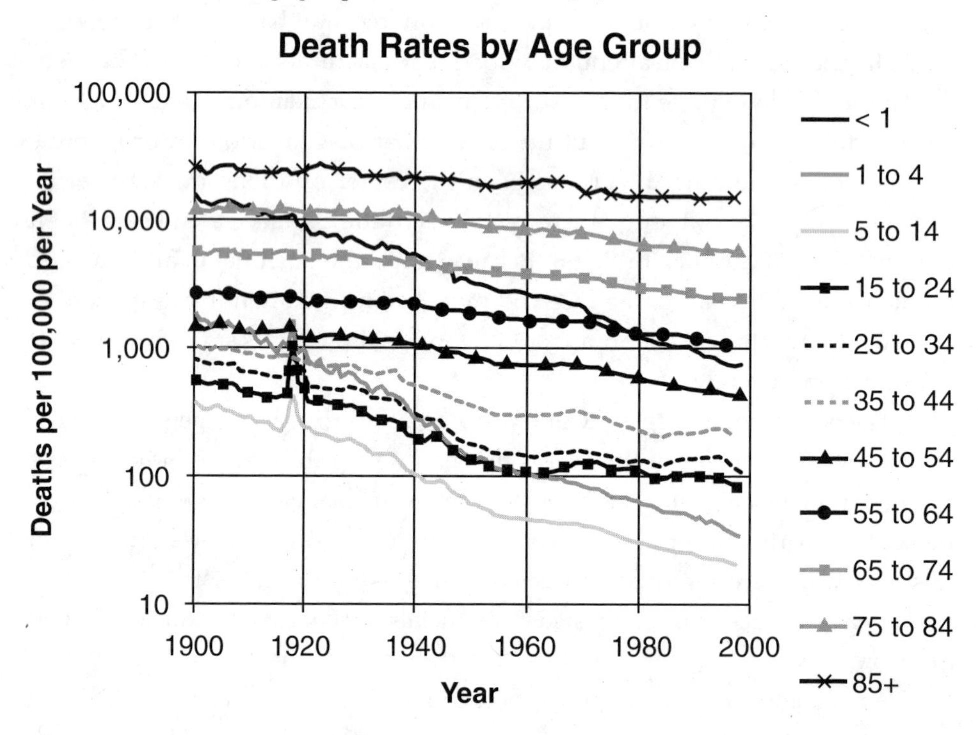

most seasonal flu, this particular strain affected healthy adults more severely than the very young or very old. The risk of a flu pandemic has only recently drawn the attention of the general public, but in the past 100 years flu has clearly had the largest single effect on global health.

8.7 ■ Economic Factors

In a world of limited resources, careful decisions must be made about the implementation of technology, and often this process starts with a risk analysis, as discussed earlier. Once the risk factors are known, economic and political factors come into play. Some economic aspects of energy, such as the concept of life cycle economic cost, have been examined previously (e.g., see Problem 2.14). In the following we briefly examine a few other important economic issues involved in making decisions about the application of new technology. A complete treatment of economics is beyond the scope of this book; the reader is referred to references [22]–[27] for further discussion of the economics involved in energy and technology applications.

Because money can be invested, a given sum of money today may have a different value in the future. Conversely, a dollar paid out in the future is worth less in today's terms because of what could be done with that dollar if it were available for investment today. If the interest rate for an investment does not change, these values can be calculated using the exponential function introduced in Chapter 1 (Equation (1.1); also see Problem 1.1). If P is the present value and F is the future value, the relation between the two is given by

$$(8.9)\ F = Pe^{rt},$$

where r is the interest rate and t is time. The inverse of this equation tells us that \$10 paid out 20 years in the future is worth $P = Fe^{-rt} = \$10e^{-.1\times 20} = \$1.35$ in today's money if the available interest rate is a constant 10%. This and the fact that interest rates do change over time are factors that help explain why conventional business planning has short time horizons; the value of money is expected to change significantly over time.

For most large construction projects such as a power-generating windmill or an array of solar cells it is necessary for investors to borrow money. An interest rate will be charged for borrowing the money, and the value of the money will also change over time. The capital recovery factor (CRF) allows us to calculate what the annual payment will be in today's dollars if we borrow P dollars at an interest rate of r for t years. The annual payment, A, is

$$(8.10) \quad A = P \times \text{CRF},$$

where

$$(8.11) \quad \text{CRF} = \frac{r(1+r)^t}{(1+r)^t - 1}.$$

So, for example, if we borrow \$20,000 for 15 years at 6% interest, our annual payment will be $A = \$20{,}000 \times \frac{0.06(1+0.06)^{15}}{(1+0.06)^{15} - 1} = \$2059.26$. This factor takes into account the time dimension to the value of borrowed money. Future dollars are worth less than current dollars because if we have that dollar today we can invest it in such a way that it will be worth more in the future. Likewise, a payment in the future must be discounted to know what it is worth in present-day dollars. Other useful economic analysis factors (which we will not use in this book) include the compound growth factor (which finds the future value of a current investment at a given interest rate), the present value factor (which finds the value of future dollars in terms of present dollars), and the uniform present value factor (which compares annual payments with a one-time payment given the discount rate of future dollars). These concepts were used to calculate the levelized cost (the yearly revenue needed to recover all expenses over the lifetime of the investment) shown in Table 5.5 for electricity generated from various renewable resources. Example 8.16 calculates the levelized cost of electricity from a home photovoltaic system built with borrowed money.

EXAMPLE 8.16

Cost of Electricity from a Small Photovoltaic Installment

The capital costs to build a photovoltaic installment include many individual components. In addition to the array itself, land must be purchased or rented, a power conditioning system to make the power usable must be installed, and an energy storage system should be included. Suppose the entire system costs \$15,000, is calculated to provide 3,000 kWh/yr, and is in operation for 20 years. What will be the cost per kilowatt-hour for electricity?

If we borrow the money at 6% interest for 20 years, the CRF is 0.0872 and the annual payment will be $1,307.77. Dividing by the power generated gives $0.44/kWh. This figure does not include operating costs, maintenance, or any profit for the owner if the electricity is to be sold. Current electrical prices are in the neighborhood of 10 cents/kWh in the United States, so it is clear that solar photovoltaic electricity cannot compete economically with conventional electricity, as was pointed out in Chapter 5.

The costs in Example 8.16 include only the price of building the system. Operating costs and profit were not included. A more detailed cost analysis would include these ongoing costs with factors to account for the changing value of money over time (see [22], [23], [25], [26], and [27] for calculations that include operating, maintenance, and other costs). In this example the fuel source is free, so ongoing costs are low and stable compared with those of a coal-fired plant, where fuel is a significant portion of the cost of generating electricity and the cost of fuel may change. The proportion of capital cost as compared to operating costs is very different for various choices of energy technology. The ratio of maintenance and fuel expenditure to capital cost might be as low as 0.15 for wind energy but as high as 2.8 for gas-powered combined heat and electricity plants [25].

One additional factor that must be included in a detailed cost analysis is the capacity factor, which is the ratio of actual power output to the rated capacity of the plant. Fossil fuel plants can be kept online as much as 90% of the time, with maintenance and other offline periods making up only 10% or 15% of the lifetime of the plant. Because of the intermittent nature of the wind and sun, windmill and solar cell plant capacity factors are generally rated at only 15% to 25%. This means that a windmill rated at 4,000 kWh/yr with a capacity factor of 25% can be relied on to supply only 1,000 kWh/yr. To supply a demand of 4,000 kWh/yr would require a plant with a peak rating of four times this number coupled with a storage system with no losses.

The cost of technology generally decreases over time. Two factors involved in this change are economies of scale and maturation of the technology. Setting up a large factory to build windmills will result in a lower cost per windmill than building them individually, an example of economies of scale. As for most technology, the energy efficiency of a particular energy conversion device tends to increase over time, meaning that the same dollar spent returns more energy. Because of improvements in technology, the cost of a kilowatt-hour of electricity generated by solar cells has decreased at a rate of about 20% per year over the past 20 years, as has the cost of wind energy. However, the cost of gas turbines, which are a more mature technology, has been decreasing at a rate of only about 10% per year in this same time span.

8.7.1 External Costs

Ongoing expenditures and capital costs are not the only economic factors involved in building an electricity-generating plant. The final price of an energy source does not always reflect the true costs or benefits of installing the resource. For example, coal plants give off carbon dioxide, which, as we saw in Chapter 7, is linked to global changes in climate. When a wind turbine is installed, the energy provided reduces the cost of importing fossil fuels and reduces the pollution from the use of these sources. In economics, expenditures or profits not directly accounted for in the construction and operation of a product or process such as environmental damage and other social costs are called externalities and are often very difficult to evaluate [27].

All energy sources have external costs in the economic sense. Mining accidents for coal, chemical waste in solar cell manufacturing, environmental and societal disruptions from the installation of hydroelectric dams, and nuclear waste from reactors are all externalities whose costs should be included in the true cost of that particular energy source. As mentioned in Section 4.6, the historical health risks associated with coal as an energy source are far greater than the risks linked to nuclear power per kilowatt-hour of electricity generation, which would indicate a higher external cost to society for coal compared with nuclear power. In addition to carbon dioxide emissions, other externalities applicable to our fossil fuel–based economy include military costs to secure access to petroleum and natural gas sources outside the United States. In general, renewable energy sources have much lower external costs than fossil fuel sources [27].

Once various externalities have been identified, they can sometimes be internalized or included in the real cost of a particular energy supply through fees, regulation, or taxes [26]. Mandatory scrubbers for smokestacks of coal-fired electric plants are an example of an externality that has been internalized by regulation. The cost of cleaning sulfur and particulates from the emissions has been included in the operating cost of the plant by the requirement that scrubbers be purchased by the operator of the plant. The successful cap-and-trade arrangement for the reduction of acid rain discussed in Chapter 1 is another example of a method of internalizing an external cost. Businesses that are required to pay these external costs often complain about taxes and regulation, but it is important to understand that capitalism works best when the price of a service or product reflects the true costs of the consumed item.

In most cases external costs are quite hard to evaluate. External costs associated with carbon dioxide emissions are very difficult to estimate, and the effects, though global, could be severe in some locations but much milder in others. Newer technologies, such as photovoltaic and wind energy, have not been in use long enough to accurately assess all the associated risks and external costs. Estimates of the external costs, including the cost of global warming, for coal range from 0.1 cent/kWh to $10/kWh of generated electricity [25]. Renewable energy fares better for external costs, but even here the estimates range from 0.001 cent/kWh to 0.1 cent/kWh for wind and photovoltaic and up to 1 cent/kWh for biomass. One way to include these externalities is by imposing a tax. In Example 8.17 the effect of a carbon tax on the cost of electricity generated from coal is calculated.

EXAMPLE 8.17

Carbon Tax

Suppose an electric plant is 35% efficient in turning coal into electricity. The coal is 70% carbon, and the cost before a carbon tax is 8 cents/kWh. What will be the increase in cost if a carbon tax of $100/mt of carbon emissions is imposed?

A metric ton of coal, of which 70% is carbon, provides about 8,200 kWh of thermal energy. Dividing by 0.35 gives the energy needed to generate 8,200 kWh of electrical energy. So we have (1 ton/8,200 kWh) × 0.7 × (1/0.35) × ($100/ton) = 2.4 cents/kWh of electrical output. This is a substantial (30%) increase in the cost of electricity.

As an alternative to taxes on sources that pollute, subsidies or tax incentives for energy technologies that are thought to have lower external costs may be effective. Tax deductions and subsidies for renewable energy sources are an attempt to account for the different external costs of these technologies compared with established sources. Example 8.18 shows how a rebate (currently in place in California) and tax deductions adjust the cost of photovoltaic sources, which are thought to have fewer externalities than electricity generated from coal.

EXAMPLE 8.18

Rebates and Tax Deductions

Suppose there is a rebate from the state for renewable energy of $2.00 per watt of energy generated and a 30% federal tax credit on the cost of the plant after rebate for the photovoltaic system in Example 8.16. The array is rated to have a peak power of 2,000 W. What will be the cost of electricity in this case?

The rebate will reduce the capital cost by 2,000 W × $2.00/W = $4,000, so you will only have to borrow $15,000 – $4,000 = $11,000. A federal tax credit of 30% reduces the capital cost by $3,300 to $7,700. With a CRF of 0.0872, this reduces the annual payment to $671.44. The cost for electricity is now $671.44/3,000 kWh = $0.22/kWh, or about half the cost without the rebate and tax advantage. Including a factor for tax-deductible interest further reduces this figure by about half, which would make solar-powered electricity economically competitive.

In addition to energy production tax credits and mandating that a percentage of power be generated from renewable sources, demand for renewable energy may be increased by direct marketing to consumers. In many cases consumers will choose renewable sources, even at a slightly higher cost, if given the choice. Tax rebates for installation of more energy-efficient appliances are also consumer-side incentives for changing energy consumption. The goal of these inducements is to create a market for conservation and renewable energy that eventually will allow economies of scale and innovation to reduce the cost so that they become competitive. Once a new technology is firmly established, the incentives can be phased out. Unfortunately, these types of incentives do not work well on a short time scale. Policies mandating production tax credits for renewable energy in the United States historically have had a lifetime of 2 years, making long-term planning and market penetration of new energy technologies difficult. There have been large fluctuations in wind turbine demand in the United States since 1992, in large part because of short-term instabilities in tax credits being offered by the government.

Several European countries have encouraged the development of renewable energies by guaranteeing a minimum price to be paid for renewable energy for some long time span. For example, France has fixed a price of 0.082 euros/kWh for renewable energy for 10 years [26]. This approach, called a feed-in tariff, levels the playing field for a significant amount of time, allowing energy suppliers and consumers to make long-term plans and investments.

As a final comment on economics, it should be pointed out that most business plans are short term. As we saw in this chapter, money changes value over time, as do government policies such as tax incentives. It is very difficult for an industry to justify an investment in developing technology that will not pay for itself until 5 or 10 years in the future. The few historic examples of businesses acting to the contrary had exceptional success, Bell Labs[5] being one famous example. But this type of research institution associated with a business is becoming rarer. One way to remedy this situation is for the government and universities to sponsor fundamental research. There are strong arguments that investment by the government and universities in basic science on a much longer time horizon than a normal business plan has significant long-term benefits to society [22]. Once a new scientific expertise has been developed and applications investigated, this information can be made available to businesses interested in developing the applications. Initially some tax incentives may be necessary to help the new technology compete equitably with mature technology, but the eventual goal is to have many entrepreneurs competing in the marketplace with new applications.

8.7.2 Cost-Benefit Analysis

Combining economics and risk assessment results in what is called a cost–benefit analysis. A cost–benefit analysis typically begins with a risk assessment and makes a value judgment as to the economic worth of a human life. Often this is expressed as the cost for preventing a year of lost life (cost per life-year) due to a given risk. Suppose you are the administrator of the highway system in your state and you want to make highway safety improvements. A number of accidental deaths have resulted from motorists striking utility poles. Should you decrease the number of utility poles from 40 to 20 poles per mile on rural roads or relocate utility poles from 8 feet to 15 feet from the edge of the highway? The economic cost for either plan can be calculated using a normal economic analysis. The risk can be estimated from historical data involving different configurations of utility poles. Suppose the cost per year of life saved is calculated to be $31,000 for decreasing the number of utility poles but $420,000 per year of life saved for relocating the poles farther from the highway [2]. We conclude that the most cost-effective measure for saving lives is to decrease the number of utility poles.

The use of a cost-per-life-saved analysis is often strongly influenced by the population affected. For example, it is estimated that the cost for screening only black newborn babies for sickle cell anemia is about $240 per year of life saved [2]. Screening every newborn for sickle cell raises the cost to $34 million for every year of life saved. This is largely because people who do not have ancestors from tropical regions of the world are very unlikely to have the genetic mutation that causes sickle cell, and there are no benefits gained by screening that population. Economically it makes great sense to screen babies with ancestors from tropical climates but not the general population.

8.8 ■ Summary

Risk is defined as probability times severity for a particular action or activity. A few elementary aspects of risk assessment have been introduced in this chapter. In particular, exposure rates, dosage, fault tree analysis, and risk perception have been discussed. Car accident rates, risk due to meteor impacts, and radiation risks were discussed in detail. Many of the techniques used for evaluating risks due to new technology originally came from the nuclear power industry. These methods have been adapted and extended to applications in the chemical and manufacturing industries and are applicable to other new technologies.

The common perception of risk is often very mistaken. There are 6 million traffic accidents in the United States every year, resulting in a death every 12 minutes and an injury every 14 seconds. This constitutes a serious threat to life and well-being, yet few drivers think seriously about changing their habits to include safer modes of travel.

Economics plays an important part in deciding how to act on the information given by a risk analysis. Most often a cost–benefit analysis combining risk and economic factors to arrive at the cost for saving a given number of lives or person-hours is the most useful approach. Economic externalities or hidden costs not reflected in the market price of a particular technology can be internalized by the use of taxes, tax incentives, rebates, and regulation.

The long-term true costs to society of a carbon-based energy system are only now becoming evident. Regulation, taxes, and government-funded research are ways to internalize these costs so that energy markets can move toward conservation and alternative energy sources with fewer environmental externalities. In the energy sector the external costs of fossil fuels are about 10 times those of most types of renewable energy [27]. Renewable resources also typically have lower fuel and maintenance costs. Many of the technical problems facing these new energy sources have been overcome, and it is mainly the capital costs that keep them from being economically viable. Small tax incentives already make wind energy economically competitive with most fossil fuel–generated electricity, as evidenced by the rapid growth of this industry. Including the external costs of carbon emissions for fossil fuels, for example by a cap-and-trade system, probably will make wind energy even more competitive without other incentives. Photovoltaic solar energy is projected to become economically competitive in the next 10 to 12 years as a result of improved efficiencies, lower manufacturing costs, and carbon emission trading. Other renewable energy sources, with a few exceptions such as geothermal in Iceland, are probably further out on the horizon in terms of economic viability.

Questions, Problems, and Projects

For all questions,

- Cite all sources you use (e.g., textbook, Wikipedia, Web site). Before using Wikipedia as a source, please read the policies of research and citing in Wikipedia: http://en.wikipedia.org/wiki/Wikipedia:Researching_with_Wikipedia and http://en.wikipedia.org/wiki/Wikipedia:Citing_Wikipedia. For Wikipedia you should verify the information in the article by reading the sources listed.
- For all sources, briefly state why you think the source you have used is a reliable source, especially Web resources.
- Write using your own words (except for direct quotes, which should be enclosed in quotation marks and properly cited).

1. Using reliable sources and your own words, give definitions of loss of life expectancy, person-days lost, working days lost, nonworking days lost, and disability-adjusted life years. Give some examples in which each is used.
2. Using reliable sources, define and comment on quality-adjusted life years (QALYs), used by economists working in the health care industry. In your opinion, should this concept be used to determine the amount of health care provided by insurance companies? Back up your opinion with data from a reliable source.
3. In 1996 a TWA passenger plane crashed in Long Island Sound. At the time someone suggested that the plane may have been hit by a meteor.

Use the following data to determine an approximate risk for a plane being hit by a meteor. Planes take off about every 5 min during daylight hours, and the number of commercial airports in the world is about 1,000. So in a 12-h day the number of takeoffs is approximately 144,000. About 3×10^3 meteors per day penetrate the atmosphere far enough to hit a plane. The vital area of a plane, the area that, if hit, might cause the plane to go down, is about 600 m^2. The area of the earth is about 5×10^{14} m^2. Complete the following steps (adapted from [28]):

a. If a typical flight is 3 h out of 24, how many planes are in the air at any given time?

b. How much total vital area does this number of planes represent in a 24-h period?

c. What fraction of the area of the earth does this vital area represent?

d. Multiply the number of meteor hits by the fraction of the earth's area covered by planes per day. If we assume 5% of these hits are deadly, we can multiply the number of hits by 5% to get the probability of a crash.

e. The probability is for a 24-h period; what is the probability of a crash in a year of travel?

f. If air traffic continues at the current volume, what is the probability of a crash in the next 50 years?

g. Was it reasonable to suggest that a meteor caused the plane to crash or totally absurd? Why?

4. In the year 2000, 553,768 people died of cancer in the United States out of a population of about 281 million. (a) What was death rate for cancer in 2000? (b) If you were going to predict the risk of cancer death for the next year based on historical trends, what risk factor would you predict? (c) Using a reliable source, find the actual cancer rate in 2001 and compare with to your calculated value. (d) Speculate on factors that might affect the cancer rate. Separate these factors into short-term and longer-term effects.

5. Using reliable sources, find reasons for the drops in automobile accidents around 1943 and 1982 shown in Figure 8.1.

6. Using reliable sources, write a summary of Robert Kubica's crash of his Formula One racing car in the Canadian Grand Prix in June 2007. Include a discussion of why he survived this 180-mph crash.

7. Use conservation of momentum to compare the acceleration of a Ford Explorer, with a weight of 4,255 lb, with a Volkswagen Beetle, with a weight of 2,762 lb, if they collide head on, each traveling at 60 mph. Assume they are identical vehicles in all respects except mass and the collision is perfectly inelastic (they are locked together after the collision). Which driver gets the larger acceleration? By how much? (*Hint*: Newton's laws say that the force on each is the same and the time of contact is the same, so the impulse is the same on each car, so $F\Delta t = \Delta p = mv_f - mv_i$ for each vehicle. Once you have the change in velocity, make an assumption about the length of time of the collision to get the acceleration for each vehicle.)
8. Using reliable sources, find a list or chart of automobile accident or death rates versus car size. What conclusions (if any) can be made about safety and vehicle size? (*Hint*: Take a look at the car safety data shown on the Safety and Fuel Economy Web site http://sites.google.com/site/safetyandfueleconomy/. Pick a few cars and follow the changes in safety and fuel economy over time. Report on any interesting trends that you find.)
9. Using reliable sources, summarize what is known about the Tunguska meteor impact.
10. Verify the energies given in Table 8.1 for various sizes of meteors by using Equation (8.1) and data given in the text.
11. Use Equation (8.2) to calculate the area destroyed for each of the classes of meteor sizes listed in Table 8.1. How do these areas compare with the size of a typical city? The area of your state?
12. Repeat the calculation of Example 8.6 for the other classes of meteors given in Table 8.3. Which size meteor is more likely to kill you?
13. What percentage of the current world population (6.7 billion) would be killed by each of the classes of meteors shown in Table 8.3?
14. Go to NASA's Near Earth Orbit Program Web page (http://neo.jpl.nasa.gov/) and pick an asteroid that is being tracked by the program. Report on what is known about the orbit, the size of the object, and when and how close it will come to the earth.
15. Use the average interval and the local death figures given in Table 8.3 to verify the column of deaths per year. Why are there fewer deaths per year for larger impacts?

16. Using reliable sources, find pictures and commentary about the collision of comet Shoemaker–Levy with Jupiter and report on what you find.
17. Given the following probabilities for the failure of various systems and components in the nuclear power plant (cooling system in figure 8.2), what is the total probability for failure of the plant (if these are the only parts involved)? P_s for excess steam pressure is 0.001, P_c for computer control failure is 0.015, P_h for the human operator being out of the room is 0.25, P_p for a pressure valve to fail is 0.001, and P_v for a backup pressure valve to fail is 0.001.
18. Using reliable sources, summarize what happened during the Chernobyl disaster (reference [13] has some information; there are others). Include a discussion of the role of operator error and comments about the design of this type of reactor compared with the design and safety features of the light water reactor, which is more commonly used in the United States. Also report on the long-term effects of the disaster.
19. Using reliable sources, summarize what happened during the Three Mile Island incident in 1979 in Pennsylvania. Report on the long-term effects of the accident.
20. Using reliable sources, locate data that support the conclusion that smoking causes cancer. Lung cancer is most directly linked to smoking, but what other cancers are implicated? What are the links between smoking and other illnesses such as heart disease?
21. Using reliable sources, investigate the accusation that the tobacco industry actively suppressed information about the health effects of smoking. Be prepared to discuss your results in class.
22. Find the decay series for radon. Which of the daughter products are also radioactive, and what is their half life? What type of decay do they undergo?
23. What would the flow rate into the room in Example 8.9 be if the measured radon level was the EPA limit of 4 pCi/l?
24. Repeat Example 8.9 for the case of a substance that has a half life of 3.8 min (this could be a different radioactive gas or an airborne substance that is removed from the air by chemically binding with other substances present). What effect does half life have on concentration levels?

25. Using reliable sources, summarize the Human Exposure/Rodent Potency (HERP) Index. What is the HERP Index used for?

26. Using reliable sources, describe the pros and cons of animal testing. What valid claims are made on both sides of this argument? List cases of the justified use of animal testing and cases in which the testing is probably unjustified.

27. Using reliable sources, summarize how a t test, a z test, and the chi-squared test work. When do you use the t test, the chi-squared test, and the z test?

28. Repeat Example 8.12 for the case of only five tumors in the group of rats that are given the chemical being tested. What is the degree of confidence?

29. Following Example 8.12, determine the effect of larger numbers of rats for both cases. Test the case in which 5 out of 100 rats have tumors and compare the results with the case given and the case of 10 out of 100. What can you conclude about the number of rats that should be used in a study?

30. For the following cases, state which of the four dose–response curves in Figure 8.3 would be more likely to apply and explain your reasoning.
    a. Pain-relieving drugs such as aspirin.
    b. Exposure to a mild carcinogen such that the body may have a chance to repair damaged DNA.
    c. Drugs for which the body gradually builds up a resistance to the effects.
    d. The skin cancer rate among lifeguards as a function of time spent in the sun.

31. The dose curves in Figure 8.3 also apply to other models of risk. Which curves are likely to apply in each the following cases? Explain your reasoning.
    a. The purchase of lottery tickets.
    b. The number of automobile accidents as a function of speed.
    c. The number automobile accidents for an individual as a function of age.

d. A nuclear power plant with a large (but finite) number of emergency backup systems.

e. A city infrastructure system (water, electrical, roads) where maintenance is not performed.

32. Suppose the risk of death from an aspirin overdose is 50% if you take 30 aspirins. (a) Assuming a linear dose model, what would be the risk if you take only 1 aspirin? (b) Using the risk for taking 1 aspirin, how many people in the United States would be expected to die as a result if everyone took an aspirin? (Use 304 million for the U.S. population.) (c) Do you think the linear dose model is accurate for aspirin? Why or why not?

33. Using reliable sources, summarize the EPA's definitions of Permissible Exposure Limit, No Observable Effect Level, Lowest Observable Adverse Effect Level, and Reference Dose.

34. The number of traffic fatalities per 10^8 vehicle miles in the United States was 1.5 for automobiles, 0.5 for buses, 34 for motorcycles, and 0.3 for planes in 2003. Find the risk of death for traveling 1,000 miles by each means of transport. Which is more dangerous?

35. Suppose you live 50 miles from the airport. Using the figures in the previous problem, compare the risk of driving to the airport to the risk of flying 500 miles (10 times as far) to your destination.

36. Convert the Environmental Protection Agency limit of 0.5 ppm for sulfur dioxide to grams per cubic meter. Using reliable sources, summarize what is known about the health effects of SO_2 for various exposures.

37. Using reliable sources, find the associated health effects of NO_2, nonmethane volatile organic compounds, and particulates. Using this information and Example 8.14, discuss the various potential energy solutions given in Table 8.4 in terms of pollution.

38. Repeat Example 8.14 for NO_2 emissions. Assume the coal plant produces NO_2 at a rate proportional to the SO_2 emissions given in Table 8.4. The mortality rate for NO_2 is 2.8 excess deaths per 44.0 $\mu g/m^2$.

39. The people of Kerala, India have a much higher background exposure to radiation (20 mSv) than the U.S. population, but they live longer on average than people in the rest of India. Initially this was thought to be a hormesis effect of radiation, but subsequent investigation has

refuted this idea. Using reliable sources, track down the details of this controversy and report on it.

40. Using reliable sources, write a report on the radiation poisoning of Russian spy Alexander Litvinenko, who died in 2006.

41. Calculate the relative risk for each of the groups in Table 8.7. What can you conclude from these results?

42. Plot the dose versus fraction of excess cancers per number of victims for the data in Table 8.7. What is the slope of the curve (cancers per sievert)? It is sometimes dangerous to extrapolate data from one situation (single large doses, in this case) to other situations (low continuous exposure). Using the slope you found for the results of a single large exposure, calculate the fraction of the population of the United States that would develop a cancer each year, assuming the average low-level exposure of 3.6 mSv/year and a linear dose model. Does the extrapolation give a believable answer? (Compare this figure with deaths by other causes, such as automobile deaths or other forms of cancer.)

43. Repeat the calculation of Example 8.15 for the average nonnatural exposure of radiation of 0.6 mSv/year.

44. Using reliable sources, repeat Example 8.15 for another country besides the U.S.

45. It is thought that the Chernobyl disaster will result in about 4,000 additional cancers in the Russian population. The current lifetime cancer risk due to all causes is about 662 per 100,000. What is the relative risk compared to cancers due to all causes in Russia?

46. Possible leakage from the Yucca Mountain storage facility (which has now been decommissioned) was estimated to be less than 0.15 mSv/year after 10,000 years in the region around the mountain. Calculate the number of additional cancers as a result of this contamination for the current population of Nevada, assuming 0.004 deaths per sievert exposure. Repeat for the Chernobyl rate of 0.45 deaths per sievert.

47. The following all have the same risk of one in a million. For each one, say whether you think the risk would probably be overestimated or

underestimated by an average person and support your answer using the factors of risk perception discussed in the text.

a. Smoking two cigarettes

b. Drinking 30 diet sodas with saccharin

c. Eating 150 (1/2-lb) charcoal-broiled steaks (aromatic hydrocarbon risk)

d. Eating 100 hundred-gram servings of shrimp (formaldehyde risk)

e. Eating 4 tablespoons of peanut butter every 10 days for person without hepatitis B1

f. Eating 350 slices of stale bread (formaldehyde risk)

g. Drinking 70 pints of beer per year (alcohol cancer risk)

h. Exposure to typical radon levels in drinking water in California for 6 months

i. Drinking water with the EPA limit of arsenic (50 ppb) for 3 days

j. Being exposed to one quarter of a typical chest X-ray

k. Nonsmoker exposed to the average U.S. radon level (1.5 pCi/l) for 1 week

l. Living in Denver (compared to Philadelphia)

m. Traveling 100 miles in a motor vehicle

n. Dying from a lightning strike in a 6-year period

48. First, without looking anything up, rank the following disasters in order of most to fewest people killed (just give it your best guess): Black Death of 1347–1351; Yellow River flood in China, 1887; World Trade Center collapse, September 11, 2001; Hiroshima and Nagasaki nuclear bomb attacks, 1945; Dresden firebombing, Germany, 1945; the 1918 Spanish flu epidemic; Chernobyl nuclear plant, 1986; Three Mile Island nuclear plant, 1978; Bhopal, India chemical plant, 1987; a large tornado in the United States; 1942 Night Club fire in Boston; the sinking of the *Titanic*, 1912; the 2010 earthquake in Haiti. Next, write a summary of what happened in each case and the number of deaths involved. How close were your guesses to the actual number of people who died?

49. For each of the following pairs of risks, decide which you think would have the higher risk to the general population and then look up the

risks (using reliable sources) and compare them. If your estimate is very different from the actual risk, explain the difference using the factors of risk perception discussed in the text.

a. Electromagnetic fields from high-tension wires or automobile accidents

b. Caffeine or radioactive waste

c. Contamination in well water from an abandoned hazardous waste site or from naturally occurring metals

d. Car accidents or train accidents

e. Food preservatives or food contaminants

50. Using reliable sources, investigate the public's reaction to the launch of the Cassini deep space probe (deep space probes use the earth's gravitational field to slingshot themselves out into deep space). The probe carried radioactive material, and there was some fear that the probe might fall into the earth's atmosphere. The risk of this occurring was calculated to be about 10^{-13}. Verify this risk calculation and compare it with other risks. Did the risk merit the public reaction that it received?

51. Many people believe (incorrectly) that modern vaccines are responsible for a rise in autism in the United States. Using reliable sources, track down the source of this belief and discuss your conclusions. Include a discussion of the benefits of various vaccines (estimated number of lives saved) and the consequences of not vaccinating.

52. In 1976 the U.S. government under President Ford reacted to a threat of a possible flu epidemic by encouraging most of the population to be vaccinated. The flu itself killed only one person, but many more were made seriously ill by the vaccine because of allergic reactions. Using reliable sources, evaluate the political aspects of this event. How many people were predicted to die from the flu if there were no vaccination program? Does this number justify the number of deaths caused by the vaccine? What decision would you have made if you were president at the time?

53. Many people believe (incorrectly) that radiation of food to kill germs will cause the food to be radioactive. Using reliable sources, track down the source of this belief and discuss your conclusions. Include a discussion of the benefits of radiating food to protect it from spoilage.

54. Using reliable sources, report on the discrepancy between the scientific evidence that cell phones do not cause cancer and public opinion about this question. (*Hint*: See http://jnci.oxfordjournals.org/cgi/content/full/93/3/166).

55. Many people believe that wealth is directly proportional to health. Using reliable sources, construct an argument to the contrary. A good article to start with is "Health and Human Society" by C. Hertzman in *American Scientist*, Nov./Dec. 2001.

56. Using reliable sources (Centers for Disease Control or Census Bureau data), list the leading causes of death in the United States. Break this down by age group (leading cause of death among 1- to 5-year-olds, 6- to 10-year-olds, etc.). Discuss your results.

57. Using reliable sources, write a report about the company 1BOG ("one block off the grid") and the projects it is carrying out in the United States.

58. Using reliable sources, write a report about Sim City (you can download and play a limited version of the game for free at http://simcity-societies.ea.com/index.php).

59. Use Equation (8.9) to determine today's value of $1 invested on the day you were born if the interest rate is 10%. How much will this dollar be worth when you are 70?

60. Use Equations (8.10) and (8.11) to determine which is better: to borrow $10,000 at 5% for 20 years or at 6% for 15 years.

61. Repeat the calculation in Example 8.16 for a windmill. You will need to find approximate costs for the windmill and estimate the electricity it is expected to produce based on figures from reliable sources. How close is your answer to the figure given in Table 5.5? Explain.

61. Repeat the calculation in Example 8.17 using other figures for rebate and tax incentives. Which has more of an effect: an increase in the rebate of 50 cents or an increase in the tax deduction of 2%?

63. Currently the corn ethanol industry is receiving tax incentives to make it competitive with other fuels. Find out exactly what these incentives are. Based on what you learned about ethanol in Chapter 5, make an argument that tax incentives are not always a good thing.

64. Summarize the main points of reference [24] on the cost of reducing greenhouse gases.

REFERENCES AND SUGGESTED READING

1. D. M. Kammen and D. M. Hassenzal, *Should We Risk It? Exploring Environmental, Health, and Technological Problem Solving* (1999, Princeton University Press, Princeton, NJ).

2. R. Wilson and E. A. C. Crouch, *Risk–Benefit Analysis* (2001, Harvard University Press, Cambridge, MA). J. Harte, *Consider a Spherical Cow* (1988, University Science Books, Herndon, VA).

3. R. Wilson, "Resource Letter: RA-1: Risk Analysis," *American Journal of Physics* 70(5) (2002):475.

4. Fatality Analysis Reporting System, developed by the National Center for Statistics and Analysis of the National Highway Traffic Safety Administration (http://www-fars.nhtsa.dot.gov/Main/index.aspx) and the National Household Travel Survey (http://nhts.ornl.gov/index.shtml).

5. National Transportation Safety Board Aviation Accident Statistics, Table 6 (http://www.ntsb.gov/aviation/Table6.htm).

6. U.S. Coast Guard Boating Safety Division report (http://www.uscgboating.org/statistics/accident_statistics.aspx).

7. Bureau of Transportation Statistics (http://www.bts.gov/data_and_statistics/) and Federal Railroad Administration Office of Safety Analysis (http://safetydata.fra.dot.gov/officeofsafety/publicsite/Query/castab.aspx).

8. T. P. Wenzel, *Safer Vehicles for People and the Planet*, Lawrence Berkeley National Laboratory publication LBNL-325E (http://repositories.cdlib.org/cgi/viewcontent.cgi?article=6832&context=lbnl).

9. National Highway Traffic Safety Administration (http://www.nhtsa.dot.gov).

10. L. Evans, "Traffic Crashes," *American Scientist* 90(3) (2002):244.

11. C. R. Chapman and D. Morrison, "Impacts on the Earth by Asteroids and Comets: Assessing the Hazard," *Nature* 367 (Jan. 1994):33.

12. NASA Near Earth Orbit Program (http://neo.jpl.nasa.gov/).

13. B. L. Cohen, "The Nuclear Reactor Accident at Chernobyl, USSR," *American Journal of Physics* 55(12) (1987):1076.

14. Radon test kits can be found by a Google Web search for "Radon Test Kit". The author has used kits from http://www.radonzone.com/ in the past.

15. Environmental Protection Agency (http://www.epa.gov/).

16. G. H. McClelland, "Working with *z*-Scores and Normal Probabilities," online *z* score calculator (http://psych-www.colorado.edu/~mcclella/java/normal/normz.html).

17. The International Energy Agency (http://www.iea.org/textbase/papers/2003/devbase.pdf).

18. D. Hafemeister, *Physics of Societal Issues* (2007, Springer Science, New York).

19. Committee to Assess Health Risks from Exposure to Low Levels of Ionizing Radiation, *Health Risks from Exposure to Low Levels of Ionizing Radiation: BEIR VII Phase 2* (2006, National Academies Press, Washington, DC) (http://www.nap.edu/catalog.php?record_id=11340).

20. E. C. Halperin, C. A. Perez, L. W. Brady, D. E. Wazer, and C. Freeman, *Perez and Brady's Principles and Practice of Radiation Oncology* (2008, Lippincott Williams & Wilkins, Philadelphia).

21. Centers for Disease Control, National Center for Health Statistics (http://www.cdc.gov/nchs/r&d/rdc.htm).

22. J. M. Deutch and R. K. Lester, *Making Technology Work* (2004, Cambridge University Press, Cambridge).

23. J. Randolph and G. M. Masters, *Energy for Sustainability* (2008, Island Press, Washington, DC).

24. J. Crets et al., "Reducing US Greenhouse Gas Emissions: How Much at What Cost?," McKinsey & Company Report, 2007 (http://mckinsey.com/clientservice/ccsi/pdf/US_ghg_final_report.pdf).

25. J. W. Tester, E. M. Drake, M. J. Driscoll, M. W. Golay, and W. A. Peters, *Sustainable Energy: Choosing among Options* (2005, MIT Press, Cambridge, MA).

26. National Academy of Science, "Electricity from Renewable Resources: Status, Prospects, and Impediments," 2009 (http://www.nap.edu/catalog.php?record_id=12619).

27. I. F. Roth and L. L. Ambs, "Incorporating Externalities into a Full Cost Approach to Electric Power Generation Life-Cycle Costing," *Energy* 29(12–15) (Oct–Dec. 2004):2125.

28. J. Harte, *Consider a Cylindrical Cow* (2001, University Science Books, Herndon, VA).

Notes

[1] The lifetime accumulated number of cancers is not exactly equal to the annual rate times 78 years because cancer rates are different at different ages and are declining.

[2] Fault trees for real reactors have hundreds of branches and multiple emergency systems.

[3] LD = lethal dose.

[4] However, reference [19] states that there is not sufficient evidence to conclude that there is a threshold dose for all types of cancers.

[5] Bell Laboratories was the research and development component of AT&T before it was sold to Alcatel–Lucent. The transistor, the laser, the UNIX operating system, and the C programming language were developed at Bell Labs, among many other innovations. Six Nobel prizes were awarded to scientists for work done there.

Conclusion

As we saw in Chapter 8, every technology has varying degrees of risk and costs associated with it. The double-edged nature of technological progress brings us full circle: We began our exploration of environmental physics by examining how scientific advancements of the last two centuries have both supported a growing, increasingly affluent global population and confronted us with unprecedented environmental challenges.

Most experts today predict that we will soon be facing a worldwide energy crisis. Globally, energy use has risen by a factor of 16 since 1800, four times faster than the population. Consumption of natural and manufactured resources has also grown faster than the population, suggesting that people today live better than they did 200 years ago. Although some parts of the world have fared better than others, on average people do live longer, eat better, have better health care, and have more leisure time, and their lives are aided much more by technology than in the past. Yet energy is a finite resource; it cannot be created or destroyed, but it can be degraded into a nonuseful form, in concurrence with the second law of thermodynamics. A continued increase in a population that enjoys today's standard of living will necessitate new energy sources or conservation.

At the same time that energy demands are going up, traditional energy supplies are showing signs of giving out. Many experts believe that the development of natural resources approximately obeys a bell-shaped curve (Hubbert's curve), with production increasing until all the resource has been located and tapering off as the resource is used up. Historically this has often been the case. For example, the world switched from whale oil to petroleum largely because overharvesting decimated the oil-bearing whale population. There is still some discussion as to whether and when this might happen with fossil fuels (petroleum, natural gas, coal), but the fact that the size of newly discovered oil fields has been declining for 40 years is cause for some concern. An increase in price will force conservation and efficiency, making the bell-shaped curve nonsymmetrical for fossil fuels as secondary and tertiary extraction methods are brought to bear on existing fields. However, once the amount of energy needed to extract the resource exceeds the energy in the extractable supply, price becomes irrelevant. Using a liter of oil to extract half a liter from the ground makes no sense, regardless of the price of oil. Clearly, new sources of energy will need to be discovered soon.

Even if supplies of fossil fuels were limitless, many believe their associated risks are too high. Rising energy use has typically meant a parallel increase in pollution, greenhouse gas in particular. Although overall emissions have decreased over the past 25 years, zinc emission is higher by a factor of five from 1900, and lead emission, mainly from the burning of leaded gasoline for transportation, is still seven times what it was 100 years ago. Because lead in the soil does not migrate very quickly, many urban areas still contain toxic amounts of it, even though lead in gasoline was phased out starting in the 1970s in the United States. The impacts are ongoing: According to the Environmental Protection Agency, approximately 9% of 1- and 2-year-olds tested in the United States in 2005 have elevated levels of lead in their blood. This exposure can contribute to learning disabilities, mental retardation, impaired visual and motor functioning, stunted growth, neurological and organ damage, and hearing loss. In a similar fashion, the adoption of any new energy source may have long-term unintended consequences.

Of all the pollutants generated by modern technology, greenhouse gases generated from the use of fossil fuels may have the farthest-reaching effects. The fact that the atmosphere now has 35% more carbon dioxide, 150% more methane, and 40% more nitrous oxide than at any time in the past 850,000 years is cause for alarm. Scientists have very good reason to think these human-caused atmospheric changes will have a significant effect on climate. And in fact we are beginning to see indications of climate change. Changes in animal migration patterns and reproduction times have been measured, as have changes in the rate of leaf maturation and plant blooming times, increased coastal erosion caused by a global sea level rise of about 23 cm has been documented, a shift in the time of peak springtime river flow has been measured, and the extent of arctic sea ice has decreased. The increase in atmospheric carbon dioxide is causing a documented change in the acidity of the ocean, affecting many forms of marine life. The impacts may not be limited to the natural world: Climate change could lead to more deaths due to heat waves, increases in vector-borne diseases, hunger caused by changing food production patterns, and populations displaced by intensifying floods and wildfire, among other consequences.

Given the potentially dire nature of the impacts of climate change and a growing realization of the finite nature of fossil fuels, society is in search of alternative energy sources. At this point, there appears to be no single source that can fully replace fossil fuels. Nuclear energy is not exempt from the Hubbert model, but the time line appears to be much longer than for fossil fuels. Many countries, such as France, Japan, China, and India, are making the nuclear choice. Hydropower is near saturation in the developed world but has great potential in the rest of the world. Wind energy, the fastest-growing renewable source, can provide a great deal of the world's energy, but because of its tie to certain geographic locations and the fact that it is intermittent, it cannot be the entire solution to the world's energy demand. Likewise, competition with food supplies and the inherently low efficiency of plant growth will limit the role of biofuels as a solution to our energy problems. Because of their lower overall potential and technical difficulties in extracting the energy, biomass, tidal, geothermal, and wave energy are unlikely to contribute more than a small percentage of the world's energy supply in the foreseeable future.

Solar energy holds the greatest hope for a one-stop world energy supply. The potential recoverable energy using existing technology is about 100 times more than any other renewable source and about 60 times what the world currently uses. The solar energy striking 2% of the surface area of the United States could supply all of that country's energy using current technology and assuming a 50% storage and transmission loss. The technical problems in storing this intermittent supply and economic factors currently limit this solution, but there appears to be moderate optimism for the future. There is enough energy in the world to replace fossil fuels, but the economic and logistical aspects of this transformation are daunting; nonhydroelectric renewable sources make up only 1% of our current energy supply.

If the astounding variety of mechanisms proposed to capture water and wind energy is any indication, humans are very creative beings. The same creativity is apparent in the array of proposed energy storage devices. Motivated by the need for lightweight, high-energy storage devices necessary in transportation applications, intense research and development are under way in battery technology, supercapacitors, compressed air, and flywheels. On a larger scale, pumped hydroelectric and large-scale solar projects are under development. Being able to store significant amounts of energy is critical to the success of intermittent energy sources.

However, it should be kept in mind that alternative energies are not necessarily environmentally benign or carbon neutral. For example, hydroelectric dams, arguably the most benign renewable energy source, emit carbon dioxide from decaying vegetation and often cause the destruction or degradation of ecosystems during construction, and large dams may cause dangerous geological shifts due to the weight of the water. All energy sources cause pollution in their initial manufacture, their daily use, or disposal of the apparatus after its normal lifetime. It is also the case that decisions made today about energy and conservation will be with us for some time. The average lifetime of a building is about 80 years, and the average power plant lasts 40 years before it is replaced or retrofitted with new technology. Clearly the long-term implications of using any new technology, including a life cycle analysis (the total energy needed to manufacture, transport, operate, and dispose of the device), should be evaluated before a proposed solution to any environmental problem is implemented.

The first step in assessing the effects of any new technology and comparing different technologies is a scientific risk assessment. Defined as probability times severity for a particular action or activity, risk may be calculated from historical or epidemiological data, animal studies, or a fault tree analysis. Accurate risk analysis often depends on understanding exposure rates and dose responses. As we saw in Chapter 8, the perception of risk is often very different from actual risk. In the United States, the average person rates smoking and driving as lower risks than expert analysts do and rates nuclear power as a much higher risk.

The final pieces of the puzzle in solving any environmental problem involve economics and politics. It was comparatively easy to ban ozone-destroying chlorofluorocarbons because there were economically viable alternatives and the political atmosphere was favorable because few people were directly affected by the switch to other chemicals. Setting up a cap-and-trade program for sulfur emissions, which cause acid rain, was more controversial and required a political compromise regarding coal mining, but again, the process did not require the direct participation of a large number of people. Capping carbon emissions, either by directly capturing emissions from fossil fuels or by switching to alternative sources, will be an enormous undertaking because so much of the world's energy supply comes from fossil fuels. The political problems are also steep because nearly everyone will be affected. Most economists believe some combination of cap and trade, tax incentives, rebates, government investment in fundamental research, and international regulation will be needed to capture the externalities associated with fossil fuels. Indeed, some European countries already have such inducements in place, and nearly every country in the world is considering policy changes to accelerate a shift away from fossil fuels.

Whatever the political and economic complications, minimizing the risks associated with energy use and modern technologies requires a basic understanding of the physical world. Without a fundamental understanding of the physical laws relating to a particular environmental problem, we cannot hope to find a satisfactory solution. This book has sought to help readers develop that core of fundamental knowledge and the skills to use it. Given the proper scientific tools and clear goals for solving these problems, the advances of the last 200 years may pale in comparison to those of the coming generations of scientists.

APPENDIX A

Useful Constants and Conversions

Dimensional Prefixes

10	deka (da)	10^{-1}	deci (d)
10^{2}	hecto (h)	10^{-2}	centi (c)
10^{3}	kilo (k)	10^{-3}	milli (m)
10^{6}	mega (M)	10^{-6}	micro (μ)
10^{9}	giga (G)	10^{-9}	nano (n)
10^{12}	tera (T)	10^{-12}	pico (p)
10^{15}	peta (P)	10^{-15}	femto (f)
10^{18}	exa (E)	10^{-18}	atto (a)

Constants

$a = e^2/hc = 1/137.036$ (fine structure constant)

$c = 2.998 \times 10^8$ m/s (speed light in a vacuum)

$e = 1.60 \times 10^{-19}$ coulomb, C (electron/proton charge)

$1/4\pi\varepsilon_0 = 9.0 \times 10^{-9}$ N m^2/C^2

$\varepsilon_0 = 8.8 \times 10^{-12}$ C^2/N m^2 (permittivity of space)

$\mu_0 = 4\pi \times 10^{-7}$ N/A^2 (permeability of space)

$G = 6.67 \times 10^{-11}$ N m^2/kg^2 (gravitational)

$g = 9.807 \times$ m/s^2 (acceleration of gravity at 45° latitude at sea level)

$h = 6.63 \times 10^{-34}$ J s $= 4.14 \times 10^{-15}$ eV s (Planck's constant)

$\hbar = h/2\pi = 1.06 \times 10^{-34}$ J sec $= 6.59 \times 10^{-14}$ eV sec

$k_B = 1.38 \times 10^{-23}$ J/K $= 8.63 \times 10^{-5}$ eV/K (Boltzmann's constant)

$k_BT = 0.26$ eV $= 1/40$ eV (at room temperature 300 K)

$m_e = 9.110 \times 10^{-31}$ kg (electron mass)

$m_p = 1.673 \times 10^{-27}$ kg = 1.6726 amu (proton mass)

$m_n = 1.675 \times 10^{-27}$ kg = 1.6749 amu (neutron mass)

$m_ec^2 = 511$ keV, $m_pc^2 = 938.3$ MeV, $m_nc^2 = 939.6$ MeV

$N_A = 6.023 \times 10^{23}$ molecules/gram mole (Avogadro's number)

volume of mole of gas at STP = 2.24 l at 2.7×10^{19} molecules/cm^3

density of water = 1,000 kg/m^3 = 62.4 lb/ft^3 (standard conditions)

density of seawater = 1,025 kg/m^3 (standard conditions)

density of dry air = 1.226 kg/m^3 (at 15°C and 1 atm)

Radiation

Quantity	Units	Definition
Rate of decay	Curie Becquerel	$1Ci = 3.7 \times 10^{10}$ decay/s 1Bq = 1 decay/s
Energy intensity	Roentgen	1R = 87 erg/g = 0.0087 J/kg
Absorbed energy	Rad Gray	1rad = 100 erg/g = 0.01 J/kg 1Gy = 1 J/kg = 100 rad
Relative biological effectiveness (RBE)	RBE	RBE = 1 for X-rays, γ-rays and β RBE = 1 – 5 for neutrons RBE = 20 for α and fission fragments
Effective absorbed dose	Rem Sievert	rem = rad × RBE 1Sv = 1 J/kg = 100 rem

The gray and the sievert are the accepted international units. rem = roentgen equivalent man.

Earth

radius R_E = 6,357 km polar and 6,378 km equatorial

area = 5.10×10^{14} km^2 (oceans 71%)

mass = 5.98×10^{24} kg

atmospheric pressure = 10^5 Pa $e^{-h/H}$, atmospheric height H = 8.1 km

atmospheric mass = 5.14×10^{18} kg with 1.3×10^{16} kg H_2O, oceanic mass = 1.4×10^{21} kg

Sun

solar flux S = 1.366 kW/m^2 = 0.13 kW/ft^2 = 2.0 cal/min cm^2 = 435 Btu/ft^{2}h

24-h average horizontal flux = 185 W/m^2 (40°N latitude); flux = 250 W/m^2 (average, all latitudes)

mass = 2×10^{30} kg

distance to Earth = 150 Mkm

Energy Conversion Factors

	Btus	Quads	Calories	kWh	MWy
Btus	1	10^{-15}	252	2.93×10^{-4}	3.35×10^{-11}
Quads	10^{15}	1	2.52×10^{17}	2.93×10^{11}	3.35×10^{4}
Calories	3.97×10^{-3}	3.97×10^{-18}	1	1.16×10^{-6}	1.33×10^{-13}
kWh	3,413	3.41×10^{12}	8.60×10^{5}	1	1.14×10^{-7}
MWy	2.99×10^{10}	2.99×10^{-5}	7.53×10^{12}	8.76×10^{6}	1
bbls oil	5.50×10^{6}	5.50×10^{-9}	1.38×10^{9}	1,612	1.84×10^{-4}
Ton oil	4.04×10^{7}	4.04×10^{-8}	1.02×10^{10}	1.18×10^{4}	1.35×10^{-3}
kg coal	2.78×10^{4}	2.78×10^{-11}	7×10^{6}	8.14	9.29×10^{-7}
Ton coal	2.78×10^{7}	2.78×10^{-8}	7×10^{9}	8,139	9.29×10^{-4}
MCF oil	10^{6}	10^{-9}	2.52×10^{8}	293	3.35×10^{-5}
Joules	9.48×10^{-4}	9.48×10^{-19}	0.239	2.78×10^{-7}	3.17×10^{-14}
EJ	9.48×10^{14}	0.948	2.39×10^{17}	2.78×10^{11}	3.17×10^{4}

	Bbls Oil Equivalent	Ton Oil Equivalent	Kg Coal Equivalent	Ton Coal Equivalent	MCF Gas Equivalent	Joules	EJ
Btus	1.82×10^{-7}	2.48×10^{-8}	3.6×10^{-5}	3.6×10^{-8}	10^{-6}	1,055	1.06×10^{-15}
Quads	1.82×10^{8}	2.48×10^{7}	3.6×10^{10}	3.6×10^{7}	10^{9}	1.06×10^{18}	1.06
Calories	7.21×10^{-10}	9.82×10^{-11}	1.43×10^{-7}	1.43×10^{-7}	3.97×10^{-9}	4.19	4.19×10^{-18}
kWh	6.20×10^{-4}	8.45×10^{-5}	0.123	1.23×10^{-4}	3.41×10^{-3}	3.6×10^{6}	3.6×10^{-12}
MWy	5,435	740	1.08×10^{6}	1,076	2.99×10^{4}	3.15×10^{13}	3.15×10^{-5}
bbls oil	1	0.136	198	0.198	5.50	5.80×10^{9}	5.80×10^{-9}
Ton oil	7.35	1	1,455	1.45	40.4	4.26×10^{10}	4.26×10^{-8}
kg coal	5.05×10^{-3}	6.88×10^{-4}	1	0.001	0.0278	2.93×10^{7}	2.93×10^{-11}
Ton coal	5.05	0.688	1,000	1	27.8	2.93×10^{9}	2.93×10^{-8}
MCF gas	0.182	0.0248	36	0.036	1	1.06×10^{9}	1.06×10^{-9}
Joules	1.72×10^{-10}	2.35×10^{-11}	3.41×10^{-8}	3.41×10^{-11}	9.48×10^{-10}	1	10^{-18}
EJ	1.72×10^{8}	2.35×10^{7}	3.41×10^{10}	3.41×10^{7}	9.48×10^{8}	10^{18}	1

For the above tables:

To convert from the first column units, multiply by the factors shown in the appropriate row.

Key: bbls = barrels; EJ = exajoule = 10^{18} J; MCF = thousand cubic feet; MWy = megawatt-year; ton = metric tons = 1,000 kg = 2,204.6 lb.; Nominal calorific values assumed for coal, oil, and gas.

REFERENCES

1. A useful online conversion engine can be found at OnlineConversion.com (http://www.onlineconversion.com/).

APPENDIX B

Error Analysis

The following is a brief introduction to the statistics involved in error analysis. The discussion is by no means complete and is meant as merely a reminder of key definitions in statistical error analysis. The reader should consult references [1], [2], [3], [4], and [5] for further details. A very useful free statistical analysis package that will perform the various statistical comparisons mentioned here is the open source program R [6].

If many measurements are taken in an experiment, we expect that there will be some error involved, no matter how good the apparatus or experimenter. In other words, not all the results will be exactly the same, but instead they will lie in some distribution around some central or average value. This is true for a set of outcomes for which the variable has discrete values such as coin or dice tosses (Chapters 3 and 8) or measurements that can have a continuous range of possible outcomes (Chapter 8).

In most cases we can predict what the distribution of data should look like and how large the fluctuations can be expected to be for a given number of data points. In other words, we often can predict the statistical model the data ought to fit. The three most common statistical models are the binomial distribution, the Poisson distribution, and the Gaussian distribution. In all three models the fluctuations are assumed to be small, the error random, and the measurements independent of each other. The most general statistical model is the binomial distribution, which applies to situations in which the outcomes are discrete. The Poisson and Gaussian distributions are approximations to the binomial distribution and can be used for continuous variables as well as discrete variables. If in a set of measurements the probabilities are small (e.g., the probability of any one atom decaying in a sample being measured for a short time period), the Poisson distribution can be used. The Gaussian or normal distribution is used to approximate the binomial distribution when the probabilities are large.

Once a statistical model is chosen, it is possible to quantify the error in a data set and to state in precise mathematical terms how closely the data set matches the predicted model. The same techniques can be used to compare two sets of experimental data to determine how closely they match each other. In the following we examine the case of continuous measurements and the Gaussian distribution, but the same ideas, with slightly different mathematics, apply to both cases. The reader is referred to the literature for further applications.

B.1 ■ Quantifying the Data

Suppose a large group of students is asked to measure the length of the hallway in a building using a meter stick. Obviously there will be slight differences in the outcomes of each measurement. Further suppose that the students act independently, and we can rule out any systematic error (e.g., a bent meter stick). For large collections of data where the errors are randomly and independently distributed, the expected outcome might look something like the data points shown in Figure B.1. Similar distributions of error will be seen for large numbers of coin or dice tosses. For several trials of tossing a coin 100 times, we do not get exactly 50 tails and 50 heads in each trial, although that is the expected outcome.

The Gaussian distribution is a statistical model that applies when the errors are expected to be random and independent. We might suspect, then, that the data in Figure B.1 will approximate a Gaussian distribution. The Gaussian curve shown in Figure B.1 does appear to fit the data, but we would like to quantify exactly how closely the data fit the curve. If the data do not fit the expected Gaussian distribution, we can conclude that the data are not random or we have chosen the wrong theoretical model. In order to accomplish this we will first define some parameters for the data and then formulate a way to compare these parameters to the Gaussian model.

First we can define the experimental mean, $\overline{x}_e$:

$$\text{(B.1)} \quad \overline{x}_e = \sum_{i=1}^{N} x_i,$$

where N is the total number of points and x_i are the individual data points. If the data form a symmetrical distribution when plotted, the mean, the median (the

FIGURE B.1

Data and a Gaussian fit to the data. Each data point represents the number of times a particular value was measured, normalized so that the area under the curve is 1. The standard deviation, σ, is 10 m in this example, and the mean is 50 m.

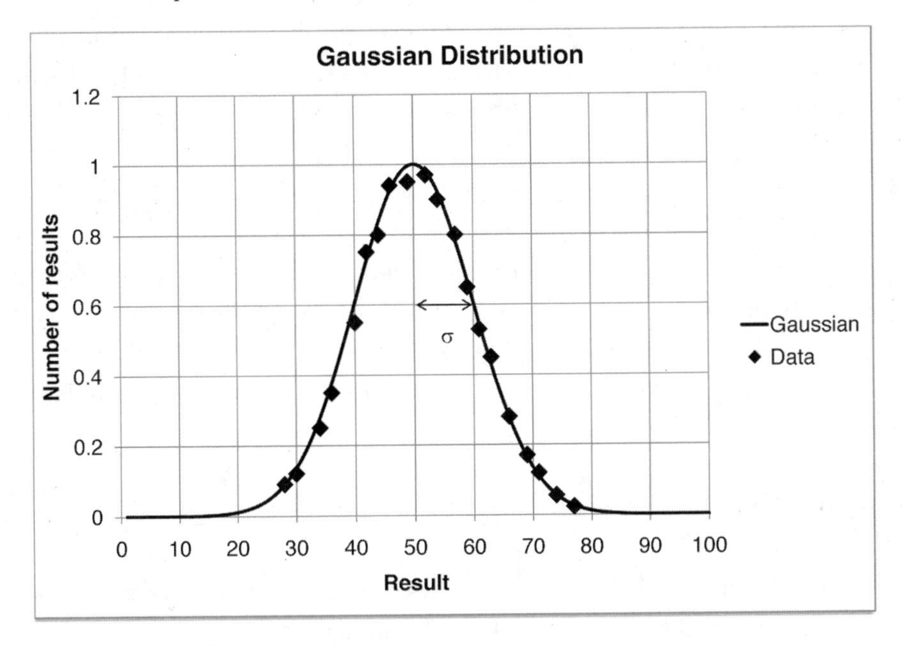

value where half of the distribution lies on either side), and the mode (the most probable value) are all equal. This is not the case for other types of distributions, nor is it always true for real data, but we will ignore this difference in the following. If the mode and the median are not equal to the mean, it indicates that the assumption of a normal distribution is false.

We may also define the distribution of the data, $F(x)$, as

$$F(x) = \frac{\text{number of results for value } x_i}{N}, \tag{B.2}$$

where x_i are the individual data points. A plot of $F(x)$ would look like the data in Figure B.1, with more results recorded near the central value and fewer as we move away from the mean. The residual[1] tells us how far away a particular data point is from the mean $d_i = x_i - \bar{x}_e$, and the sum of the residuals is called the sample variance, s^2:

$$s^2 = \sum_{x=0}^{\infty} (x - \bar{x})^2 F(x) = \frac{1}{N-1} \sum_{i=1}^{N} (x_i - \bar{x}_e)^2, \tag{B.3}$$

where $\bar{x}$ is the true mean, assuming an infinite number of data points could be taken (or, equivalently, that the actual distribution is known). Notice that in the second expression the sample variance is seen to be the average deviation from the experimental mean. Equations B.2 and B.3 link an experimental distribution function with a continuous set of variables to the (necessarily) finite set of experimental data points.

These quantities that characterize the data can be compared with a statistical model, such as the Gaussian distribution. The Gaussian probability distribution for random error (the curve shown in Figure B.1) is given by

$$\text{(B.4)} \quad P(x) = \frac{1}{\sqrt{2\pi\bar{x}}} \exp\left[-\frac{(x-\bar{x})^2}{2\bar{x}}\right]$$

and calculates the probability that a particular measurement has the value x. When plotted, $P(x)$ for the statistical model should have the same shape as $F(x)$ for the real data if the Gaussian distribution really does model the data. The predicted variance for the Gaussian distribution, σ^2, is defined in a similar way as the sample variance:

$$\text{(B.5)} \quad \sigma^2 \equiv \sum_{x=0}^{\infty} (x-\bar{x})^2 \; P(x).$$

Substituting Equation (B.4) into (B.5) allows us to calculate $\sigma^2 = \bar{x}$ for a Gaussian distribution (other distributions will have other expressions for T^2).

To see whether the data actually match the statistical model (the Gaussian distribution in this case), we assume the true mean (for an infinite number of data points) equals the experimental mean, $\bar{x} = \bar{x}_e$, and check to see whether the predicted variance equals the sample variance. It should be the case that $\sqrt{s^2} = \sqrt{\sigma^2} = \sigma$ if the data fit the model. The parameter σ is called the standard deviation and may refer to either the square root of the predicted variance or the sample variance.[2]

If the standard deviation of the data is small, we can say the data are more precise because the average variance is small. A second way to think about σ is if we make a plot of the Gaussian distribution, Equation (B.1), as in Figure B.1. If σ is small, the curve will be taller and more narrow, which indicates that the error or spread of the data is smaller. For the Gaussian model the height of the curve at the maximum can be shown to equal $(\sigma\sqrt{2\pi})^{-1}$, and the height at $x = \pm\sigma$ is $1/\sqrt{e} = 0.607$, where σ is the model standard deviation. If we integrate the area under the curve from $-\sigma$ to $+\sigma$ we get 0.68; in other words, any data point is expected to have a 68% chance that it lies between $-\sigma$ and $+\sigma$ around the mean, assuming the Gaussian distribution models the data.

Because σ quantifies the error in the data, it is often reported as the experimental precision.[3] In other words, the results of a set of length measurements might be given as 50.39 m ± 0.02 m, where 50.39 m is the mean and $\sigma = 0.02$ m. However, because the range $\pm\sigma$ covers only 68% of the data, 32% of the collected data points will lie beyond $\pm\sigma$ from the mean. Integrating from -2σ to $+2\sigma$ gives 95% and from -3σ to $+3\sigma$ gives 99.7% of the data points (Equation (B.4) is normalized, which means integrating from $-\infty$ to $+\infty$ equals 1 or 100%). This means there is a 95% probability that a given measurement will lie within $\pm2\sigma$ of the mean and a 99.7% chance that it will be within $\pm3\sigma$ of the mean or a 0.3% chance the measurement will lie outside the range $\pm3\sigma$.

One way to quantify the match between experimental data and the statistical model is to define the variable chi-squared as

$$\text{(B.6)}\quad \chi^2 = \frac{1}{\bar{x}_e}\sum_{i=1}^{N}(x_i - \bar{x}_e)^2 = \frac{(N-1)s^2}{\bar{x}_e}.$$

If the fit is good, the sample variance equals the predicted variance, and $s \cong \sigma$. For the Gaussian model, $\sigma = \sqrt{\bar{x}} = \sqrt{\bar{x}_e}$, so we expect $\chi^2/(N-1) \approx 1$ in the case that the data fit the model. The further χ^2 is from 1, the worse the data fit the model. This is usually quantified by a table or computer program that returns the probability that a random sample from the predicted distribution has a larger value of χ^2 than the actual value calculated for χ^2 from the data.

In many references on statistics Equation (B.4) is converted into a simpler form using the dimensionless variable $z = (x - \bar{x}_e)/\sigma$. The standard form of the Gaussian distribution then becomes

$$\text{(B.7)}\quad P(z) = \frac{1}{\sqrt{2\pi}}\exp\left[-\left(\frac{z}{2}\right)^2\right].$$

The variable $z = (x - \bar{x}_e)/\sigma$ in the Gaussian distribution in standard form relates the variance (how far away a particular data point, x, is from the mean, $\bar{x}_e$) to the standard deviation (the spread in the data) and is sometimes called the z score. The closer the data point is to the mean, the smaller the z score. Notice that a chi-squared value (Equation B.6) is equivalent to a sum of z scores for a Gaussian model.

The z score can be used to test how closely an individual data point (or in some cases a theoretical prediction) fits with a given data distribution, assumed to be Gaussian. Suppose we have a theory that gives us an answer of x' and we want to know how well this fits with a given set of experimental data. First we calculate $z = (x' - \bar{x}_e)/\sigma$ using the mean and standard deviation from the data.[4] Obviously, if the theoretical point is equal to the mean, the z score is zero and the theoretical hypothesis is true with 100% confidence. For other values of z we can consult tables or a computer program that tells us the probability that the theoretical point lies within $\pm z\sigma$ of the mean.[5] From a standard table of z scores or a z score calculator [7] we see that a z score greater than 1.96 corresponds to a value of 0.95, so the area from -1.96σ to $+1.96\sigma$ includes 95% of the data. Only 5% of the data lies beyond a z score of 1.96, so a z score higher than 1.96 indicates that the hypothesized value has only a 5% chance of being related to the data. Other values of the z score give the percentage of confidence for the hypothesized or theoretical point, x'.

The z test can also be used to indicate the accuracy of a set of measurements by substituting the true value for x' if the true value is known. For example, if students measuring a hallway use a bent meter stick for all their measurements, the results might be precise (small standard deviation) but not accurate (large z score when the data are compared with the true value).

Modern statistical programs such as R give the probability directly without calculating a z score [6]. The user inputs the data and the test data point (theoretical, experimental, or true value), and the program returns the probability that the point is correlated to the data, the mean, and the standard deviation or standard error.

B.2 ■ Hypothesis Testing

In the previous section we demonstrated how the chi-squared test can tell us how well a set of data matches a proposed statistical model using the Gaussian distribution as the predicted model. The exact same procedure can be used to compare a data set to other statistical models, such as the Poisson or binomial distributions. The standard deviation of the data set gives an idea of the precision of the data. We also saw that the z test is one way to test how an individual data point (theoretical or experimental) is related to a set of data that is assumed to fit some statistical model. The z test can also be used to indicate the accuracy of a data set if the true value is known. These are all examples of hypothesis testing. In this section we briefly describe how we can use the same ideas to compare two different sets of experimental data and give an example using the open source software R.

Suppose we want to know whether a new drug is significantly more effective than another drug or a placebo for the treatment of high blood pressure. We might give the new drug to one group of individuals and a placebo to another group and measure the change in blood pressure after some period of time. If we can assume the distribution of changes of blood pressure is random, we expect to find a Gaussian distribution of blood pressure changes around some mean value, with a standard deviation for each of the two groups. Suppose the control group has an average change in systolic blood pressure of 3 mm Hg (millimeters of mercury), with a standard deviation of ±1 mm Hg due to random changes in the health of the patients in this group. Suppose the group receiving the drug has a mean change of 12 mm Hg, with a standard deviation of ±2 mm Hg. The 95% confidence level ($\pm 2\sigma$) for the control group is ±2 mm Hg and ±4 mmHg for the group receiving the drug. Notice that the two results, 3±2 mm Hg and 12±4 mm Hg, do not overlap at the 95% confidence level. We can conclude with 95% confidence that the effects of the drug are significant compared to the null hypothesis (no drug).

Obviously, in this example if the standard deviations do overlap significantly, it will be more difficult to compare the two results. However, the chi-squared can still be used to find how well the two data sets match each other. In fact, the chi-squared test can also be used to compare other types of data sets besides Gaussian distributions. For example, if a set of data is expected to fit a straight line, the theoretical values would be the points generated by the equation $y = mx + b$, where m, the slope of the line, and b, the y-intercept, are determined from the data. The chi-squared test then measures the confidence with which the N data points, x_i, actually fit the theoretical line.

A chi-squared test is used when sets of large numbers of continuous variables with randomly distributed error are being compared. Several other statistical analyses are used under different conditions. Student's t test assumes a distribution slightly different from the Gaussian distribution, which is more accurate for smaller numbers of data points. In the chi-squared and t test the data in both sets being compared are assumed to have the same distributions (either Gaussian, binomial, Poisson, or some other distribution). The F test evaluates the variances in each of the data sets being compared so that not only is the overlap between sets measured but also the type or shape of the distribution. In principle, the F test could distinguish between a data set that was a straight line with zero slope from a sinusoidal data distribution, whereas the chi-squared test might show the two distributions as statistically equivalent if there were sufficient scatter in the data. All these tests can be used to compare a theoretical distribution to a set of real data or two sets of experimental data.

The easiest way to perform these tests is to use a computer program, such as R, which calculates probabilities (or p values) directly from the data. The following excerpt is taken from an R session based on the R-manual, which shows the ease with which these statistical tests can be performed using this open source software.

EXAMPLE B1

Using R to Compare Two Data Sets

Suppose you have two data sets, measured by two different methods, and you want to know whether the two methods are equivalent. Use the free open source program R to find the statistics for each set and then compare the two sets statistically.

```
Method A: 79.98 80.04 80.02 80.04 80.03 80.03 80.04 79.97 80.05 80.03 80.02 80.00 80.02
Method B: 80.02 79.94 79.98 79.97 79.97 80.03 79.95 79.97
```

First, enter the two data sets into R as arrays (hit "Enter" after each line):

```
> A <- scan()
1: 79.98 80.04 80.02 80.04 80.03 80.03 80.04 79.97 80.05 80.03 80.02 80.00 80.02
14:
Read 13 items
> B<- scan()
1: 80.02 79.94 79.98 79.97 79.97 80.03 79.95 79.97
9:
Read 8 items
```

To t test each set separately,

```
> t.test(A)

        One Sample t-test

data: A
t = 12038.79, df = 12, p-value < 2.2e-16
alternative hypothesis: true mean is not equal to 0
95 percent confidence interval:
 80.00629 80.03525
sample estimates:
mean of x
 80.02077
```

```
> t.test(B)

        One Sample t-test

data: B
t = 7211.703, df = 7, p-value < 2.2e-16
alternative hypothesis: true mean is not equal to 0
95 percent confidence interval:
 79.95253 80.00497
sample estimates:
mean of x
 79.97875
```

Here the value *df* is the degrees of freedom ($N-1$), and the very small p value is telling us that the mean is definitely not equal to zero. The value for t is a statistical parameter that indicates fit in a way similar to the chi-squared parameter.

To find the standard deviation of each set type,

```
> sd(A)
[1] 0.02396579
> sd(B)
[1] 0.03136764
```

To compare the two sets using a t test, type

```
> t.test(A, B)

        Welch Two Sample t-test

data:  A and B
t = 3.2499, df = 12.027, p-value = 0.00694
alternative hypothesis: true difference in means is not equal to 0
95 percent confidence interval:
 0.01385526 0.07018320
 sample estimates:
 mean of x mean of y
 80.02077  79.97875
```

The low p value tells us that the two samples have a low probability of being equivalent; a value of 1 would indicate that they are identical. The smaller the p value, the more confident we are that the two sample means really are different numbers. When statistics are calculated by hand, the t value is calculated first from the means of the two sets and the standard deviation and compared to a table to get the p

value. The program R calculates the p value directly, making the t irrelevant.

We can also do an F test to see whether the variances of the two samples are equivalent:

```
> var.test(A, B)

        F test to compare two variances

data:  A and B
F = 0.5837, num df = 12, denom df =  7, p-value = 0.3938
alternative hypothesis: true ratio of variances is not equal to 1
95 percent confidence interval:
0.1251097 2.1052687
sample estimates:
ratio of variances
0.5837405
```

Again, the low p value tells us that the variances are not equivalent, although the match is better than that of the sample sets as a whole.

REFERENCES AND SUGGESTED READING

1. P. R. Bevinton and D. K. Robinson, *Data Reduction and Error Analysis for the Physical Sciences* (1992, McGraw-Hill, New York).

2. L. Lyons, *A Practical Guide to Data Analysis for Physics Science Students* (1991, Cambridge University Press, Cambridge).

3. *Electronic Statistics Textbook* (2007, StatSoft, Inc., Tulsa, OK) (http://www.statsoft.com/textbook/stathome.html).

4. D. M. Kammen and D. M. Hassenzal, *Should We Risk It? Exploring Environmental, Health, and Technological Problem Solving* (1999, Princeton University Press, Princeton, NJ).

5. G. F. Knoll, *Radiation Detection and Measurements*, 3rd ed. (2000, Wiley, New York).

6. The R Project for Statistical Computing (http://www.r-project.org/).

7. G. H. McClelland, "Working with *z*-Scores and Normal Probabilities," online *z* score calculator (http://psych-www.colorado.edu/~mcclella/java/normal/normz.html).

Notes

[1] The difference between a data point and the true value, if known, is called the deviation. The residual is more convenient because the mean can be found directly from the data.

[2] This causes a certain amount of confusion in the literature, but in most cases we are interested in the standard deviation of the real data.

[3] Many student laboratory manuals use other measures of precision, such as discrepancy or relative error, because the number of data points is typically very small, so that a Gaussian distribution cannot

be assumed. The standard error, $\sigma / \sqrt{N}$, where N is the number of data points, is also commonly used as a measure of precision in the case of a small number of data points.

[4] Some references use $\sigma / \sqrt{N}$, the standard error, instead of σ in the denominator, depending on the number of data points in the original data. For other distributions, such as the binomial distribution, the definition for z is slightly different, but it can be used in the same way.

[5] Use z tables with caution. For some tables the z score returns the area under the normal distribution from the center out to $z\sigma$. Other references give the percentage of the area under the Gaussian from minus infinity out to $z\sigma$, and negative values of z are used for areas to the left of the peak. In many other tables (e.g., references [1] and [2]) the area given is $\bar{x}_e \pm z\sigma$.

Index

Note: page numbers followed by n refer to endnotes.

Island Press | Board of Directors